MINNESOTA STUDIES IN THE PHILOSOPHY OF SCIENCE

Minnesota Studies in the
PHILOSOPHY OF SCIENCE

HERBERT FEIGL AND GROVER MAXWELL, GENERAL EDITORS

VOLUME VIII

Foundations of Space-Time Theories

EDITED BY

JOHN EARMAN, CLARK GLYMOUR, AND JOHN STACHEL

FOR THE MINNESOTA CENTER FOR PHILOSOPHY OF SCIENCE

UNIVERSITY OF MINNESOTA PRESS, MINNEAPOLIS

Published by the University of Minnesota Press, 2037 University
Avenue Southeast, Minneapolis, Minnesota 55455, and
published in Canada by Burns & MacEachern Limited,
Don Mills, Ontario

Library of Congress Catalog Card Number 77-083503
ISBN 0-8166-0807-5

Preface

This book arose from the proceedings of two small conferences: one held in Minneapolis under the auspices of the Minnesota Center for the Philosophy of Science, the other held in Andover under the auspices of the Boston University Institute of Relativity Studies. The similarity of topics and the overlap of participants decided the editors to join forces and include the material from both conferences in a single volume. Indeed, one paper in this volume, the first by Howard Stein, was presented and discussed at both conferences. Further details about the conferences and a list of the participants will be found in the Appendix to this Preface. We would like to thank all the participants for their presence and their contributions. Special thanks are due the speakers for their patience in awaiting the printed version of the proceedings.

J.E.

C.G.

J.S.

Notes on the Andover Conference

The Boston University conference was held to mark the issuance of the second, considerably revised and expanded edition of Adolf Grünbaum's seminal book, *Philosophical Problems of Space and Time*, in the Boston Studies series.[1] I would like to again express my gratitude to Adolf Grünbaum for consenting to this meeting and for his contributions to the proceedings, both formal and informal. In addition to the papers of Stein and Grünbaum, Peter Bergmann's contribution was presented and discussed. Wesley Salmon made his paper available to the participants, but unfortunately time prevented discussion of it. Jürgen Ehlers also gave a paper entitled "Structures on Space-time," which was discussed at length. The material he presented is already partly available in his contribution to the Dirac Festschrift,[2] "The Nature and Structure of Spacetime"; the rest will ultimately appear in a volume of the Pittsburgh Studies.[3] I very much regret that considerations of space made it impossible to reprint Ehlers' paper and urge readers interested in an up-to-date discussion of "The Nature and Structure of Spacetime," as treated in current physical theories, to read it.

Now let me (ab-?) use an editor's privilege to make comments on some of the papers, primarily those discussed at Andover.

In connection with Howard Stein's historical survey, I should just like to propose the addition of Euler to the space-time pantheon. His is a name too much neglected I feel—not in the absolute sense, of course, but relative to the discussion of the space-time problem. *The Encyclopedia of Philosophy* and the *Dictionary of Scientific Biography*, for example, are silent on this work, as indeed is Adolf Grünbaum's book. Yet, as Arnold Koslow has noted[4] in his introduction to a translation of excerpts from Euler's *Reflections on Space and Time* (a translation, by the way, which seems badly in need of repair at certain places):

Euler, according to an insight of Ernst Casirer [sic][5] was one of the first to reverse a traditional relation between physics and philosophy. Before his

vii

work, it was customary to think that no concept was to be used in physics or science generally unless it met certain standards of metaphysics. . . . Euler's reverse consisted in his assertion that the laws of physics, by which he meant Newton's laws of motion, were so well confirmed that no one would be right to deny them. These laws not only state truths about the world, but they also explain why things happen; that is, they give the causes of certain kinds of events. Euler argued that if one followed philosophers like Leibniz and used a relational theory of space, two consequences would follow: (1) substituting Leibniz's concept of space for Newton's transforms the true laws of Newton into demonstrably false statements, and (2) we cannot explain why certain kinds of events occur, if we employ Leibniz's concept of space rather than Newton's. Granted, then, that we do have true laws and explanations, the conclusion is, clearly, that both Leibniz and Descartes did not have the proper understanding of space and time.

Euler's argument seems to show that the Newtonian concepts are certainly preferable to those of Leibniz, if we believe that we have true laws and explanations which involve space and time. . . . The major thrust of Euler's study lies in its suggestion that the ideas and principles of metaphysics ought to be regulated and determined by the knowledge which physics has established, and not the other way around.

I shall not follow Howard Stein's example, however tempting, and add an appendix of translations from Euler; [6] but I cannot refrain from quoting a few lines from the above-mentioned essay:

. . . since it is metaphysics which is concerned in investigating the nature and properties of bodies, the knowledge of these truths of mechanics is capable of serving as a guide in these intricate researches (of metaphysics). For one would be right in rejecting in this science (of metaphysics) all the reasons and all the ideas, however well founded they may otherwise be, which lead to conclusions contrary to these truths (of mechanics); and one would be warranted in not admitting any such principles which cannot agree with these same truths.

This is surely a sentiment in which Howard Stein will join (as will we all); and it is perhaps sobering to reflect that it dates from 1748. At any rate, I think this essay marks an important watershed in the evolution of the space-time problem; one the spirit of which is very much in accord with the thrust of Howard Stein's arguments, not only in his historical paper but (if I understand him correctly) also in his letter to Adolf Grünbaum.

On the Great Debate between Grünbaum and Stein my inclination is rather to be taciturn than prolix (mainly on the ground that "bócca fermato è mezzo salvato," which could be loosely translated as "keep your trap shut and you're half saved"). However, I would like to let Einstein have a word, since he seems to come down in this quotation rather on the side of Stein on certain of the issues. This will perhaps serve to balance Adolf Grünbaum's epigraph from Einstein, to which Howard Stein chivalrously refrained from responding. It comes from an unpublished letter to Lincoln Barnett, [7] and since the English letter actually sent varies somewhat from the German draft, I shall reproduce both:

I do not agree with the idea that the general theory of relativity is geometrizing Physics or the gravitational field. The concepts of Physics have always been geometrical concepts and I cannot see why the g_{ik} field should be called more geometrical than f.[or] i.[nstance] the elctromagnetic field or the distance of bodies in Newtonian Mechanics. The notion comes probably from the fact that the mathematical origin of the g_{ik} field is the Gauss-Riemann theory of the metrical continuum which we are wont to look at as a part of geometry. I am convinced, however, that the distinction between geometrical and other kinds of fields is not logically founded.

Ich kann nicht mit der weit verbreiteten Auffassung übereinstimmen, dass die allgemeine Relativitätstheorie die Physik 'geometrisiere.' Die Begriffe der Physik sind natürlich von jeher 'geometrisch' gewesen, und ich kann nicht sehen, warum das g_{ik}-Feld 'geometrischer' sein soll als das elektromagnetische Feld oder die Distanz von Körpern in Newtons Mechanik. Wahrscheinlich stammt die Ausdrucksweise aus dem Umstand, dass das g_{ik}-Feld seinem mathematischen Ursprung (Gauss, Riemann) Begriffen entstammt, die man als 'geometrisch' zu betrachten gewohnt ist. Genauere Überlegung zeigt aber, dass die Unterscheidung zwischen geometrischen und anderen Feldbegriffen sich nicht objeetiv begründen lässt.

(In the last sentence, the phrase "aufrecht erhalten lässt" has been crossed out and replaced by "objectiv begründen lässt.")

While we are on the subject of Einstein's views, let me try to summarize what I think to be his outlook on the role of rods and clocks (and by extension, all of Grünbaum's family of concordant standards or FCS) in general relativity. I see this to be a three-step affair:

(1) One provisionally accepts the existence of rods and clocks as foundational elements to give a physical interpretation to the tensor field $g_{\mu\nu}$ as the (I shudder to utter it!) metric tensor.

(2) One then develops the theory of this field, and any generalization of it or other fields that one may need to introduce, to get a theory of the structure of matter.

(3) It is now incumbent on such a theory to allow the *construction* of entities which can be *proved* to show the behavior *postulated* in step 1 for rods and clocks (to sufficient approximation).

At least, this seems to me the correct construing of the following passage from Einstein's "Autobiographical Notes":[8]

One is struck [by the fact] that the theory (except for the four-dimensional space) introduces two kinds of physical things, i.e., (1) measuring rods and clocks, (2) all other things, e.g., the electromagnetic field, the material point, etc. This, in a certain sense, is inconsistent; strictly speaking measuring rods and clocks would have to be represented as solutions of the basic equations (objects consisting of moving atomic configurations), not, as it were, as theoretically self-sufficient entities. However, the procedure justifies itself because it was clear from the very beginning that the postulates of the theory are not strong enough to deduce from them sufficiently complete equations for physical events sufficiently free from arbitrariness, in order to base upon such a foundation a theory of measuring rods and clocks. If one did not wish to forgo a physical interpretation of the co-ordinates in general (something which, in itself, would be possible), it was better to permit such inconsistency—with the obligation, however, of eliminating it at a later stage of the theory. But one must not legalize the mentioned sin so far as to imagine that intervals are physical entities of a special type, intrinsically different from other physical variables ("reducing physics to geometry," etc.).

Indeed, it was the inability to complete step two of this program within classical general relativity ("the postulates of the theory are not strong enough") which constituted one of the main—if not *the* main—spurs to Einstein's search for a unified field theory.

At any rate, if we accept my construction of the Einstein program for relativistic field theories, might we not see the work of Grünbaum—as well as many others, including some in this volume—as primarily concerned with step one of the program, in which rods, clocks—and whatever other FCS's—are treated as entities not fully explicable within the theory as it has been so far developed? Then it would not be surprising that the considerations of these authors cannot have more than a heuristic significance in discussing the development of the program. Whereas many of Howard Stein's comments (though by no means all) would refer to a state of the fully developed (step three) field theory, in which any singling out of intervals—or indeed of any element of the theory—as primarily geometrical[9] in nature would have only that historical justification which Stein—and Einstein—have emphasized.

If this analysis has any merit, we could understand some of the puzzlement of and mutual incomprehension between the two interlocutors—in spite of the evident mutual respect and good will—as owing to the fact that one point of view is based upon a level of theoretical development which by its very nature must deal with certain concepts heuristically, while the other is based upon the assumption of the existence of a fully developed theoretical framework which does not now exist—nor is even accepted as possible by most physicists (unified field theory)!

The trouble is, of course, that we never know how the story is going to end. A science keeps developing, so that what "a theory" tells us at any stage of its development—perhaps it would even be better to switch terms to the currently more fashionable phrase: "research program"—is never quite unproblematic; [10] even a completely superseded theory may offer some surprises when rediscussed from a more current viewpoint, as witness Howard Stein's work on Newtonian theory.

Just to give an example, which happened to come to my attention while working on this preface, illustrating that the current status of general relativity does not allow any final judgments about the role of the "geometrical" component, let me quote from a recent (excellent) article by two working physicists on the problem of setting up and interpreting a quantum theory of gravity. [11]

Suppose . . . that one obtained a formalism involving a wave function on the space of possible classical configurations of the gravitational field. How is such a wave function to be interpreted physically? What does 'a probability distribution for the results of a measurement of the geometry' mean when the measuring instruments themselves are immersed in that geometry?

The point is that as long as any research program is developing, there is no clear and final line of demarcation between the heuristic and the established theoretical concepts. Indeed, it has been argued that it is just the tension between these two elements that creates the space in which epistemological considerations may validly establish themselves. [12] No doubt I have gone too far already and provided by these remarks a demonstration *a contrario* about the wisdom of keeping one's trap shut.

I shall make no lengthy comments on the discussion of Robb's work, running through a number of the papers, but merely point out that the significance of the work would not have remained so long unknown to us all had we just carefully read the works of Hermann Weyl—something to be recommended for many other reasons, I hardly need add! In the first of his 1922 lectures on the space problem,[13] he gave a concise characterization of the mathematical essence of the issue and of Robb's contribution.

1. *The group of homothetic mappings is the intersection of the projective group and the group of conformal transformations [Möbiusschen Kugelverwandtschaften].*

2. *The former may be characterized by the concept of straight line, the latter by the concept of null element.*

If one . . . understands by a mapping onto the number space a single-valued invertible continuous mapping of the complete infinite space-time [Welt] onto the entire number space, then it is determined by the first condition alone, that each straight line must go over into another straight line; space-time is fixed not merely as a projective, but as an affine space. For the only projective mappings which take finite points into finite points and points at infinity into points at infinity are the affine ones. On the other hand, the only conformal transformations which leave all finite points as finite points are the homothetic mappings. Therefore one can, if one works with the complete space-time, construct the Minkowski geometry completely on the basis of the *single* concept of null element; the straight line is dispensable. This was carried through a few years ago in a certainly rather circumstantial and artificial axiom system by the English mathematician Robb.

I should like to thank Ms. Barbara Rahke, the Conference Secretary, for the care and expertise she contributed to the smooth running of the meeting.

Were it not a vain illusion to think that anyone reads prefaces, I should close by apologizing for having held back the reader so long from the main business of this book.

John Stachel

Notes

1. Adolf Grünbaum, *Philosophical Problems of Space and Time*, 2nd enlarged edition, being Volume XII of *Boston Studies in the Philosophy of Science* (Dordrecht/Boston: Reidel, 1974).

2. Jagdish Mehra (ed), *The Physicists Conception of Nature* (Dordrecht/Boston: Reidel, 1973).

3. Robert Colodny (ed), *University of Pittsburgh Series in the Philosophy of Science.*

4. Arnold Koslow (ed), *The Changeless Order: The Physics of Space, Time and Motion* (New York: George Braziller, 1967), pp. 115–16.

5. The reference to Ernst Cassirer probably refers to the opening pages of his book, *Zur*

Einstein'schen Relativitätstheorie, translated as the Supplement in *Substance and Function and Einstein's Theory of Relativity* (Chicago and London: Open Court Publishing Co., 1923). Cassirer points out the influence of Euler's work on the precritical Kant, who in his 1763 "Attempt to Introduce the Concept of Negative Magnitudes into Philosophy," states:

> The mathematical consideration of motion in connection with knowledge of space furnishes many data to guide the metaphysical speculations of the times in the track of truth. The celebrated Euler, among others, has given some opportunity for this, but it seems more comfortable to remain with obscure abstractions, which are hard to test, than to enter into connections with a science which possesses only intelligible and obvious insights.

The above is quoted from pp. 351–52 of the English translation of Cassirer's book; the original is Immanuel Kant, "Versuch, den Begriff der negativen Grössen in die Weltweisheit einzuführen," in *Immanuel Kants Werke*, Vorkritische Schriften Bd. II (Berlin: Bruno Cassirer, 1922).

6. Another relevant Englished excerpt will be found in Milič Čapek (ed), *The Concepts of Space and Time: Their Structure and Development*, being Volume XXII of *Boston Studies in the Philosophy of Science* (Dordrecht/Boston: Reidel, 1976).

7. This extract from the letter and draft of the letter of June 19, 1948, is reproduced with the kind permission of the Trustees of the Einstein Estate. It may be found on Reel 6 of the Einstein Papers, available at the Princeton University Library. Incidentally, I wish that I had had this quotation when I wrote my paper on "The Rise and Fall of Geometrodynamics" [in K. E. Schaffner and R. S. Cohen, eds, *PSA 1972: Proceedings of the 1972 Biennial Meeting of the Philosophy of Science Association*, being Volume XX of *Boston Studies in the Philosophy of Science* (Dordrecht/Boston: Reidel, 1974)]. Its first sentence would have provided much stronger evidence for the claimed dissociation of Einstein from Wheeler's geometrodynamic program than I was able to give in my note 1a.

8. Pp. 59–61 in Paul Arthur Schilpp (ed), *Albert Einstein: Philosopher-Scientist*, (*The Library of Living Philosophers*) (LaSalle, Ill.: Open Court Publishing Co., 1970).

9. Incidentally, perhaps I should go on record again, although it may be well known by now, stating that while rods and clocks are postulated to respond directly to the $g_{\mu\nu}$ field, massive structureless test particles are postulated to respond directly to the affine structure; and massless test particles ("photons")—or better, null wave fronts—to the conformal structure of space-time. It is only in a theory, such as general relativity, that postulates the *concordance* of all these structures, that conclusions about metric structure (interval) may be drawn from test particle behavior. For further details, see the paper by Ehlers mentioned in note 2.

10. I have made some brief comments on the danger of "freezing" the discussion of a theory at one stage of its development in "A Note on the Concept of Scientific Practice," in R. S. Cohen, J. Stachel, M. Wartofsky (eds.), *For Dirk Struik*, being Volume XV of *Boston Studies in the Philosophy of Science* (Dordrecht/Boston: Reidel, 1974).

11. A. Ashtekar and R. Geroch, "Quantum Theory of Gravitation," in *Rep. Prog. Phys.* 37, 1211 (1974), p. 1215. Note that they are not discussing Einstein's research program—to subsume general relativity within a wider classical field theory which would explain the structure of matter—but about a different one: to "find some physical theory which encompasses the principles of both quantum mechanics and general relativity" (p. 1213).

12. See Jean Toussaint Desanti, *La philosophie silencieuse ou critique des philosophies de la science* (Paris: Editions du Seuil, 1975), especially the essays "Qu'est-ce qu'un problème épistémologique?" and "L'épistémologie et son statut," for an attempt to give a rigorous definition of what an epistemological problem is and how epistemology differs from traditional philosophical approaches to the sciences.

13. First issued in Catalan, and then in German as *Mathematische Analyse des Raumproblems* (Berlin: J. Springer Verlag, 1923), pp. 7–8. In Appendix I (pp. 62–64) he gives a short mathematical exposition of the ideas expressed in this quotation.

Conference on the Foundations
of Space-Time Theories

Held in Minneapolis, May 9–11, 1974

Participants

David Bantz	Adolf Grünbaum	Alan Shapiro
Peter Bowman	Paul Horwich	Lawrence Sklar
John Earman	Roger Jones	John Stachel
Arthur Fine	David Malament	Howard Stein
Michael Friedman	Grover Maxwell	Roger Stuewer
Robert Geroch	Hilary Putnam	John Winnie
Clark Glymour	Richard Schlegel	

Conference on Absolute and Relational
Theories of Space and Space-Time

Held at Osgood Hill Conference Center of Boston University,
N. Andover, Mass., June 3–5, 1974

Participants

Peter G. Bergmann	Donald S. Hockney	F. David Peat
Robert S. Cohen	Clifford A. Hooker	Wesley C. Salmon
Stanley Deser	Allen Janis	Abner Shimony
John Earman	Arthur Komar	Gerrit J. Smith
Jürgen Ehlers	David Malament	John Stachel
Arthur Fine	Ellen Meadors	Howard Stein
Adolf Grünbaum	Andre Mirabelli	Caroline Whitbeck
Peter Havas	Charles Misner	Steven Wolfson
Robert Hermann	Thomas Pascoe	

Contents

Contents

MINNESOTA STUDIES IN THE PHILOSOPHY OF SCIENCE

Some Philosophical Prehistory
of General Relativity

As history, my remarks will form rather a medley. If they can claim any sort of unity (apart from a general concern with space and time), it derives from two or three philosophical motifs: the notion—metaphysical, if you will—of *structure in the world*, or *vera causa*; [1] the epistemological principle of the *primacy of experience*, as touchstone of both the content and the admissibility of knowledge claims; and a somewhat delicate issue of scientific method that arises from the confrontation of that notion and that principle. The historical figures to be touched on are Leibniz, Huygens, Newton; glancingly, Kant; Mach, Helmholtz, and Riemann. The heroes of my story are Newton and Riemann, who seem to me to have expressed (although laconically) the clearest and the deepest views of the matters concerned. The story has no villains; but certain attributions often made to the credit of Leibniz and of Mach will come under criticism.

I

It is well known that Leibniz denied, in some sense, to space the status of a *vera causa*. In what precise sense he intended the denial is perhaps less well known; indeed, as I shall soon explain, I myself consider that sense in some respects difficult if not impossible to determine. The fact that Leibniz characterizes space as not "real" but "ideal," or as an "entity of reason" or abstraction, by itself decides nothing; for he also tells us that the structure thus abstracted is the structure—or, as he puts it, the "order"—of the situations of coexistent things; furthermore, of the situations of all actual *or possible* coexistent things; or, again (in the fourth letter to Clarke), that space "does not depend upon such or such a situation of bodies; but it is that order, which renders bodies capable of being

NOTE: Acknowledgment is due the John Simon Guggenheim Memorial Foundation, during the tenure of a Fellowship from whom a part of the work on this paper was done.

situated, and by which they have a situation among themselves when they exist together." It is abundantly clear from this and many other explicit statements, as well as from his scientific practice, that Leibniz regarded the attribution to bodies at an instant of ordinary geometrical relations—distances, angles, etc.—as having objective significance; that he held these relations to be subject to all the principles of Euclidean geometry; that he regarded geometrical distinctions (i.e., nonsimilar arrangements) as in principle discernible; that he considered these distinctions legitimate to invoke in laws of nature; and thus that the Euclidean spatial structure of the world at an instant was, for him, in the only sense I am concerned with, a *vera causa.*[2]

It is, then—of course—the *connection through time*, the problem of *motion*, that is seriously at issue. But on this issue certain of Leibniz's statements seem to face in several different directions. Let us consider the one that seems most radical. In his treatise on dynamics (*Dynamica de Potentia et Legibus Naturae corporeae*, Part II, Sec. III, Prop. 19) Leibniz states *as a theorem* what appears to be a general principle of relativity. He employs the phrase "equivalence of hypotheses"—evidently derived (and generalized) from the usage in astronomy, where the Ptolemaic, Copernican, and Tychonian "hypotheses" were in question: so "hypothesis" means "hypothesis about motion," or more precisely "choice of a reference body to be regarded as at rest." The proposition asserts the dynamical equivalence of hypotheses for any closed system of interacting bodies, "not only in rectilinear motion (as we have already shown), but universally." Unfortunately, this last expression is obscure: does "not only in rectilinear motion" refer to the reference body—which would give us our "general relativity"; or does it refer to the *interacting* bodies, and generalize from what Leibniz had "already shown," namely for *impacts*, to arbitrary interactions, allowing also *curvilinear trajectories*? The evidence offered by this and related texts seems to me to weigh almost equally for each alternative. A thorough discussion would hardly be appropriate here; but let me sketch some of the salient points.

First, one naturally wants to know how this remarkable theorem is proved. There are in fact two proofs, both very short. The first cites two previous propositions: the analogous result, already mentioned, "for rectilinear motions"; and a proposition stating that *all motions are composed of rectilinear uniform ones*. From these our proposition is said by Leibniz to follow. Now, the exact meaning of the second premise is not immediately clear—not even from its own proof; but it *is* quite plain that, *whatever* it means, it cannot justify an inference from Galilean to "general" relativity. A possible clue comes from a number of passages—in the *Dynamica* and in other writings—in which Leibniz asserts the nonexistence in the world of any true "solidity" or "cohesion," or anything in the nature of a real "cord" or "string." For he says, in several of these places, that if there were any such bond *it would not be true that all motions are composed of uniform rectilinear ones*: rather, there would be real *circular* motions. On the other hand, he says, in actuality all cohesion results from the "crowding in" of particles by an ambient medium; and when a body rotates, its particles not only strive to go off on the tangent, but *actually begin* to go off, and are then turned aside by the medium. Whether Leibniz means this "begin" in the sense of a minute finite segment, or of some kind of infinitesimal, seems uncertain; but in any case it is plausible to conclude that, in saying that all motions are composed of uniform rectilinear ones, Leibniz essentially means that *all interactions are by impact*; and this at least would justify the inference from impact to interaction in general.

Leibniz's second proof makes no appeal to any previous dynamical result at all. It rests upon the claim that a motion by its nature consists in nothing but a change in the geometric relations of bodies; hence "hypotheses" describing the same antecedent relative motions of the bodies are indistinguishable, and their results must be likewise indistinguishable. This does certainly look like an argument for a general principle of relativity. Of course, the argument is philosophical, not dynamical: that is, if one accepts it, one accepts a *constraint* or *condition of acceptability* for any proposed system of dynamics. Does Leibniz's own dynamics satisfy this constraint? Of course it does not! It is in fact very hard to conceive what the structure of a dynamical theory satisfying this constraint could be like. Let me try to be precise about the difficulty.

We may start from space-time—its structure and its automorphisms. Clearly, for Leibniz, *simultaneity* ("coexistence") has objective meaning; so, we may assume, does *ratio of time-intervals*; and I have already remarked that the Euclidean structure on space at each instant is to be presupposed. Leibniz's claim about the purely relative nature of motion seems to imply that this is all the objective structure there is; although I think we may concede (without now questioning its grounds) the *differentiable* structure of space-time as well. Now consider any smooth map

5

of space-time onto itself that preserves simultaneity and ratio of time-intervals, and that restricts, on each instantaneous space, to a Euclidean automorphism.[3] Such a map is an automorphism of the entire Leibnizian structure, and should therefore be a symmetry of the dynamics: i.e., should carry dynamically possible systems of world-lines to dynamically possible systems of world-lines. This immediately leads to the crucial difficulty: take any time-interval $[t_1, t_2]$, and any time t outside this interval; then there exists a mapping of the sort described that is the identity map within $[t_1, t_2]$ but not at t. It follows that the dynamics cannot be such as to *determine systems of world-lines on the basis of initial data* (for the automorphism just characterized preserves all data during a whole time-interval—which may even be supposed to be infinite in one direction, say to include "the whole past"—but changes world-lines outside that interval). It must not be thought that this argument demonstrates the impossibility of a deterministic Leibnizian dynamics; the situation is, rather, that all of the systems of world-lines that arise from one another by automorphisms have to be regarded as objectively equivalent (i.e., as representing what is physically one and the same actual history). But the argument does show that a Leibnizian dynamics cannot take the form of a system of "differential equations of motion," for such equations precisely do determine the world-lines from initial data. Or to put what is essentially the same point in a more sophisticated way: in Leibnizian space-time the "phase," or instantaneous state of motion, of a system of particles cannot be represented by an assignment of 4-vectors to the world-points of the particles at that instant. In short, the basic conceptual apparatus for a cogent formulation of a dynamics satisfying this version of "Leibnizian relativity" would have to be significantly different from the structural framework we are used to. So far as I know, the appropriate concepts have never been defined. The principles of his dynamics, as Leibniz actually formulates them, uncritically employ the standard kinematical notions; Leibniz not only states that a moving body tends to continue its motion *uniformly in a straight line*, but emphatically declares that for a body free from contact with others to move continually in a circle would be a miracle, because contrary to the nature of body. It follows that *in his scientific practice* Leibniz treats the distinction between uniform rectilinear motions and all other motions as a *vera causa*, and so presupposes—in conflict with at least some of his statements of principle, and with the usual interpretation of his views—that there is in the world an objective

6

structure that supports this distinction.[4] (According to Reichenbach, Leibniz regarded the inertial structure of the world as the result of an interaction of bodies with the ether, and he cites as evidence Leibniz's discussion of rotation in the *Dynamica*. That suggestion is really not coherent: the ether, for Leibniz, acts dynamically; hence its actions themselves involve the inertial structure. And of the passage in the *Dynamica*, although it is very obscure and would call for a much fuller discussion than I can give here, one thing at least is plain: the effect it attributes to the ether is not any *inertial* behavior of bodies, but their cohesion.)[5]

II

Not only in contrast with Leibniz's obscurity, but by any standard, the dynamical writings of Huygens are models of clarity. In these writings, from quite an early date, relativistic considerations play a major role—most notably in his work on impact, where the principle of Galilean relativity is stated as one of the axioms and employed to far-reaching effect. As to the philosophical question of the nature of motion, we have, for instance, this statement, in August 1669: "According to me rest and motion can only be considered relatively, and the same body that one calls at rest with respect to certain bodies may be said to move with respect to others; and [I say] even [that] there is no more reality of motion in the one than in the other."[6] Again, at about the same time, he wrote that "the motion of a body may be at the same time truly equal and truly accelerated according as one refers its motion to other different bodies."[7] This statement clearly asserts the relativity of acceleration, as well as that of uniform motion.

But one must ask: what did Huygens mean by the relativity of acceleration? In his treatise on centrifugal force, Huygens calculates the tension of a cord by which a body is held at the edge of a rotating disk, by imagining an observer fixed to the disk and holding the cord. He remarks that if the cord were cut the body would move off with uniform velocity on the tangent, and calculates from this the trajectory as it would appear to the observer on the disk: this trajectory, one easily sees, is the *involute* of the circle bounding the disk, described by *uniform unwinding*; and Huygens shows that the involute, so described, leaves the disk at right angles, with a velocity of zero at the initial point and initial acceleration v^2/r (v being the circumferential velocity of the rotating disk, r its radius). He concludes that the tension of the cord that prevents this acceleration must be the same that would, e.g., support the same body on an incline down

7

which it would otherwise slide with the same acceleration. This beautiful argument—which implicitly exhibits, in the later parts of the trajectory of the body, the effect not only of centrifugal but of Coriolis force—may be considered an application of the relativistic idea just quoted: the same body is in the one sense in uniform motion on the tangent, and, at the same time, in the other sense, accelerating from rest to traverse its curved path. But this species of relativistic consideration is not an example of the "equivalence of hypotheses," or invariance of the laws of nature. And Huygens seems to have been quite clear about this; for in one of his notes from nearly the same time as our passage on the relativity of acceleration we find this: "Straight motion is only relative among several bodies, circular motion is another matter and has its κριτηριον which the straight does not have at all."

This last statement expresses a view which, in the joint opinions of Huygens and Leibniz—in their correspondence in 1694—agrees with that of Newton in the *Principia*.[8] (In his *Dynamica*, by the way, Leibniz enigmatically remarks that that view *would be right* if there were any real solidity or real cords in nature.) But Huygens and Leibniz also agree, in that exchange, that the view in question, which Huygens himself had long held, is wrong; and Huygens says that he has only recently—within two or three years—arrived at a more correct understanding of the matter.[9]

What this more correct understanding was cannot be inferred from the correspondence. In particular, Huygens makes no explicit reference there—as Leibniz does—to the "equivalence of hypotheses." But there is a set of late manuscript notes by Huygens on the nature of motion that shed some light on the question. Unfortunately, there is a problem of dates that creates some uncertainty whether we do have here Huygens's final view: that, from the statement to Leibniz, should date from around 1691, whereas the position described in the manuscripts seems to go back three years further, to 1688.[10] On the other hand, Huygens remarks more than once in these fragments that he "had long considered that we have in circular motion a κριτηριον of true motion from centrifugal force," but considers so no longer; and this appears to be just the change of view he avows to Leibniz. In any case, what Huygens says in these several notes amounts to the following: *Only relative, not absolute, motion is real; but there can be relative motion with no change in the relative positions of bodies.* In particular, he says, this is the case in a rigid rotation: all

8

distances are preserved, and yet, e.g., the opposite ends of a diameter are in relative motion.

At first sight, this seems a very lame and disappointing argument; even a naïve blunder, confusing the proper concept of *relative motion*—as, say, change of coordinates with respect to a system of reference (or, more generally and more fundamentally, as *change of geometrical relations*)—with that of *velocity difference*. And I think this impression is not entirely wrong: insofar, namely, as Huygens believes that his new analysis squares with the epistemological argument for relativity of motion—the argument that the empirical content of the concept of motion can extend no further than to change of observed, hence relative, position.[11] But there is another aspect to Huygens's analysis, which seems to me to show remarkable—and instructive—insight.

In discussing *how circular motion is known*—is recognized in experience—Huygens says there are two ways. First, "by reference to bodies that are round about and relatively at rest among themselves and free"; and he adds that this is the case with the fixed stars—"unless they are fastened to a solid sphere as some used to believe"—so we know from observation of the stars that the earth rotates. Second, even if there are no such surrounding bodies, circular motion can be known by the centrifugal projection of bodies—as, he remarks, the observations of the behavior of pendulum clocks on voyages have revealed that "the earth flings bodies harder near the equator." We see, then, that no purist dogma is here maintained by him, that motion can "mean" only change of observed position; rather, the concept of motion is allowed to claim, as "empirical content," whatever accrues to it by virtue of the application to experience of a theory in which it appears. We also see that Huygens has by no means embraced the equivalence of the Copernican and Tychonian hypotheses, and by no means abandoned centrifugal force as a criterion of circular motion. But then, in what has his view changed at all? Is not the discussion just reviewed of criteria of the earth's rotation exceedingly close to the discussion in Newton's scholium on space, time, place, and motion?

The answer is yes, the parallel between the two discussions is very close. There is only one difference: Newton believes that the circumstances reviewed prove the need for a concept of *absolute place* and *absolute motion*. Huygens says this is not so; he says that although centrifugal force *is* a criterion of circular motion, circular motion itself is *not* a criterion of "absolute" motion, *in the sense of change of absolute place*. Indeed, the concept of absolute motion, or absolute velocity, Huygens dismisses as a "chimera"; but he admits the objective significance of what he calls "relative motion," which amounts to (absolute) difference-of-velocities, even though this may be attended by no change in the relative positions of bodies. This, as we can see very clearly today, is precisely the right conclusion about the structure of the world as it is represented in Newtonian dynamics. In Newtonian space-time, a state of uniform motion is represented by a (nonspatial) *direction* in space-time; and all such directions are equivalent: for any two, there is an automorphism of space-time taking the one to the other; there is no "absolute motion" or "absolute rest." But a pair of such states, or directions, have what is analogous to an *angle* between them, and this can be measured by a (spatial) vector, the "velocity-difference." Of course neither Newton nor Huygens had available the mathematical language to express this as I have just done. It seems to me quite remarkably penetrating of Huygens to have grasped, and expressed in the language available to him (in which it sounds paradoxical), the heart of the matter: velocity a "chimera," velocity-difference real. And I think his insight not only remarkable, but instructive, when one considers how he arrived at it: not by standing upon philosophical dogmas, whether empiricist or metaphysical, about admissible concepts and theories; but by considering, for a theory known to have fruitful application, exactly how its concepts bear upon experience.

III

I said at the outset that Newton was a hero of the story. Clearly Huygens merits consideration: on the issue of the space-time structure of the world, I think he does see farther than Newton. Not that the distance between them is so very great: we know that Huygens had held Newton's view; we have seen how closely his late discussion parallels part of Newton's scholium. In point of fact, that late discussion of Huygens's was directly *stimulated* by Newton's *Principia*. In the latter, too, we find a clear statement of the principle of Galilean relativity for dynamical systems—and from this it follows, if we regard dynamics as the fundamental theory of physical interactions, that neither absolute place nor absolute motion is a *vera causa*. So when Huygens tells Leibniz that he is eager to see whether Newton will not revise his views on motion in a new edition of the *Principia*, we can take this as a reasonable expectation based upon a perceived affinity of views. But Newton did not revise his opinion, and Huygens on this point stands alone until the threshold of the twentieth century. [12]

I do not, of course, wish to have my use of the word "hero" taken seriously; nonetheless, I do intend a serious point in continuing to sing Newton's special praise as a philosopher—and the point concerns the twin issues of "structure" and "knowledge-source." Newton has often been represented as rather crude in his philosophical views: his epistemology a crude empiricism; his metaphysics dogmatic and crudely realistic (with theological overtones); the epistemology and the metaphysics unresolvedly incompatible with one another. I believe this to be an erroneous representation, and have attempted on previous occasions to offer correctives. [13] What I wish to emphasize at present is that Newton's empiricism was indeed a radical one: he considered *all* knowledge of the world to be grounded in, and in principle corrigible by, experience. Thus he presents the laws of motion as propositions long accepted, and "confirmed by abundance of experiments"; he tells us that geometry itself is "founded in mechanical practice"; and when he offers theological remarks, he claims for them no *a priori* metaphysical status, and no sort whatever of justificatory force to ground propositions of natural science—quite the contrary, he concludes one set of such remarks with the words, "And thus much concerning God; to discourse of whom *from the appearances of things*, does certainly belong to Natural Philosophy."

What has made this radical empiricism appear crude and shallow is, I think, Newton's failure to discuss in detail the methods by which claims to knowledge are to be empirically grounded and tested, or the difficulties that are faced by an empirical "justification" of knowledge. Yet he does take up both these issues, in several brief but pregnant passages. On the first, for instance, the "Rules of Philosophizing" in Book III of the *Principia*, although they are certainly not definitively satisfying, do constitute a substantial statement; and they have the inestimable virtue of being immediately put into detailed application, in an argument upon which they provide a commentary, and which provides a substantial exemplification—and thus data for the explication—of them in turn. On the second issue—the *difficulties* of empirical justification—we have such a (terse) statement as this in the *Opticks*: "[A]lthough the arguing from Experiments and Observations by Induction be no Demonstration of general Conclusions; yet it is the best way of arguing which the Nature of

11

Things admits of. . . ." This antedates Hume. But Hume's critique of induction is in effect an explication of Newton's first clause; and all attacks upon Hume's problem are in effect attempts to explicate the second. The statement itself, once again, does not *finally* satisfy; yet I do not think it unreasonable to call it farseeing and remarkably balanced—or deep and clear.

But, as I have already suggested, the aspect of Newton's empiricism that I consider most remarkable and most philosophically meritorious is its dialectical relationship to his views about the fundamental structure of the world. The notion of such a structure he took very seriously indeed; and he put forward, for the guidance of physical investigation, a general conception of the kind of structure to be sought—a general program for physical theory or explanation. Huygens and Leibniz, too, had such conceptions or programs—closely similar to one another, since they were variants of the prevalent view of the time, that all physical action is "mechanical" action by contact, through pressure or impact. Leibniz considers this conception to be a necessary consequence of the nature of body;[14] and Huygens repeatedly refers to it as defining *our only hope to achieve understanding in physics.*[15] Both Huygens and Leibniz rejected universal gravitation—that most famous of *verae causae* discovered by Newton—because they saw no way to accommodate it to their general conceptions of physical action: "[I]t cannot," says Leibniz, "be explained by the nature of bodies."[16] This mode of argument—the dogmatic refutative use, against a proposed theory, of structural or "metaphysical" preconceptions—is precisely what Newton objects to in his well-known deprecation of the appeal to "hypotheses" in experimental philosophy. Just because experience is (on the one hand) our only authoritative guide, and cannot (on the other hand) demonstratively establish general conclusions, we must *always* be prepared to *modify* any conclusion, including our most general conceptions about the structure of the world. In a certain sense, this methodological principle plays, for Newton, a role analogous to that of the structural principle of mechanical causation for Huygens: it is the basis of all our hopes for understanding in natural philosophy. To violate it—to "evade the argument of induction by hypotheses"—is to risk the stultification of inquiry.[17] I believe the case can be made, in historical detail, that Newton was in practice consistently faithful to this principle: that in all the controversies he engaged in, he *never argued against* a rival physical theory on the grounds of what could

12

be called a "hypothesis."[18] And his expositions of his own deepest physical conceptions are also characteristically in the spirit of the same methodological principle. I shall return to this point later, and shall cite two examples.

IV

Before advancing now from the seventeenth to the nineteenth century, I should like to make one slight remark about Newton and space-time—a remark that really bears more upon the technique of historical interpretation than upon Newton's theory. I have found that some historians are suspicious of the very use of a term like "space-time" in explicating Newton's thought, on the grounds that such a use is anachronistic. On one occasion, I cited a statement made by Newton in one of his most metaphysical-theological passages: "Since every particle of space is *always*, and every indivisible moment of duration is *every where*, certainly the Maker and Lord of all things cannot be *never* and *no where*"; and I suggested that to say that every particle of space *is always* and every moment of duration *is everywhere* is exactly to identify the "particles of space" or points with certain one-dimensional submanifolds of space-time, and the "moments of duration" or instants with certain three-dimensional slices or submanifolds. The suggestion was considered extravagant. It was, therefore, with a certain shock of delight that I recently noticed exactly the same reading of the same passage by no less a figure than Kant—to suspect whom of harboring such a notion as space-time might otherwise itself seem anachronistic. The Kantian statement occurs in a footnote in section 14 of Kant's inaugural dissertation (a footnote, by the way, to the striking remark that *simultaneity* is the most important concept following from that of time). In the midst of this note, Kant says:

Although time is of only one dimension, yet the *ubiquity* of time (to speak with Newton) . . . adds to the quantum of what is actual another dimension. . . . For if one designates time by a straight line produced to infinity, and the simultaneous in any point of time by ordinate lines, the surface so generated will represent *the phenomenal world*, in respect of substance as well as accidents.

I am not, of course, suggesting that Kant's reading of Newton is to be taken as necessarily correct—any more than my own is necessarily so; yet I think that Kant's words, written within sixty years of Newton's, carry some persuasive force. But to put my point about interpretation and

13

anachronism in a general form, it is this: For the avoidance of anachronism, it is *neither necessary nor sufficient* to restrict one's conceptual vocabulary to that of the period under discussion. To impose such a restriction is to inhibit flexibility of thought *without any important compensating guarantee against error*. It is an intellectual stratagem analogous to that of the shallow empiricism in science, that seeks security in rules for the construction of concepts, and achieves only a hobbling of theory.

V

In Mach, of course, we have a classic case of this abusive empiricism. It is a case that also exemplifies a characteristic tendency, a kind of Nemesis, of what we might call "hypercritical" philosophic theories—theories that lay down methodological standards or criteria which are actually impossible to practice. The tendency is to lose, at crucial junctures, basic critical control of the conceptual process. Such critical failure is to be seen in Mach whenever he engages on phenomenalistic grounds in polemic against a physical theory. For instance, Mach's opposition to the kinetic-molecular theory is based upon the fact that, as he puts it, atoms are "mental artifices." But what about perfectly ordinary objects? "Ordinary matter," Mach says, is a "highly natural, unconsciously constructed mental symbol for a relatively stable complex of sensational elements"; the only distinction he finds to the disadvantage of atoms is that of the "natural unconscious construction" versus the "artificial hypothetical" one.[19] To conclude, as Mach does, on the basis of this distinction, that *atomic theories* should eventually be replaced by some "more naturally attained" substitute[20] is very strange: not only is the argument at right angles to Mach's view of the "economic" objective of science, it actually accords a preference to the instinctive and unconscious over the conceptual and deliberate mental processes. And it in no way makes philosophically plausible the claim that atoms are more "artificial" than, for example, thermodynamic potentials.

In Mach's discussion of the issues of space, time, and motion, such loss of critical control occurs in an especially acute form. Let us consider Mach's critique of Newton's inference from the water-bucket experiment. Newton laid stress upon four stages of the experiment: two in which the rotational velocity of the water relative to the bucket was zero, and two others in which it was maximal. In each of these pairs, there was a case in which the water surface was plane, and a case in which it was concave: namely, whenever the water was at rest *relative to the earth* its surface was plane, and whenever it was rotating relative to the earth its surface was concave. Newton argued that these observations indicate that not relative but absolute rotation is responsible for the dynamical phenomenon.

Prima facie, one may wonder why Newton put such stress on the relative motion of the water *with respect to the bucket*: do not the observed facts of the experiment naturally suggest that rotation with respect to the earth is the relevant circumstance? The commentators with whom I am acquainted have failed to notice that Newton's emphasis here is motivated, not by a mistaken estimate of the demonstrative force of this argument, but by a dialectical enagagement with *Descartes's* theory of motion—according to which, as it is presented in Descartes's *Principia Philosophiae*, the "true" or "absolute" motion of a body is that relative to *the bodies immediately touching* the given one. *This* notion the bucket experiment quite convincingly shows not to be the appropriate one to serve as the basis for dynamical theory. On the other hand, the question remains: does the bucket experiment show that dynamics requires an "absolute" notion of rotation? What rules out rotation relative to the earth as the appropriate fundamental concept?

Now, it is really quite plain that the bucket experiment does *not* establish this point. Yet this question, a natural one not only from Mach's point of view but intrinsically, is not the question Mach raises: he proceeds at once to the stars: "Try to fix Newton's bucket and rotate the heaven of fixed stars and then prove the absence of centrifugal forces." But why the fixed stars? Is this not—from the phenomenalist point of view especially—a very artificial view of the situation? (Doubtless the stars were not visible at all to Newton during the experiment!) Mach says, in a famous and true remark, that the world is given to us only once, and he concludes that it is "not permitted to us to say how things would be" if that world were other than it is; so, he says, we know only that centrifugal forces accompany those states of motion which are rotations relative to the fixed stars. But Mach does not make it a general rule for science that in every statement based upon experience there should appear a list of all the circumstances over which we have no control (the universe being given only once), in order to avoid seeming to claim that we know that the statement would continue to be true even if these things were otherwise.

15

Such a rule would not only grievously violate Mach's "economy of thought," it would make science impossible. So, again: if we need not bring the stars in everywhere, why here? And why the stars rather than the earth?

The answer to the second question is obvious: we know—essentially from Newton's dynamical analysis of the solar system—that, although the bucket experiment is insufficiently sensitive to show it, rotation relative to the earth will not do as the basic dynamical concept; for the earth itself has to be regarded as rotating. So Mach appeals to the stars because they provide his only recourse: Mach's epistemology teaches him—or he thinks it does—that the only differences among states of motion that can be taken for *verae causae* are differences among *relative* motions; and the only relative motion he finds that can be assigned responsibility for centrifugal force is that relative to the stars. This argument, however, is not really epistemological: it is metaphysical, and is quite akin to Mach's rejection of atoms: namely, it is a rejection of the *possible* reality of certain structures, although they are suggested by and serve for the systematic characterization of phenomena, on the grounds that they are not in some sense properly *constituted out of* phenomenal "elements."

But Mach's confusion on this issue can be documented more fully, and not just in the philosophy but in the physics. One avenue of approach to the point is this: Mach says that we are "not permitted to say" how things would be if the universe were differently arranged. But scientific theories ordinarily do permit all sorts of inferences about situations different from the actual one. Does Mach have a way to formulate a theory of motion that will not permit such inferences? More especially, does he have a dynamical theory that erases the distinction between a rotating universe and a nonrotating one? Or, again—to put the same question another way—it is agreed that we have *evidence*, within the frame of Newton's theory, that the earth is in rotation; does Mach exclude the possibility of finding analogous evidence about the system of the visible stars and galaxies?

To the first, most general question, the answer is a simple negative: Mach suggests no way to formulate a theory in which inferences about states contrary to fact cannot be made; and the very idea seems to contradict our usual notion of what we mean by a "theory." But on the more special issues about motion and the stars, the situation is more complicated. It has not, so far as I know, been noticed that in his discussion of Newton's theory of motion in *The Science of Mechanics*, in the space of

16

about four pages, Mach sketches or suggests *three entirely different phys-ical theories*, each of which puts the subject on a footing satisfactory to him, but which are (as he does not seem to realize) quite incompatible with one another.

The most elaborately presented of these three theories, and the least satisfying, is essentially an attempt to make precise and general the idea of taking the cosmic masses as defining the basic dynamical reference frame. I shall not attempt to present this Machian theory, which is given in chap. ii, sec. vi, parag. 7 of the book, in any formal detail, but shall only comment upon the trouble with it. Mach's formulation is sketchy and loose, and his exact meaning a little hard to determine. The basic idea, however, is, not to refer explicitly to the fixed stars (or galaxies), but just to take successive determinations of the average relative positions and motions of *all* the surrounding bodies over various solid angles and out to increasingly great distances, and to use the limits of sequences so ob-tained to define the dynamical variables.[21] Difficulties of formulation aside, this strategy has a quite basic defect. First, it *cannot* be made general: that the sequences involved converge, and that the limits taken in different directions fit together into a coherent geometrical and kinematical framework, are special assumptions about the cosmic geog-raphy. If these assumptions fail, the theory just collapses. Second, the special assumptions involved, since they *are* cosmological (or cosmo-graphic), go far beyond anything for which convincing empirical evidence is available. Finally, the theory really rests upon the assumption that Newtonian dynamics itself is correct; it is entirely parasitic upon the latter theory, and is merely an attempt to *reformulate* it in a way that refers only to relative motion. Since the device by which this is to be accomplished is the one I have described—in effect, the positing and then the exploita-tion of a very special cosmic geography—the theory does indeed imply that there is no difference between a universe globally at rest and a rotating one; but this is achieved, not through any theory of inertial structure as an effect of interaction, but only by the special assumptions I have mentioned—which are tantamount, so far as this question is con-cerned and from the wider Newtonian point of view, to the mere arbitrary exclusion of an average rotation of the universe from the range of envis-aged possibilities. It is hard to see in this theory propounded by Mach an exemplification of anything like what has subsequently been discussed under the rubric of "Mach's Principle." And, what is ironical above all, in the interest of purging Newtonian dynamics of an allegedly nonempirical component, Mach has been led to put forward a theory which must be regarded as on an *empirically weaker footing* than Newton's own—since Mach's theory is equivalent to the *conjunction* of Newton's and of special cosmological assumptions. In short, I submit that this is a clear case of ideology out of control.

But in quite sharp contrast to the foregoing theory is a recurring remark of Mach's, to the effect that the true and complete principle of inertial reference systems is contained in Newton's fifth Corollary to the Laws of Motion—i.e., the principle of Galilean relativity.[22] This remark is a little vague; but it clearly represents a very different point of view from the one I have just been discussing—in particular, it makes no reference to the cosmic masses, and is perfectly compatible with a Newtonian rotational state for these (or, for that matter, with there being no global steady state at all). The most natural interpretation of this Machian view is the following: The laws of mechanics are to be construed as asserting that the relationships they express hold *for some kinematical reference frame*; Corollary V tells us how all such frames may be obtained from any one of them; beyond this, no identifying or "individuating" mark of a distinguished reference system is given.[23] This point of view is precisely the appropriate one for Newtonian dynamics; and it rests, as Mach entirely fails to notice, not indeed upon absolute space, but nonetheless upon "absolute uniform motion" as a *vera causa*—*not* explicated through phenomena of relative motion.

Mach's third theory—or rather suggestion; for it does not amount to a theory, or even to the sketch of a theory—is presented in a series of detached remarks, all of which point out not only that (according to Mach's general caveat) *we have no way of knowing, and ought not to presume*, how things would be if the universe were differently arranged, but that, positively, *we must be prepared for possible surprises in novel circumstances*. The most striking remark of this sort is the famous one in which Mach says we cannot know how the bucket experiment would turn out if the sides of the bucket were several leagues thick.[24] This is a *very* different thing from the analogous comment about putting the stars into rotation about the bucket—because in this case, unlike the one of the stars, we are not obviously beyond the hope of obtaining information to decide the issue. So this third point of view, in contrast to the other two, raises the prospect of a *possible revision* of physical theory, on the basis of

18

new discoveries; although it neither suggests the actual content of a revised theory, nor in any way provides evidence for the belief that new discoveries will in fact lead to revision: it simply points out the possibility that this may happen. What has made it appear to readers of Mach (most notably, to Einstein) that evidence has been given that it is somehow "right" or "desirable" for a rotating hollow mass to induce centrifugal forces on stationary bodies inside it, is the occurrence of this suggestive physical speculation in the confused context of the conflated epistemology and metaphysics I have just been describing.

In summary, then: of the three views, the first implies that there is no difference between rotation of the stars about the earth and rotation of the earth beneath the stars. It is equivalent to the statement that the universe is Newtonian, with the stars, on the average, in an inertial state; and would be defeated by evidence against this assertion. The second implies that there *is* a difference between those two cases, and leaves the issue of rotation of the stars open to empirical investigation. The third suggests that there *may be* no difference between the two cases, because *moving masses may induce "inertial" effects*. This is of course the suggestion that caught the imagination of Einstein.

In connection with the second of the Machian positions we have considered, a question arises: Why does Mach single out Newton's fifth Corollary for praise? The *sixth* Corollary to the Laws of Motion goes further: it states that the internal motions of a system of bodies are unaffected by any *accelerated* motion shared by the bodies of the system. Why does Mach not take this up, as a more far-reaching embodiment of the sole relevance of *relative* motion? I am unable to answer this—unless by echoing Mach's own phrase about Newton, that he "was correctly led by the tact of the natural investigator" (or of the good interpreter). For despite Corollary VI, absolute acceleration is in Newton's theory a *vera causa*. But how is it so, despite Corollary VI? A superficial answer is that, for Newton, an acceleration requires a force to produce it, and conversely an unbalanced force requires an acceleration. To this, the proper objection can be made that Newton's dynamics gives us no independent general criterion of the presence or absence of force; thus it is merely glib to cite force as "the criterion" of acceleration. The solution of the puzzle, however, is the following: Newton indeed does not provide us with what a naïve empiricism used to demand—something like an "operational definition" of force or of acceleration. But his dynamics imposes upon dynami-

19

cal systems the general condition that *all the forces in such a system occur in action-reaction pairs*. A shared acceleration, common to all the bodies of a system, cannot come from such forces: it can come, if at all, only from some source outside the system in question. Thus Newton's theory affords a way of assigning kinematical states up to a Galilean transformation, *on condition that one has succeeded in accounting completely for the relative motions by a system of action-reaction pairs*, and on the further condition that there is no reason to suspect the system in question to be subject to an outside influence imparting equal accelerations to all its members. Newton had the good luck to find such a system: namely, the solar system; and the skill to effect its thorough dynamical explication. That explication is the explicit basis upon which Newton rests his determination of the "true" or "absolute" motions.[25] Mach, when he places his emphasis upon the fixed stars as reference bodies, has failed to notice that this choice in no way suffices to decide—on the basis of seventeenth-century data— whether the earth's center, or the sun's, or some other point, is to be described as fixed;[26] Newton, however, was in a position to assert that *neither* earth nor sun can be at rest, but that the center of mass of the solar system—which is never far from the sun—is "either at rest or moving uniformly in a right line."[27]

Newton's Corollary VI, however, is not in the *Principia* for mere ornament. Newton needs it to establish that, to a first approximation, the system of a planet and its satellites can be treated as an *isolated* gravitating system if one ignores the shared orbital motion around the sun. Thus we have a case—treating the sun's gravitational field in the region of the system in question as essentially homogeneous—of what Einstein was to call the "principle of equivalence." Indeed, Newton's sixth Corollary (which deals with a homogeneous field of acceleration), and Huygens's discussion of centrifugal force (which deals with an inhomogeneous one), together adumbrate the principle, exploited so fruitfully by Einstein, that the dynamical states and behavior of bodies in no way distinguish between, on the one hand, a certain kinematical state, and, on the other, a second kinematical state implying the same distances and rates of change of distance, together with a suitable applied field of force. It is a little surprising that Mach, with his relativistic view of motion and his interest in seventeenth-century mechanics, did not at all notice these things.

I said, however, at the outset, that my story has no villains; and I particularly do not wish to give the impression of using Mach as a whipping-boy. I believe he deserves better. I do think that his epistemology was faulty, and his application of it confused. But—despite what I have earlier characterized as "ideology"—the honesty of his mind seems to me beyond dispute; his critique of basic concepts, however defective, has been stimulating for philosophical analysis, for historical interpretation, and—not least—for physical theory; and the same can be said of the sheer physical speculation which, in the guise of critique, appears in his suggestion that a whirling container might induce a field of centrifugal force. I should like to close this lengthy and rather critical section on Mach with two quotations that should leave a pleasanter flavor. The first is from Einstein: "I see Mach's greatness," he wrote, "in his incorruptible skepticism and independence." The second is from Mach, commenting upon experiments designed to detect possible induced centrifugal force fields (about which he was in fact quite skeptical): "[W]e must not," he said, "underestimate even experimental ideas like those of Friedländer and Föppl, even if we do not yet see any immediate result from them. Although the investigator gropes with joy after what he can immediately reach, a glance from time to time into the depths of what is uninvestigated cannot hurt him."

VI

I shall have to abbreviate my discussion of Helmholtz and Riemann; let me try to reduce it to bare essentials. Helmholtz was evidently led to questions about the foundations of geometry by way of his studies in the physiology of visual perception, and the associated questions about the genesis of our perceptions and conceptions of space; and then he was further stimulated by the posthumous publication, in 1867, of Riemann's *Habilitationsvortrag*.[28] The work that resulted was of substantial philosophical interest, and of rather deep mathematical interest. In its latter aspect, as completed and improved some eighteen years later by Sophus Lie, it led to the celebrated group-theoretical foundation of Euclidean and non-Euclidean geometry. Helmholtz's leading idea was that *all our knowledge of space comes from observation of the properties of* (approximately) *rigid bodies*, and therefore that the general properties of space should be deducible from the conditions that must be satisfied by bodies in motion if they are to qualify as "rigid." These conditions can be expressed more advantageously in terms of the motions themselves, directly; that is—since Helmholtz assumes that any rigid motion can be extended, conceptually, through all of space—as assumptions about the group of those mappings of space upon itself that take each figure to a congruent figure: the group of "congruence transformations." The execution of this program led to a twofold result, in relation to the theory of space presented in Riemann's essay: first, the "Pythagorean" metric, postulated by Riemann in an avowedly arbitrary way as just the first and simplest case to consider, is *derived* by Helmholtz from his basic assumptions; second, Helmholtz is led to a far more drastic restriction: for of all the structures that Riemann's theory allows, only those of uniform curvature satisfy Helmholtz's postulates.[29]

Helmholtz's view of his contribution, in its relation to Riemann's theory, is sharply expressed in the title of his basic paper: Riemann's essay, of course, was called, "On the Hypotheses Which Lie at the Basis of Geometry"; Helmholtz's is, "On the Facts That Lie at the Basis of Geometry." Riemann says, in his introduction, that the postulates of Euclidean geometry, as well as those of the (ordinary) non-Euclidean geometries, are "not necessary, but only of empirical certainty—they are hypotheses; one can thus investigate their probability—which, of course, within the limits of observation, is very great—and subsequently judge the reliability of their extension beyond the limits of observation, both on the side of the immeasurably great and on the side of the immeasurably small." The implication of Helmholtz, in substituting "facts" for "hypotheses," is that by reducing the theory to its more fundamental presuppositions he had narrowed the range of open possibilities, and in particular had eliminated the "hypothetical" character of the postulates of "Pythagorean" metric and constant curvature.

Helmholtz, of course, was wrong. It is childishly easy—after the fact—to point out the source of error. It is true that we arrive at our notions about congruence, distance, etc., from manipulations and observations of bodies; and that the general notions we form are connected with the behavior of *rigid* bodies, which we do conceive as conforming to Helmholtz's postulates. But it does not follow from this—either from the psychological facts of the genesis of our spatial notions, or from the mathematical relationship of Helmholtz's postulates to metric geometry—that these postulates, or the geometry they entail, must hold strictly in the world. It follows at most that they hold in some sense "approximately" (as Riemann says: they have a high degree of assuredness or probability, within the limits of observation). The question then arises, *does it make sense to ask what the "exact" state of affairs is?* One may guess that Helmholtz thought something like this: that since the very notion of length is based upon that of congruence, and congruence is based upon the motion of rigid bodies, *either* the basic spatial relationships of one of the geometries of constant curvature must hold, *or* geometry will just break down altogether and our question about the "exact state of affairs" will be devoid of meaning.

VII

Riemann's *Habilitationsvortrag* is one of the most marvelous documents in the history of the human intellect. It is about fifteen pages long; contains almost no formulas; is singularly lucid, and yet so dense with ideas that I am tempted to say that to understand it is to be a wise and a learned man. It is primarily a work of mathematics, and the richness of its mathematical content would merit a very extensive commentary. But I have to confine myself to two or three points that bear upon physics and the philosophy of physics; and I shall first take up Riemann's discussion of the matter just adumbrated in connection with Helmholtz. Here is what he says—lightly paraphrased: [30]

If one presupposes that bodies exist independently of place, then the measure of curvature is everywhere constant [—this of course is Helmholtz's point of view]; and it follows from the astronomical measurements that [that measure cannot be much different from zero]. But if such an independence of bodies from place does not obtain, then one cannot infer the measure-relations in the infinitely small from those in the large; in that case the measure of curvature can have an arbitrary value at each point in three directions, if only the total curvature of every measurable part of space does not differ noticeably from zero. . . . Now the empirical concepts in which the spatial measure-determinations are grounded—the concept of the solid body and of the light-ray—appear to lose their validity in the infinitely small; it is therefore very well conceivable that the measure-relations of space in the infinitely small are not in accord with the presuppositions of [ordinary] geometry—and one would in fact have to adopt this assumption, as soon as the phenomena were found to admit of simpler explanation by this means.

Notice that Riemann has made a remark that one would not have expected Helmholtz, of all people, to ignore: our basic source of spatial information is not bodies only; light also plays a fundamental role. But both of these physical structures do in fact break down in the small. When

Riemann says this about solid bodies, he undoubtedly has in mind the atomic constitution of matter—which he is thus not disposed to dismiss as an "artifice" or as irrelevant. When he says the same about the light-ray, he is of course referring to diffraction. We know a great deal more now about the breakdown of ordinary physical conceptions in the very small; and clearly we have not yet learned all there is to know about it. Riemann's statement of the case was quite remarkably on target.[31]

But what about the objection that if the empirical basis of our geometric knowledge gives way, we are left with no sensible conceptions at all—no concept of a structure to be investigated?

Riemann's stand on this issue is the following—expressed by him in almost lapidary prose, without circumlocution, with the greatest simplicity; yet, for some reason, apparently not often appreciated at its true worth: Our conceptions about spatial structure, and most particularly about structure *in the small*, are essentially bound up with our whole theoretical understanding of physical interaction. "Upon the exactness with which we pursue the phenomena into the infinitely small," he says, "essentially rests our knowledge of their causal connection. . . . The questions about the measure-relations of space in the immeasurably small therefore are not idle questions." From this position it follows, (1) that we must not take a too narrow view of what empirical sources may be used to obtain spatial information: if the ordinary sources of such information break down in the very small, this is an indication that we should be prepared for something possibly quite new and surprising; and empirical information bearing upon possible spatial revision may come from *any* source relevant to fundamental physical theory itself. It also follows (2) that it is appropriate to think of space in a way that has, after all, something in common with the point of view of Leibniz: not with the somewhat rigid dogma that spatial properties are nothing but relations among bodies, but with the broader view that our spatial notions, insofar as they are brought to bear in physics, have their significance only as structural aspects of a more embracing structure: that of physical interaction itself. This is how I understand one most important remark of Riemann's that does seem to me obscurely phrased: his famous statement that if "the reality that lies at the basis of space" is not a discrete manifold, then "the ground of its measure-relations must be sought . . . in binding forces that act upon it." Setting aside the issue of discrete versus continuous, the essential point seems to me this: By "the reality that underlies space,"

Riemann means that aspect of the real structure of the world which we express in terms of spatial concepts. That the ground of the measure-relations is to be sought in *binding forces* expresses partly the general principle that the full physical meaning of the spatial structure comes from its role in physical interaction, but also the more special point that our ordinary middle-sized spatial knowledge does derive from our experience of solid bodies, and that *solid bodies have to be understood, on a more fundamental view, as equilibrium configurations.* This is the same point that Einstein makes when he says that it is wrong, in a fundamental sense, simply to postulate that objects of some kind "adjust" to the metric field—that, roughly, the metric is what we read from meter-sticks—because how an object behaves in relation to the field ought to be derived in the theory *from the object's physical constitution.* Notice that this applies to Newtonian physics quite as well as to any other; and nothing in Riemann's statement implies that it will in fact be necessary to give up Newton's physics—he implies only that it *may* be necessary to do so.

This brings me to the third main consequence of Riemann's view. It is that, since the possible sources of information that might bear upon spatial structure are as wide as physics itself, one cannot hope to foresee in detail what will eventually prove relevant. What one can do—specifically, what the mathematician (I think one should say, the philosophical mathematician) can do—is to explore as well as he can the conceptual possibilities. Here is how he puts it:

> The decision of these questions can only be found by proceeding from the traditional and empirically confirmed conception of the phenomena, of which Newton has laid the foundation, and gradually revising this, driven by facts that do not admit of explanation by it; such investigations as, like that conducted here, proceed from general concepts, can serve only to ensure that this work shall not be hindered by a narrowness of conceptions, and that progress in the knowledge of the connections of things shall not be hampered by traditional prejudices.

VIII

It is with this last passage from Riemann that I should like to compare two of Newton's characterizations of his own most basic physical results, and his program for the development of physics. The first occurs near the end of the *Opticks*, after a summary statement of Newton's fundamental principles: the laws of motion, and the notion of *forces of nature*, among which he mentions "that of Gravity, and that which causes Fermentation, and the Cohesion of Bodies." "These Principles," he says, "I consider, not as occult Qualities, . . . but as general Laws of Nature, . . . their Truth appearing to us by Phaenomena, though their Causes be not yet discover'd. For these are manifest Qualities, and their Causes only are occult. . . . To tell us that every Species of Things is endow'd with an occult specifick Quality by which it acts . . . is to tell us nothing: But to derive two or three general Principles of Motion from Phaenomena, and afterwards to tell us how the Properties and Actions of all corporeal Things follow from these manifest Principles, would be a very great step in Philosophy, though the Causes of those Principles were not yet discover'd: And therefore I scruple not to propose the Principles of Motion above-mentioned, they being of very general Extent, and leave their Causes to be found out."

The second Newtonian characterization appears in the Preface to the *Principia*. Having explained what the work—in particular, the third book—accomplishes, namely the establishment of the theory of gravity and the derivation from it of "the motions of the Planets, the Comets, the Moon, and the Sea," he continues thus: "I wish we could derive the rest of the phaenomena of Nature by the same kind of reasoning from mechanical principles. For I am induced by many reasons to suspect that they may all depend upon certain forces by which the particles of bodies, by some causes hitherto unknown, are either mutually impelled towards each other and cohere in regular figures, or are repelled and recede from each other; which forces being unknown, Philosophers have hitherto attempted the search of Nature in vain. But I hope the principles here laid down will afford some light either to that, or some truer, method of Philosophy."

IX

I have been discussing a few of the strands which—with others as well—were drawn together by Einstein in creating the general theory of relativity. A philosophically satisfying account of the process by which Einstein accomplished that great work has, I believe, never been given. In taking up these pieces of prehistory, I hope to have helped to clarify some of the issues—and some of the confusions and obscurities—that formed the background of Einstein's achievement. I have tried also to suggest—I hope persuasively—a certain general philosophical attitude toward science and its history. Of science, it is the view that, through all the complexities and perplexities of epistemological analysis, all of science is most fruitfully thought of as having one great subject: the structure of the natural world; that to understand a scientific theory is to understand what it says about that structure; but that the touchstone of the genuineness of a structural claim is its connection with experience. Of the history of science, my view is correspondingly old-fashioned: where some see "incommensurable" theories, I see, from the seventeenth century up to today, a profound community of concerns, and a progressive development that has involved both cumulative growth and deepening structural understanding. It is partly a matter of emphasis. Einstein unquestionably effected a conceptual revolution. Such a revolution—I will not say that it was foreseen, but its possibility was foreseen by Riemann. And this work of Riemann's and of Einstein's was entirely in the spirit of Newton's hopes for physics. Indeed, I doubt that a history of three hundred years has ever more gloriously crowned the wishes of a man than the past three hundred years have crowned the expressed hope of Newton: that his principles might afford some light, either to his own or some truer method of Philosophy.

Notes

1. The term *vera causa* is to be taken here, not as bearing a technical sense, but rather as presystematic (a "commonplace"). This term appears to have entered the literature on scientific method in Newton's first "Rule of Philosophizing" (*Principia*, Book III): "We ought to admit no more causes of natural things, than such as are both true and sufficient to explain their phenomena." Newton's phrase was taken up and commented upon, with animadversions variously favorable and unfavorable, by (for example) Whewell, Mill, and Peirce. The sense in which it has been generally understood is expressed as follows by Dewey (*Logic: the Theory of Inquiry* [New York: H. Holt, 1938], p. 3): "Whatever is offered as a hypothesis must . . . be of the nature of a *vera causa*. Being a *vera causa* does not mean, of course, that it is a *true* hypothesis, for if it were that, it would be more than a hypothesis. It means that whatever is offered as the ground of a theory must possess the property of verifiable existence in *some* domain. . . . It has no standing if it is drawn from the void and proffered simply *ad hoc*." (The word "standing" is particularly suggestive here; I have wondered whether there might not have been a prior usage of the term *vera causa* in jurisprudence, to signify a *case* with standing in court—that is, with a prima facie claim upon the court's attention.)

It should be noted, however, that my own emphasis is principally upon a connection the reverse of that asserted by Dewey: I am concerned primarily, not with "true causality" as a credential for admission into a theory, but with the inferences concerning what are and are not "true causes" that may be drawn from a theory taken as already adopted.

2. To relate this sense to the metaphysics professed by Leibniz is a formidable task; but some remarks bearing upon the problem are perhaps in order. First, in Leibniz's ontology the only true "beings" ("substances"; "beings capable of action") are the *monads*, whose states are states of *perception*, whose (exclusively internal) processes are governed by *appetition*, and which together—all mutually adjusted in a harmony pre-established through God's

will—constitute the *kingdom of final causes* or *of grace* (cf. the *Monadology*—e.g., in Gottfried Wilhelm Leibniz, *Philosophical Papers and Letters*, ed. Leroy E. Loemker [2nd ed., Dordrecht-Holland: Reidel, 1970], pp. 643ff., paragraphs 14, 15, 18, 19, 79, 87; and *The Principles of Nature and of Grace, Based on Reason, ibid.*, pp. 636ff., paragraph 1). At the same time, in a sense which it is not at all easy to explicate, each monad is associated with a particular body (cf. *The Principles of Nature and of Grace, Based on Reason*, paragraph 4); and Leibniz sometimes refers to the monad and its body together as a "corporeal substance" (see letter to Bierling, 12 August 1711, quoted by A. G. Langley in his edition of Leibniz's *New Essays Concerning Human Understanding* [3rd ed., La Salle, Ill.: Open Court, 1949], p. 722). Bodies themselves are not substances, and thus not, strictly speaking, capable of action; they are only *phenomena*. But they are none the less real; Leibniz dismisses Berkeley with some scorn (letter to des Bosses, 15 March 1715; Loemker, p. 609): "We rightly regard bodies as things, for phenomena too are real. . . . The Irishman who attacks the reality of bodies seems neither to offer suitable reasons nor to explain his position sufficiently. I suspect that he belongs to the class of men who want to be known for their paradoxes."

The realm of "well-founded phenomena" comprising bodies and bodily processes is the *kingdom of efficient causes* or *of (corporeal) nature*. Notwithstanding the incapacity of bodies for "action," this kingdom can be considered as if it were autonomous; and its autonomous concordance with the kingdom of grace is a manifestation, according to Leibniz, of the pre-established harmony of the monads. In the words of the *Monadology* (paragraphs 79 and 81):

> Souls act according to the laws of final causes through their appetitions, ends, and means. Bodies act according to the laws of efficient causes or the laws of motion. And the two kingdoms, that of efficient and that of final causes, are in harmony with each other.
>
> In this system bodies act as if there were no souls . . . , and souls act as if there were no bodies, and both act as if each influenced the other.

There are in this striking analogies to Kant (whose metaphysical schooling was of course of Leibnizian derivation), even in terminology (e.g., "kingdom of final causes" = "kingdom of ends"); in particular, Leibniz's "real" or "well-founded" corporeal phenomena are precisely equivalent to Kant's "objects of (external) experience," and constitute, just as with Kant, the subject matter of natural science. The "substantial" foundation of these phenomena involves most crucially, according to Leibniz, the notion of *active force* or *power*; he accordingly coined the word "dynamics," i.e., the theory of power, for the basic science of corporeal nature; and he gave to his major treatise on the subject the title "Dynamica de Potentia et Legibus Naturae corporeae"—that is, "Dynamics, on Power and the Laws of Corporeal Nature."

The central difficulties in an attempt to find a coherent formulation of Leibniz's metaphysical principles and of their connection with the principles of his physics can, then, be summarized as follows:

(1) How are we to understand the relation of bodies to the monads?
(2) What role and what status have space and time in this system?
(3) How can Leibniz's concept of force perform the function he claims for it as the metaphysical foundation of bodily phenomena—i.e., their foundation *in the kingdom of final causes* or realm of monads? How, that is, can a conceptual transition be effected between the two realms?

A sketch of answers to these questions, with brief indications of textual evidence, will be essayed here.

To begin with point (2): Space and time are (to use Leibniz's term) "orders"—i.e., systems of structural relations—of *everything that exists*; in particular (and here we have a very sharp divergence between Leibniz and Kant) of the (*noumena* or) monads—even including God.

28

(That spatio-temporal relations affect monads is, e.g., explicit in the letter to de Volder of 20 June 1703; Loemker, p. 531: "I had said that extension is the order of possible coexistents and that time is the order of possible inconsistents. If this is so, you say you wonder how time enters into all things, spiritual as well as corporeal, while extension enters only into corporeal things. I reply that . . . every change, spiritual as well as material, has its own seat, so to speak, in the order of time, as well as its own location in the order of coexistents, or in space. For although monads are not extended, they nevertheless have a certain kind of situation in extension, that is, they have a certain ordered relation of coexistence with others, namely, through the machine which they control." This passage, to be sure, is restricted by the next sentence to *finite* substances; but the application to God is assured by, e.g., the discussion of God's "immensity" and "eternity" in the fifth letter to Clarke, paragraph 106; Loemker, p. 714: "These attributes signify . . . , in respect to these two orders of things, that God would be present and coexistent with all the things that should exist.") Because they are systems of relations, space and time are "beings of reason"; the reality that underlies them is a certain collection of affections of the monads taken severally (cf. letter to des Bosses of 21 April 1714; Loemker, p. 609: "My judgment about relations is that paternity in David is one thing, sonship in Solomon another, but that the relation common to both is a merely mental thing whose basis is the modifications of the individuals"). But what modifications of the individual monads constitute the "basis" of the spatial relations? The answer, presumably (since the only states of monads are perceptive states), is: *their own spatial perceptions.* The pre-established harmony must then ensure a concordance among the spatial perceptions of all the monads, which can be expressed in the formula that *the monads have perceptions of their spatial relationship to one another.*

If this is correct so far, it makes prima facie good sense to identify bodies either with (a) certain *collections of monads* (not necessarily, for "the same body," the same monads at all times), or with (b) certain *correlated perceptions*—appearances or phenomena—of (i.e., internal to) all the monads (taken severally). For each of these identifications will endow bodies with spatial relations; and the first has some conformity with Leibniz's assertion that each monad "has a body," the second with his characterization of bodies as "phenomena." The second identification is in effect that made by Kant: bodies, as objects of experience, are constituted by a connection of perceptions "in consciousness in general." The first contains, by contrast, what Kant would call a "transcendent"—and therefore an illegitimate—judgment. But for Leibniz *both* of these identifications are possible, for one can appeal to the harmony to guarantee compatibility between them; and the double identification agrees very well with a number of passages in which Leibniz speaks almost interchangeably of "aggregates" and "phenomena." For instance, to de Volder, 20 June 1703 (Loemker, pp. 530–531): "[I]n appearances composed of aggregates, which are certainly nothing but phenomena (though well founded and regulated), no one will deny collision and impact." And: "[S]ince only simple things are true things, and the rest are beings by aggregation and therefore phenomena [etc.]." And again: "[A]n internal tendency to change is essential to finite substance, and no change can arise naturally in the monads in any other way. But in phenomena or aggregates every new change arises from an impact according to laws prescribed partly by metaphysics, partly by geometry. . . ."

We have, thus, a hypothesis in answer to question (1) above. As to question (3), a general answer is easy to give (although not an answer that will relieve Leibniz's doctrine of obscurity, or defend it against Kant's criticism of all pretensions to "transcendent" knowledge). Leibniz seeks, among the descriptive parameters of corporeal phenomena (motions), one that is suitable as a *measure* of "power" or "force"; and he argues for the suitability of the quantity he calls "living force" (*vis viva*)—or, over an interval of time, of a quantity he calls "moving action," which is in effect the time-integral of living force. Unfortunately, the connection of these physical quantities (and of the laws he proposes for them) with the monadic metaphysical realm is merely claimed, rather than convincingly explicated (much less established), by Leibniz. (For the best account of the notions themselves, and of the conservation laws suggested by Leibniz, see the French paper "Essay de Dynamique sur les

Loix du Mouvement," in Leibniz. *Mathematische Schriften*, ed. C. I. Gerhardt [Halle: H. W. Schmidt, 1860; photographically reprinted, Hildesheim: Georg Olms Verlagsbuch-handlung, 1962], vol. VI, pp. 215ff.; English translation as Appendix V in Langley's version of the *New Essays* cited above. This paper contains an impressive anticipation of the general principle of the conservation of energy, including some discussion of the problem of inelastic impact. Cf. also, on the same general subject and for some account of the metaphysical connections as Leibniz views them, the essay *Specimen Dynamicum*; English translation in Loemker, pp. 435ff.) Leibniz does, however, make one pregnant suggestion: inspired by Fermat's derivation of the optical law of refraction from a principle of *minimality*, he suggests that the basic laws of the "kingdom of efficient causes" may be derivable from a principle of "final causes" in the form of an extremality principle ("principle of the optimum"). (See *Specimen Dynamicum*; p. 442 in Loemker; and *Tentamen Anagogicum: an Anagogical Essay in the Investigation of Causes*; Loemker, pp. 477ff. It should be noted that Leibniz did not—so far as I know—apply the principle of extremality to his quantity of "moving action": this step was apparently first made by Maupertuis, and first put in a satisfactory form by Euler.)

If this rough account of Leibniz's general doctrine is accepted, it will be seen that within it the spatial structure occupies a critical position, and in some critical ways an ambiguous one. This structure may be said to manifest itself on three levels (which, to press the Kantian analogue, correspond respectively to the Transcendental Aesthetic, the Transcendental Analytic, and the Transcendental Dialectic): first, spatial qualities belong, "subjectively" in Kant's sense, to the perceptions of each monad—they are *phenomena* (and it is of this aspect that Leibniz is presumably speaking when he says: "I can demonstrate that not merely light, heat, color, and similar qualities are apparent but also motion, figure, and extension"—see "On the Method of Distinguishing Real from Imaginary Phenomena"; Loemker, p. 365). Second, spatial relations hold *objectively* among bodies (and, despite Leibniz's assimilation, in the passage just quoted, of extension, figure, and motion to the "secondary qualities," he remained committed to the Cartesian view that only motion can make a real difference among bodies—cf. Appendix II, pp. 44 and 45). It is in view of this circumstance that Leibniz sometimes characterizes space and time, like bodies, as "well-founded phenomena." (That he does so is a remark for which I am indebted to Arthur Fine; cf. Leibniz's letter to Arnauld of 9 October 1687; Loemker, p. 343: "[M]atter . . . is only a phenomenon or a well-founded appearance, as are space and time also.") Finally, as we have seen, spatial relations hold, according to Leibniz, among the monads themselves. Yet it should be noted that this is so only in an indirect or derivative sense. Leibniz's monads, as he puts it, "have no windows": each is a world to itself, characterized in a fundamental sense only by what is internal to it, its "perceptions" and its "appetition." Only through the harmony can the monads be said to perceive, or be related to, one another; and the harmony is a harmony among the perceptions, i.e., the phenomena. Since this is just what constitutes the realm of corporeal nature, it is after all in this realm that space or extension primarily occurs; and accordingly, in the letter to de Volder quoted from above, monads are said to have spatial relations to one another "through the machine which they control." So, in the letter to des Bosses of 16 June 1712 (Loemker, p. 604), Leibniz can say the following (italics added here): "I consider the explanation of all phenomena solely through the perceptions of monads functioning in harmony with each other, with corporeal substances rejected, to be useful for a fundamental investigation of things. In this way of explaining things, space is the order of coexisting *phenomena*, . . . and *there is no spatial or absolute nearness or distance between monads*. And to say that they are crowded together in a point or disseminated in space is to use certain fictions of our mind when we seek to visualize freely what can only be understood."

We come thus to the conclusion that in their *central* sense, for Leibniz, *spatial relations are objective relations among bodies*; and that whatever is to be regarded as "real" or objective about space must be expressible in terms of such relations. However, a critical problem remains; for this view of what is objective about space is hardly reconcilable with

Leibniz's claim that "force" has absolute metaphysical standing, and that the phenomenal manifestation of this absolute force is *vis viva*. The conceptual incoherence shows itself, for example, in the manifest ambivalence of Leibniz's position on absolute versus relative motion in his correspondence with Huygens (see Appendix I below, and cf. also notes 4 and 5 below).

What I wish most to emphasize, in concluding this rather lengthy but hardly adequate note, is that the substance of the discussion in the main text does *not* depend upon the resolution of these intricate issues of the metaphysical interpretation of Leibniz. In particular, what Leibniz's physics implies about the "real structure of the world" is a question that can be approached through a philosophical analysis of that physics; and the compatibility of the result with his metaphysical principles—or with some proposed interpretation of those principles—can be taken as one measure of the success of (the proposed interpretation or of) Leibniz's philosophical program. This point of view, although it is not without its own hazards, seems to me fruitfully applicable to many philosophers who have given serious attention to natural science (and seems to me to have been, on the whole, neglected by historians of philosophy and of science).

3. It is not quite clear whether "automorphism" (for the Euclidean structure, what is usually called "similarity") or "isometry" ("congruence") is the more appropriate notion here. From a general point of view, it should certainly be the former, since isomorphic structures are structurally indiscernible. But Leibniz seems to have conceived of Euclidean metric relations—and not just ratios of distances, but distances themselves—as having a kind of absolute (monadic) basis (cf. *The Metaphysical Foundations of Mathematics*; Loemker, pp. 666–667: "In each of the two orders—that of time and that of space—we can judge relations of *nearer to* and *farther from* between its terms, according as *more* or [*fewer*] middle terms are required to understand the order between them"); and this suggests that Leibniz would choose congruence.

4. "In conflict with *some* of his statements of principle"—namely, those asserting the "equipollence of hypotheses" for more than just uniform rectilinear motions. (I have suggested that Leibniz may have intended the generality here to apply only to the interacting bodies, not to the reference systems. In this case, however, his strictures against Newton—so far as the dynamical issue goes, setting the metaphysics aside—are unwarranted: for, as Leibniz recognizes, Newton admits the dynamical principle of Galilean relativity; and this is absolutely general so far as the interacting bodies are concerned.)

At the opposite extreme, we have a collection of statements in which Leibniz maintains that motion—"or, rather, force" (i.e., *vis viva*—kinetic energy)—must have a determinate subject. *These* statements of principle are quite compatible with the view that rectilinear motion is distinguished; but, so far as physics is concerned, they are essentially equivalent to the theory of absolute space, and are hard to reconcile with *any* version of the "equipollence of hypotheses." The editors of vol. XVI of the *Oeuvres complètes de Christiaan Huygens*, published by the Société hollandaise des Sciences (The Hague, 1929), suggest a resolution of this problem by cutting the knot (p. 199, n. 8 of the cited volume): "Remarquons que chez Leibniz il faut toujours faire une distinction entre le point de vue du métaphysicien et celui du physicien. La force absolue et le 'mouvement absolu véritable' . . . peuvent exister sans que (suivant Leibniz) le physicien puisse les apercevoir." To me this seems a counsel of despair, making nonsense of Leibniz's philosophy. After all, for Leibniz the identity of indiscernibles is a fundamental *metaphysical* principle; how, then, for him, can physically indistinguishable hypotheses be metaphysically distinct?

5. Reichenbach tells us that Leibniz "would argue . . . that the appearance of centrifugal forces on a disk isolated in space proves its rotation relative to the ether and not relative to empty space" (*The Philosophy of Space and Time*, trans. Maria Reichenbach and John Freund [New York: Dover, 1958], p. 212); and then comments (*ibid.*, n. 2): "This view is not precisely formulated by Leibniz, but it may legitimately be extrapolated from a passage in his *Dynamics* (Gerhardt-Pertz, *Leibnizens mathematische Schriften*, VI, 1860, p. 197) and also from his defense of the relativity of motion in the exchange of letters with

Howard Stein

Clarke." I have been unable to find in the letters to Clarke any passage bearing such a construction as Reichenbach suggests; and the page cited in Gerhardt is in the midst of an article on the cause of gravity—an article that bases itself upon Huygens's theory of weight as *an effect of the centrifugal force of a type of vortical motion of the ether,* which is surely not to the point! I should guess that Reichenbach meant to refer either to the *Dynamica* or to the *Specimen Dynamicum,* both of which address the issue of apparent deviations from the "equipollence of hypotheses." In both these works, such apparent deviations are attributed to hidden interactions of some kind with an ambient medium (see, e.g., the last sentence of the long first paragraph under Proposition 19 of the passage from the *Dynamica* given below in Appendix II, p. 42). In the cited passage from the *Dynamica* it is clear beyond a doubt, from the immediate context and from the following Proposition 20, that the hidden interactions in question are those responsible for cohesion. A careful reading of the closely related passage in the *Specimen Dynamicum* (see Loemker, pp. 449–450) reveals the same; except that in this passage gravity is cited as another effect of hidden interaction, and the inhomogeneity—in particular, the varying direction—of the gravitational field is named as the reason why projectile motion on a moving ship is not, when considered with extreme precision, "phenomenologically" the same as on a ship at rest.

The connection Leibniz repeatedly asserts between the equivalence of hypotheses and the nature of cohesion (cf., besides the passages already referred to, the last two sentences of the selection from a letter to Huygens given in Appendix I(d), p. 41) is surely a crux for any interpretation of his dynamical relativism. What seems to me the melancholy truth of the matter is that *no* interpretation resolves this crux: that it stands as evidence of a fundamental confusion in Leibniz's thought on this subject. I have suggested, as alternative constructions of Leibniz's "general" principle of equivalence, that the generality may apply either to the *interactions* or to the *reference systems;* and I have suggested that the former (and more restricted) version has the merit of making sense of Leibniz's first argument for this principle. Supporting evidence for this reading of Leibniz is found in a passage in the *Specimen Dynamicum* (Loemker, p. 445) in which, having praised Descartes for maintaining the purely relative character of motion, Leibniz criticizes him for failing to infer from this principle "that *the equivalence of hypotheses is not changed by the impact of bodies upon each other* and that such rules of motion must be set up that the relative nature of motion is saved, that is, so that the phenomena resulting from the collision provide no basis for determining where there was rest or determinate absolute motion before the collision." Here the statement that the equivalence "is not changed by the impact" clearly means only that *Galilean* relativity—the freedom of choice as to "where there was rest or . . . motion before the collision"—is not broken by collisions. (Descartes's laws of impact grossly violated Galilean relativity.) If this passage is compared with Leibniz's phrase to Huygens (Appendix I(b) below, p. 40) that he has reasons to believe that nothing, including circular motion, "breaks the general law of equivalence," the reading under discussion gains considerable plausibility. *But this reading shatters upon our crux.* For it is explicitly demonstrated by Newton in the *Principia* that Newtonian dynamics satisfies Galilean relativity without qualification; and forces of attraction, or inextensibility of cords, or perfect rigidity, make no difficulties in the matter. So if the reading in question is the right one, Leibniz's statements about cohesion and solidity can only be taken to show that he had failed to understand some of the basic elementary arguments and theorems of Newton's mechanics.

The second interpretation—that Leibniz professed an honest principle of "general relativity"—seems in better accord with his philosophy, and (as we have seen) with his second argument for the equivalence of hypotheses. On the other hand, the difficulties with this interpretation are formidable, as we have also seen; and it, too, fails to solve our crux. I have attempted to put the notion of general relativity and Leibniz's conception of cohesion from impact by ambient particles into intelligible connection in the following way, which may be of some interest although I cannot claim that it finally succeeds: Leibniz says (Appendix II below, p. 42) that if there were true circular motion (as ordinarily conceived) the general equivalence of hypotheses would be violated by the occurrence of

32

centrifugal force on a rotating disk. Could he have believed that, on his view of the nature of cohesion, "rotating reference systems" are without strict physical meaning—because, in that view, there are *in principle* no rigid bodies? Of course, this is hardly a tenable position—for whether the earth, for example, is or is not rigid, one can adopt it as a kinematical reference body. Could, then (to approach something more cogent), Leibniz have reasoned as follows?—If the earth, for instance, or Newton's famous bucket, were a strictly rigid body, then the two systems, earth at rest and earth rotating (or bucket at rest and bucket spinning), would be kinematically equivalent but (as the phenomena of centrifugal force, or Newton's experiment of putting water in the bucket, show) dynamically inequivalent; so in this case the general principle of equipollence of hypotheses would indeed be broken by circular motion. In fact, however, because there is no perfect rigidity, *the two systems in question here are not even kinematically equivalent*: the rotating earth has—necessarily—*a different shape*, that is a different internal geometrical configuration, from that of a nonrotating earth; and the same will be true of the bucket *or of any other body*. Therefore, rotating systems are not examples of dynamically inequivalent *but kinematically equivalent* ones, and the principle of equipollence is unbroken. It seems to me indeed possible that Leibniz's reasoning was something like this; it is, at any rate, as close as I can come to a solution of the crux I have posed. Yet it is a highly problematic solution. On the basis of the texts known to me, it is (to use Reichenbach's term) a rather fargoing "extrapolation." It leaves unresolved the other difficulties that have been discussed, and therefore fails to absolve Leibniz of serious confusion. And so far as the crucial issue itself is concerned, it falls short in two ways: First, although this line of argument depends upon there being no true *rigidity*, it does not in any clear way exclude some sort of true *elasticity* of bodies. And second, for a viable theory to be developed along such lines, it would be necessary to give rules for *determining kinematically*, on the basis of the configurations and changes of configuration of bodies and their parts, those states that are to be taken as "inertial"; but of the existence of such a problem—to say nothing of its solution—there is not a word in the writings of Leibniz. These two defects are not necessarily fatal to the proposal, as an interpretation of Leibniz, for we have found no way to avoid attributing to him a defective theory; but it can claim nothing like the satisfactory status argued for in section II below, in the interpretation there offered of the position of Huygens. We are left, for Leibniz, in an ambiguity of confusions.

6. *Oeuvres complètes de Huygens*, vol. VI, 1895, p. 481.

7. *Ibid.*, p. 327.

8. It will be seen presently that some qualification is required here. The quoted statement of Huygens is not altogether explicit; it is argued below that that statement, taken quite literally, continued to "express Huygens's view" even when his view changed and no longer agreed with Newton's.

9. See Appendix I below for the pertinent passages of the Huygens-Leibniz correspondence. I have previously commented on these matters, and especially on Huygens's theory of relative motion, in my paper "Newtonian Space-Time"; see Robert Palter, ed., *The Annus Mirabilis of Sir Isaac Newton* (Cambridge, Mass.: MIT Press, 1970), pp. 258ff.

10. Appendix III below contains translations of three of these notes of Huygens, from the *Oeuvres complètes*, vol. XVI. The piece published as No. III of this set (*ibid.*, pp. 222–223) is assigned by the editors to the year 1688, since, itself undated, it occurs in a manuscript between pages dated 27 March 1688 and 8 November 1688 respectively. It is tempting (and not impossible) to construct an argument suggesting that No. III is an earlier note than the others here translated, and that their contents involve a significant advance, which may have been made at a date that accords with the statement to Leibniz; but this would be skating on rather thin ice. It is at least as plausible that Huygens, in writing to Leibniz, used the phrase "2 ou 3 ans" in the imprecise sense of "a few years." The best reason for believing that these notes represent the view Huygens speaks of in his letters to Leibniz is that they quite certainly do represent a late change in his conception of motion; that their contents are fully compatible with his statements in the letters; and that it is somewhat unlikely that he should have experienced a further fundamental change of views before the letters to Leibniz and

yet not have included in them a hint of the nature of the change—but unlikely to the second order that, if this *were* true, there should remain no manuscript evidence of it. (It was clearly Huygens's habit to ruminate, in his private notebooks, over his theoretical conceptions, and to preserve his notes. For instance, the editors of vol. XVI of the *Oeuvres complètes* tell us that they are printing only a representative selection of the papers dealing with the same new conception of relative motion; and they print eight such pieces.)

11. Cf., e.g., below, Appendix III(a), last sentence (p. 46), and Appendix III(b), pp. 47–48.

12. I had originally written "until the twentieth century"; but I have learned from an article of Jürgen Ehlers—"The Nature and Structure of Spacetime," in J. Mehra, ed., *The Physicist's Conception of Nature* (Dordrecht-Holland: Reidel, 1973), p. 75—that Ludwig Lange advanced essentially this view in 1885; see his paper "Ueber das Beharrungsgesetz," *Berichte über die Verhandlungen der königlich sächsischen Gesellschaft der Wissenschaften zu Leipzig*, Mathematisch-physische Klasse, vol. XXXVII (1885), pp. 333ff.

13. See my paper, already cited (n. 7 above), "Newtonian Space-Time"; and "On the Notion of Field in Newton, Maxwell, and Beyond," in Roger H. Stuewer, ed., *Historical and Philosophical Perspectives of Science*, Minnesota Studies in the Philosophy of Science, vol. V (Minneapolis: University of Minnesota Press, 1970), pp. 264ff.

14. Cf., e.g., the following (already quoted in part), from the letter to de Volder of 20 June 1703 (Loemker, p. 531): "[I]n phenomena or aggregates every new change arises from an impact according to laws prescribed partly by metaphysics, partly by geometry. . . . So any body whatever, taken by itself, is understood to strive in the direction of a tangent, though its continuous motion in a curve may follow from the impressions of other bodies." Also the third letter to Clarke, paragraph 17 (Loemker, p. 684): "I maintain that the attraction of bodies, properly so called, is a miraculous thing, since it cannot be explained by the nature of bodies." (Cf. also the fourth letter to Clarke, paragraph 45; and the fifth letter, paragraphs 112 *et seq.*)

15. See, e.g., his *Traité de la Lumière*, chap. 1, p. 3 of the original edition; *Oeuvres complètes*, vol. XIX (1937), p. 461: "[In considering the production of light and its effects, one finds everywhere] that which assuredly manifests motion [of some matter]; at least in the true Philosophy, in which one conceives the cause of all natural effects by mechanical reasons. Which it is necessary to do in my opinion, or else to renounce all hope of ever understanding anything in Physics."

16. Cited above, n. 14.

17. For the quoted phrase, see *Principia*, Book III, fourth "Rule of Philosophizing."

Of the several passages in which Newton expresses his views about "hypotheses," the one that I consider the clearest and most rounded occurs in a relatively obscure place in his early optical correspondence (letter to Oldenburg for Pardies, 10 June 1672; H. W. Turnbull, ed., *The Correspondence of Isaac Newton*, vol. I [Cambridge: University Press, 1959], p. 164): "The best and safest way of philosophizing seems to be, that we first diligently investigate the properties of things, and establish them by experiments, and then more slowly strive towards Hypotheses for their explanation. For *Hypotheses* should only be fitted to the properties that call for explanation, not made use of for determining them—except so far as they may suggest experiments. And if one were to guess at the truth of things from the mere possibility of *Hypotheses*, I do not see how it would be possible to determine any settled agreement in any science; since it would be always allowable to think up further and further Hypotheses, which will be seen to furnish new difficulties."

18. This is not, of course, to say that Newton was never wrong, or that he was never influenced in his judgment by his theoretical leanings. He was, for example, wrong about the law of double refraction and wrong in asserting the impossibility of correcting the chromatic aberration of lenses; and both these errors arose from too hastily adopted theoretical conclusions. My contention is, rather, that Newton never argued against a rival theory either on the grounds that the phenomena might be otherwise explained, or on the grounds that the theory itself did not admit of "explanation" in some suitable sense—as (for the first part) Hooke attacked Newton's theory of light on the grounds that an explanation in

terms of waves was also possible, Huygens his theory of "mutual" gravitation on the grounds that pressure by an ether might produce gravitation towards a center with no reciprocal action upon that center; and (for the second part) Huygens was reserved towards Newton's theory of the composition of white light, Huygens and Leibniz very adverse to the theory of universal gravitation, on the grounds in each case that a "mechanical" explanation seemed impossible.

19. Ernst Mach, *The Science of Mechanics*, 6th English ed. (La Salle, Ill.: Open Court, 1960), pp. 588–589; *The Analysis of Sensations*, revised English ed. (New York: Dover, 1959), p. 311.

20. *Mechanics*, pp. 588–589: "[T]he mental artifice atom . . . is a product especially devised for the purpose in view. . . . However well fitted atomic theories may be to reproduce certain groups of facts, the physical inquirer who has laid to heart Newton's rules will only admit those theories as *provisional* helps, and will strive to attain, in some more natural way, a satisfactory substitute."

21. See *Mechanics*, pp. 286–287. (The third sentence of paragraph 7 is mistranslated; it should say that the alterations of the mutual distances of remote bodies are proportional *to one another* [not "to those distances"]—i.e., if in two time-intervals, I and II, bodies A and B have distance-changes Δr_I and Δr_{II}, and if bodies C and D have distance-changes Δs_I and Δs_{II} in these same intervals, and if these bodies are all "remote from one another," then $\Delta r_I : \Delta r_{II} = \Delta s_I : \Delta s_{II}$.)

22. See *Mechanics*, pp. 284–285, 293, 340; and Preface to the Seventh German Edition, p. xxviii. (This view seems to be the one Mach held to most firmly; note that it is, for example, restated in the last sentence of the Preface to his last edition.)

23. That this is Mach's intention seems plain from a passage added in a late edition of the *Mechanics* (p. 339), in which Mach suggests that, rather than "refer the laws of motion to absolute space," we may "enunciate them in a perfectly abstract form; that is to say, without specific mention of any system of reference." The import of this last phrase may seem obscure; but the sequel makes clear that it implies in effect an *existential quantification* over reference systems: Mach says that this course "is unprecarious and even practical; for in treating special cases every student of mechanics looks for some serviceable system of reference."

24. *Mechanics*, p. 284. (Thus all three views appear in pp. 284–287.)

25. For a discussion of this point, see my paper (already cited—n. 7) "Newtonian Space-Time."

26. Two astronomical phenomena provide optical evidence of the earth's annual motion relative to the fixed stars: the aberration of starlight and stellar parallax. Aberration was discovered in 1725 by James Bradley, and was explained by him in 1727 (the year of Newton's death). Stellar parallax, sought for since (at least) the time of Tycho Brahe, was not detected until 1838 (by F. W. Bessel).

27. Proposition XI of Book III of the *Principia*, which states "that the common center of gravity of the earth, the sun, and all the planets, is immovable," is asserted under the hypothesis—expressly so labeled—"that the center of the system of the world [i.e., solar system] is immovable": in other words, that there exists a point, definable by its geometrical relations to the earth, sun, and planets, which is immovable. Newton's argument for Proposition XI is that, by the Laws of Motion, the center of gravity must be either at rest or in uniform rectilinear motion; but if the center of gravity has a constant velocity different from zero, there clearly cannot be any fixed point in the system at all; so the hypothesis requires that the center of gravity be at rest. He immediately infers (Proposition XII) that the sun itself is in continual motion. Of course, the whole argument assumes that the solar system is dynamically closed. On this point Newton comments further, in his subsequent, less formal work *The System of the World*, section 8 (see *Sir Isaac Newton's Mathematical Principles of Natural Philosophy and His System of the World*, ed. Florian Cajori [Berkeley, California: University of California Press, 1946], p. 558): "It may be alleged that the sun and planets are impelled by some other force equally and in the direction of parallel lines; but by such a

force (by Cor. VI of the Laws of Motion) no change would happen in the situation of the planets one to another, nor any sensible effect follow: but our business is with the causes of sensible effects. Let us, therefore, neglect every such force as imaginary and precarious, and of no use in the phenomena of the heavens. . . ."

28. See the introductory paragraphs of Helmholtz's paper "Ueber die Thatsachen, die der Geometrie zum Grunde liegen," in his *Wissenschaftliche Abhandlungen*, vol. II (Leipzig: Johann Ambrosius Barth, 1883), pp. 618ff.

29. Some further comments about the theorem of Helmholtz and Lie seem to be called for:

(1) It was known to Riemann, and clearly stated in his *Habilitationsvortrag*, that "free mobility" of figures "without distension" is possible, in a connected manifold with Riemannian metric, if and only if its curvature is uniform (but see (3) below for a qualification of this statement). What is distinctive in the result of Helmholtz is, therefore, not the *sufficiency* of the condition "Riemannian metric of uniform curvature," but the *necessity* of the condition "Riemannian metric," for such free mobility.

(2) However, the sense of this "necessity" requires elucidation (for I have found, in discussing the point with colleagues, that the mathematical facts are less well known than I should have supposed). The work of Lie and his successors has shown that there are several theorems, not just one, "of Helmholtz type"; what they have in common is the following: One defines a certain condition, *C*, upon a manifold, involving a class of mappings (to be called "rigid displacements"); and proves that if the condition *C* is satisfied, then *there is an essentially unique Riemannian structure on the manifold for which the already postulated "rigid displacements" are precisely the isometries of the Riemannian metric.* ("Essentially unique" here means unique up to a constant positive factor—i.e., up to the choice of a "unit of length.") The clause in italics is, therefore, a necessary condition for the condition *C* to hold.

I have been asked whether this result truly establishes "necessity" that a manifold with condition *C be* Riemannian in structure—or whether it merely shows that such a manifold "admits" a suitable Riemannian structure. The issue is perhaps in part merely verbal; but these points should be noted: (a) The *essential uniqueness* stated in the theorem means, in effect, that if one is "given" a manifold *M* with "rigid displacements" satisfying the condition *C*, one is *ipso facto* "given" a Riemannian structure upon that manifold. (b) In ordinary Euclidean geometry, axiomatized with the help of the relation of "congruence," a metrical structure is determined in no other sense than this: the length or distance, satisfying the Pythagorean theorem, is *definable* (up to the choice of a unit) *in terms of the axiomatized relations*, in precise analogy to the Helmholtz-Lie situation. (c) What is perhaps the central thing, from the point of view of Helmholtz, is this: A careful reflection on the proof(s) of the Helmholtz-Lie theorem(s) will show that the Riemannian structure whose existence and (essential) uniqueness is thereby established is *the* metrical structure that is obtained through the "ordinary, standard procedures of measurement" using "freely mobile rigid bodies" as measuring instruments. (d) It might be asked, beyond this, whether there exist, under the stipulated conditions, besides the (essentially unique) Riemannian metric, still other—non-Riemannian—metric structures for which the "rigid displacements" are the isometries. The answer to this question depends in part upon the generality allowed to the notion of a "metric." If one only demands satisfaction of the standard "metric space" axioms for a distance-function, then there are indeed non-Riemannian metrics admitting the same isometries as the Riemannian one (and, in the Euclidean case, *there exist "non-Pythagorean" metrics whose isometries are exactly the Euclidean congruences*). This follows easily from the fact that if *d* is a distance-function on a set, and if *f* is an arbitrary *monotonically increasing* and *concave* real-valued function on the non-negative real numbers, with $f(0) = 0$, then the composite function $f \circ d$ is again a distance-function (and has the same isometries as *d*). But in the spirit of differential geometry, a natural specialization of the notion of a metric is obtained as follows: First, for an *arbitrary* metric space, one can define the notions of a "rectifiable curve" and of the "length" of such a curve in a straightforward way. It is then always true that the length of a rectifiable curve from p to q is greater than or equal to the distance from p to q. Let us call a *metric* "rectifiable" if for each pair of points p and q there exists a rectifiable curve from p to q whose length is as close as one likes to the distance from p to q—i.e., if the distance from p to q is the *greatest lower bound* of the lengths of curves from p to q. (Thus "rectifiability of the metric" entails that the metric space in question is pathwise connected.) Now we can assert that, under "Helmholtz-Lie conditions" C on a connected manifold, *the only* RECTIFIABLE *metrics whose isometries are the "rigid displacements" are the Riemannian metrics, unique up to a unit of length, given by the theorem of Helmholtz and Lie.*

(3) In the older literature of geometry, authors were notoriously careless of the distinction between "local" and "global" structures; and this leads to some difficulties in our present context. Thus in the work of Helmholtz and Lie—and, indeed, in all the later work known to me on the "Helmholtz-Lie problem"—the notion of "rigid displacement" is taken to mean a certain class of mappings *of the whole manifold onto itself*. But if the notion is so construed, it is not true that uniform Riemannian curvature is a sufficient condition for free mobility under rigid displacement: there are required, in addition, certain "global" conditions on the manifold (finiteness of the fundamental group, and metric completeness). Moreover, from the quasi-epistemological point of view from which Helmholtz began, based upon considerations of spatial *measurement*, this "global" construction of the notion of rigid displacement is inappropriate: for purposes of measurement, one does not carry around the whole space—one carries around only a small measuring body. It therefore seems worthwhile to give here a strictly local form of the Helmholtz-Lie theorem (which does in fact single out all and only the Riemannian manifolds of constant curvature).

We suppose given, then, a connected n-dimensional differentiable manifold M, and a class R of mappings (the "rigid displacements"), each of which is *a diffeomorphism into M, defined on a domain* (i.e., connected open set) in M. These are required to satisfy the following conditions:

(a) "Local character" of rigid displacement: A diffeomorphism f defined on a domain U is in R if and only if for each p in U there is a neighborhood V of p such that for any domain W contained in V the restriction of f to W is in R.

(Intuitively: a diffeomorphism f is a rigid displacement on U if and only if it is a rigid displacement on all small enough parts of U.)

(b) "Group-like properties" of the set of rigid displacements: If f, defined on U, is in R, then f^{-1} (defined on the domain $f(U)$) is in R; if f, defined on U, and g, defined on $f(U)$, are both in R, then so is the composite mapping $g \circ f$ (defined, again, on U).

To facilitate formulation of the requirement of "free mobility," it is convenient to introduce the notion of a "flag" on the manifold M: namely, a sequence $a = (a_0, a_1, \ldots, a_{n-1})$, with a_0 a point ("0-dimensional subspace") of M, a_1 a 1-dimensional subspace of the tangent space to M at a_0, a_2 a 2-dimensional subspace of that tangent space *containing* the 1-dimensional subspace a_1, and so forth up to the $(n-1)$-dimensional subspace a_{n-1} (which contains all its predecessors). The flag $a = (a_0, a_1, \ldots, a_{n-1})$ is called a flag "at the point a_0." The set of all flags on M has itself a natural manifold-structure (of $n(n+1)/2$ dimensions), and the set of all flags at a fixed point a_0 is a submanifold (of $n(n-1)/2$ dimensions).

(c) "Condition of free mobility": For every pair of points (p, q) of M, there exist domains U, V—neighborhoods of p and q respectively—such that:
(i) "Freedom" and "rigidity": For every flag a at p and every flag b at a point in V, there exists a unique $f_{(a, b)}$ in R, defined on U, under which a is carried to b.

(Intuitively: *some* ["small enough"] body at p [namely, U] can be moved freely everywhere near q, and rotated every which way; but given the "where" and the "way"— specified by the flag b—the displacement is rigidly determined.)

Howard Stein

(ii) "Continuity": For a fixed flag a at p, the mapping $(b, v) \rightarrow f_{(a, b)*}(v)$ (with b a flag at a point in V, v a tangent vector at a point in U, and $f_{(a, b)*}$ the "induced" mapping on the tangent bundle) is continuous.

(In effect, this can be construed as saying that if b' is near b, the rigid displacement that takes a to b' is near the one that takes a to b, where "nearness" of the differentiable maps is measured by their effect both upon points and upon tangent vectors.)

This completes the exposition of our "local Helmholtz-Lie condition C"; the conclusion now follows, as already stated in (2): that there exists an essentially unique Riemannian metric on M, such that a diffeomorphism f of a domain U into M is in R if and only if f is a Riemannian isometry.

The original proof of this theorem—in its weaker, therefore *easier*, "global" version—proceeded by a rather formidable induction on the dimension of M, using properties of the projective spaces. With more modern techniques, however a conceptually rather straightforward proof is possible, and a sketch of the proof will now be given: We first choose a fixed point p, and take $q = p$ in (c) above. Fixing also a flag a at p, we write g_b instead of $f_{(a, b)*}$; then g_b is a nonsingular linear transformation on the tangent space at p, and by (c) (i) and (ii) the mapping $b \rightarrow g_b$ is a 1-1 continuous map of the flag-manifold at p into the Lie group of all such transformations. Since the flag-manifold at a point is compact, so is its image set in the linear group. But using (a) and (b), one easily shows that the image set is a *subgroup* of the linear group, and is therefore itself a *compact Lie group*. By the standard technique of "averaging" or "integration" over the action of the group, one can then define a *positive-definite quadratic form on the tangent space at p, invariant under all "rigid rotations"*; and it follows easily from (c) (i) that this form is unique up to a constant positive factor. Finally, the rigid displacements "from p to q" can be used to "transport" this form over the entire manifold, giving *a Riemannian structure invariant under all the (local) rigid displacements*; and with this, since "essential uniqueness" follows at once from the construction, the proof is complete.

(4) Our formulation of the condition of free mobility in (3) depended essentially upon the *differentiable* structure of the manifold M; for only this structure gives us the notions of "tangent space" and "flag." Lie gave a second version of the (global) Helmholtz theorem, in which free mobility is expressed without such an "infinitesimal" construction. Nevertheless, Lie's second theorem also involves assumptions of differentiability, because these were presupposed in the very foundations of his theory of "continuous groups of transformations," which provided the tools for his attack upon Helmholtz's problem.

The motive of eliminating all explicit differentiability conditions from the Helmholtz-Lie foundations of geometry has played a noteworthy role in the subsequent history of mathematics; for it was the direct inspiration of the celebrated "Fifth Problem" of Hilbert—the problem *to what extent assumptions of differentiability can be dispensed with in the theory of Lie groups.* This was fifth in the list of twenty-three problems posed by Hilbert at the International Congress of Mathematicians in Paris in 1900; its complete solution—showing that the differentiability assumptions in question can be dispensed with *entirely*—was obtained (after a number of preliminary advances by a number of illustrious mathematicians) in 1952 by Andrew M. Gleason, Deane Montgomery, and Leo Zippin. Hilbert himself introduced this problem in explicit connection with the Helmholtz-Lie foundations of geometry (see his *Gesammelte Abhandlungen*, 2nd ed. [Berlin: Springer-Verlag, 1970], vol. III, p. 304); and when, in 1902, in a paper "Über die Grundlagen der Geometrie" (*Mathematische Annalen* 56 (1903), pp. 381–422; also published as Anhang IV in his book, *Grundlagen der Geometrie*, 7th ed. (Leipzig: B. G. Teubner, 1930), Hilbert gave an axiomatization of geometry in two dimensions on the basis of the group of motions without any assumptions of differentiability, he expressed the view that this work "answers, for the special case of the group of motions in the plane, a general question concerning group geometry, which I have posed in my lecture 'Mathematical Problems,' Göttinger Nachrichten 1900, Problem 5." The generalization of this geometrical result to n dimensions finally became possible as a consequence of the results of Gleason, Montgomery, and Zippin—cf. the article of Hans

38

Freudenthal, "Neuere Fassungen des Riemann-Helmholtz-Lieschen Raumproblems," *Mathematische Zeitschrift* 63 (1955–56); so the solution of Hilbert's group-theoretic problem did lead to the geometrical result he had hoped for.

30. Departures from strict translation occur only in the material within brackets.

31. But it should be noted that in one important point Riemann's anticipation has not proved correct: despite great efforts to account for the structure of the physical world "in the small" with the help of the space-time curvatures, no satisfactory account of this kind has been achieved; i.e., we have no evidence of strong fluctuations of curvature in regions of microscopic scale, averaging out on the scale of ordinary bodies, such as Riemann foresaw. Instead it is after all "in the large" that we have come to know phenomena which "admit of simpler explanation" through the assumption that "the measure-relations . . . are not in accord with the assumptions of [ordinary] geometry."

Appendix

For convenience of reference, there are added here translations of passages from the writings of Huygens and of Leibniz which may not otherwise be easily accessible.

I. From the Correspondence of Huygens and Leibniz.

(a) From a letter of Huygens to Leibniz, dated 29 May 1694:

I shall not touch this time on our question of the void and of atoms, having already been too lengthy, against my intention. I shall only say to you that I have noticed in your notes on des Cartes that you believe it *to be discordant that no real motion is given, but only relative*. Yet I hold this to be very sure, and am not checked by the argument and experiments of Mr. Newton in his Principles of Philosophy, which I know to be in error; and I am eager to see whether he will not make a retraction in the new edition of this book, which David Gregorius is to procure. Des Cartes did not sufficiently understand this matter.

(b) From a letter of Leibniz to Huygens, dated $\frac{12}{22}$ June 1694:

As to the difference between absolute and relative motion, I believe that if motion, or rather the moving force of bodies, is something real, as it seems one must acknowledge, it is quite necessary that it have a *subject*. For, *a* and *b* moving towards one another, I maintain that all the phenomena will occur in the same way, in whichever of them one posits motion or rest; and if there were 1000 bodies, I remain convinced that the phenomena could not furnish to us (nor even to the angels) an infallible criterion for determining the subject of motion or its degree; and that any of them could be considered by itself as being at rest; and this I believe is all that you ask. But (I believe) you will not deny that in truth each has a certain degree of motion—or, if you will, of force—notwithstanding the equivalence of hypotheses. It is true that I infer this consequence, that there is in nature some other thing than what Geometry can there determine. And among many arguments of which I make use to prove that, besides extension and its variations (which are purely geometrical things), it is necessary to recognize something higher, namely force, this one is not the least. Mr. Newton recognizes the equivalence of hypotheses in the case of rectilinear motions; but in respect of the circular ones, he believes that the effort of circulating bodies to increase their distance from the center or axis of circulation manifests their absolute motion. But I have reasons that make me believe that nothing breaks the general law of equivalence. It seems to me nevertheless that you yourself, Monsieur, were formerly of the sentiment of Mr. Newton with respect to circular motion.

(c) From a letter of Huygens to Leibniz, dated 24 August 1694:

. . . As to what concerns absolute and relative motion, I am amazed at your memory—that you recall that I used to be of Mr. Newton's opinion in regard to circular motion. Which is so, and it is only 2 or 3 years since I have found what is truer—from which it seems that you too are now not far, except that you would have it, when several bodies are in mutual relative motion, that they have each a certain degree of veritable motion, or of force; in which I am not at all of your opinion.

(d) From a letter of Leibniz to Huygens, dated $\frac{4}{14}$ September 1694:

. . . When I told you one day in Paris that one would be hard put to it to know the veritable subject of motion, you answered me that this was possible by means of circular motion, which gave me pause; and I recalled it in reading almost the same thing in the book of Mr. Newton; but this was when I already believed that I saw circular motion to have no privilege in this respect. And I see that you are of the same opinion. I hold therefore that all hypotheses are equivalent, and when I assign certain motions to certain bodies, I neither have nor can have any other reason than the simplicity of the hypothesis, believing that one may take the simplest (all things considered) for the true one. Having thus no other criterion, I believe that the difference between us is only in the manner of speaking, which I seek to accommodate to common usage as much as I can, *salva veritate*. I am even not far from your own, and in a little paper that I sent to Mr. Viviani and which seemed to me suited to persuade Messrs. of Rome to license the opinion of Copernicus, I accommodated myself to it. Nevertheless, if you have these opinions about the reality of motion, I imagine that you must have opinions about the nature of bodies different from the customary ones. I have on this subject very singular views, which seem to me demonstrated. . . .

II. From Part II, Section 4 of Leibniz's *Dynamica*.

Proposition 19.

The Law of Nature that we have established of the equipollence of hypotheses—that a Hypothesis once corresponding to present phenomena will then always correspond to subsequent phenomena—is true not only in rectilinear motions (as we have already shown), but universally: no matter how the bodies act among themselves; but provided that the system of bodies does not communicate with others, i.e., that no external agent supervenes.

[Note: The explicative material that follows the colon might belong instead to the clause set off by dashes: i.e., it may either (as put above) amplify "universally," or further explain the notion of "equipollence of hypotheses" itself; the Latin is entirely ambiguous on this point.]

This is demonstrated from prop. 16 [note: there is no Proposition 16(!) —Proposition 17 is evidently intended; or rather, all propositions printed by Gerhardt with numbers greater than 16 should have their numbers reduced by 1, so that the present one should be Proposition 18], namely that all motions are composed of rectilinear uniform ones, for which the thing is so by prop. 14. But the same is demonstrated in another way from the general Axiom, that of those things whose determinants cannot be distinguished, the determinates cannot be distinguished either. And so, since in the cause or antecedent state the diverse hypotheses cannot be distinguished, namely insofar as the bodies are carried by free rectilinear motions, they clearly cannot be distinguished either, in any way, in the effects or subsequent states; nor, therefore, in

41

collisions or any other events, even if some motions are perhaps converted from rectilinear to circular through the cohesion or solidity of bodies, or through restraining cords. Since, therefore, all motions—even circular or other curvilinear ones—can arise from preceding rectilinear uniform motions, changed into curvilinear ones perhaps by thrown cords; and since a motion once given, no matter how it was first produced, ought now to have the same outcome as another that is in all ways like it, even though otherwise produced; therefore in general Hypotheses can be distinguished in mathematical rigor by no phenomena ever. Universally, when motion occurs, we find nothing in bodies by which it could be determined except change of situation, which always consists in relationship. Therefore motion by its nature is relative. And these things are understood with Mathematical rigor. However, we ascribe motion to bodies according to those hypotheses by which they are most aptly explained; nor is a hypothesis true in any other sense than that of aptness. Thus, when a ship is borne on the sea in full sail, it is possible to explain all the phenomena exactly, by supposing the ship to be at rest and devising for all the bodies of the Universe motions agreeing with this hypothesis. But although no mathematical demonstration could refute this, it would still be inept. I remember, indeed, that a certain illustrious man formerly considered that the seat or subject of motion cannot (to be sure) be discerned on the basis of rectilinear motions, but that it can on the basis of curvilinear ones, because the things that are truly moved tend to recede from the center of their motion. And I acknowledge that these things would be so, if there were anything in the nature of a cord or of solidity, and therefore of circular motion as it is commonly conceived. [But] in truth, if all things are considered exactly, it is found that circular motions are nothing but compositions of rectilinear ones, and that there are in Nature no other cords than these laws of motion themselves. And therefore if ever the equipollence of hypotheses is not apparent to us it is because sometimes all events are not apparent, on account of the imperceptibility of the ambient bodies; and often some system of bodies seems not to be communicating with others, although the contrary is the case.

Moreover (what is worth mentioning), from this single principle, that motion by its nature is relative and therefore all hypotheses that once agree produce always the same effects, it would have been possible to demonstrate the other laws of Nature expounded so far.

Proposition 20.

The solidity or cohesion of the parts of bodies arises from the motion or tendency of striking of one body against another.

For (by prop. 17) all motions are rectilinear uniform ones compounded together. But if the solidity of bodies comes from anything but composition of motions, rotation too will derive from something other than composition, namely from that very necessity by which it follows from the hypothesis of solidity. And so indeed if a straight line that is corporeal or endowed with density and is solid, LM, is struck simultaneously in its extremities L and M, with equal respective force of contrary motions AL, BM, by bodies A and B, it is necessarily, by the advance of the bodies, put in rotation about its midpoint N; but in this way matter near L or M tending to recede from the center N will be retained solely by the solidity of the body, not by contrary impressed motion; and, therefore, this circular motion does not consist in a composition of rectilinear ones, unless we explain that solidity by a certain motion of pressing. The same is shown from prop. 19, which we have demonstrated not only from prop. 17 but from another different ground; and from this conversely prop. 17, together with the present 20, would (in a certain regress) be demonstrated from prop. 19 in another way than above. Doubtless, since it is shown in prop. 19 through the relative nature of motion that hypotheses are indiscernible, it cannot be known whether some particular body is rotated; but if we posit solidity, and therefore rotations not derived from the composition of rectilinear motions, a criterion to discern absolute motion from rest is given. Indeed, let body ACB rotate about its center C, near the row of points ADB [which points are themselves disposed in a circle about C as center], and now suppose the solidity of the body to be dissolved so that its extreme part A is separated by the breach of the connection: it will go along the straight [tangent] line towards E, if the motion of the body was a true one [note: "*versus*," in Gerhardt's text here, is an obvious error for "*verus*"]; if it was merely apparent, part A will stay with the remainder of the body ACB, notwithstanding the dissolution of the connection. And so we should possess a necessary ground of discerning true motion from apparent, against prop. 19. And this will not be avoided, unless the solidity of the body ACB arises from a pressing in of the bodies around it. Since, then, in this way all motions are rectilinear, and no other rotation has come about than a certain determinate composition of rectilinear motions; and since in purely rectilinear motions, speaking absolutely and of geometrical necessity, hypotheses cannot be discerned from one another (by prop. 19); it follows that they cannot be discerned in rotations either. But let us show more distinctly in what way a certain rotation about a center and a pressing in of bodies would arise from the sole impression of rectilinear tendencies. Indeed, let the *mobile* A be going in the direction and with the speed represented by the indefinitely small elementary straight line $_1A_1\alpha$; but let the tendency of the surrounding bodies be continually driving the *mobile* A towards the center C, so that it always keeps the same distance from the latter (namely because otherwise the present motion of the surrounding bodies is disturbed), . . . [there follows here a straightforward account (only slightly obscured by some notational errors in the letters referring to Leibniz's diagram) of uniform circular motion as the result of a suitable combination of tangential velocity and a continual "pressing in" (treated as a sequence of very small impulsive forces) towards the center; concluding:] And so from motion that is *per se* uniform rectilinear, but changed into circular by an added *tendency towards a center*, there arises a circulation also uniform; which is noteworthy, and agrees with experiments. We have therefore explained the conversion of rectilinear motion into circular by compositions of rectilinear tendencies—on which basis alone the equipollence of Hypotheses can be satisfied.

It is certain that the cause of cohesion is to be explained from these things that we understand of bodies—such as are magnitude, figure, rest or motion. But besides motion there is nothing that makes a boundary in a thing.

For let there be a body ABC, whose part AB, struck by a blow coming in the line DE, does not leave BC in its former place but moves with it; the reason of this *dragging* is sought. And for instance if we wish to reduce it to *pushing* by conceiving certain hooks of the body AB to be inserted in handles of the other body BC, or if we imagine certain ropes or fibrous webs or other tangled textures, we have accomplished nothing; because it is asked in return what, then, connects the parts of the fibers or hooklets. But contact alone, or rest of one beside another, or common motion, surely does not suffice; for it cannot be understood why one body drags another from this alone, that it touches it. And universally we understand no reason why a body is moved except this, that two bodies cannot be in the same place, and hence if one is moved then those others also must be

44

moved into whose place it enters. We have demonstrated the same in this place from the laws of Nature. And just as, from the law that change cannot be by a jump, we have shown all bodies to be flexible, or Atoms not to be given; so from the posited general law of Nature, that phenomena must proceed in the same way whatever hypothesis is made concerning the subject of motion, we have shown solidity to arise in no other way than from the composition of motions. If indeed some derive the solidity of bodies from the pressure of the air or ether, on the analogy of two polished tablets which are separated with difficulty, then although this is in some ways true, yet it does not explain the first origins of solidity or cohesion; for there remains the question of the very solidity or cohesion of the tablets. Since, therefore, a mass of matter cannot be discriminated except by motion, it is manifest that the ultimate grounds of the solidity of both the larger and the smaller ones must be sought in this alone.

III. From the Notes of Huygens on the Nature of Motion.

The following are among the notes published by the Société hollandaise des Sciences, in vol. XVI of the *Oeuvres complètes de Christiaan Huygens*, under the heading of "Pièces et Fragments concernant la Question de l'Existence et de la Perceptibilité de 'Mouvement Absolu'."

(a) No. III, assigned to the year 1688 on the basis of its position in the manuscript; from the Latin:

All motion and rest of bodies is relative. Nor without mutual reference of bodies can something be said or understood to be moved or to rest.

For they err who imagine certain spaces unmoved and fixed in the infinitely extended world—whereas that immobility cannot be conceived except with reference to a resting thing.

But the parts of a body can be moved with reference to one another (which is called whirling motion), preserving their distance on account of a bond or an obstacle: on account of a bond, as in the case of a top or the composite of two bodies connected by a cord; on account of an obstacle, as in the case of water swirled in a round vessel.

In this motion the parts tend to recede from one another or from a point defined with reference to them, and this with the greater force the greater is their relative motion. Whence, moreover, judgment can be made of the quality of this relative motion, when it cannot be made from change of distance.

Bodies which are moved with reference to one another are moved truly.

Between two bodies motion is produced by impelling either of them. And the same motion can be produced, whichever of the two is impelled; even though a smaller force is needed if the smaller of the two is impelled.

Any body continues its once received speed with reference to others, which are regarded as at rest, uniformly and along a straight line with reference to those other bodies.

Of rest we have no idea except through relation of bodies.

(b) No. IV, no date determined; from the French:

It must therefore be understood that one knows that bodies are mutually at rest, when being free to move separately, and in no way bound or held together, they maintain their mutual position. Thus if several balls are put on a smooth table and if each remains motionless in its place on the table, then they are at rest among themselves and with respect to that table. I have said that they must be free to move separately because they might maintain their place, being bound together or attached to the table, and yet be in motion among themselves—which may seem strange; but it is in this that the nature of circular motion consists, which occurs when two or more bodies, or the different parts of a single body, are impelled to move in different directions, and their separation is prevented by the bond that holds them together—so that it is relative motion among these bodies or among the parts of a single body, with continual change of direction, but with constancy of distance on account of the bond.

As when two balls A and B, held together by the thread AB, and being mutually at rest (which is judged, according to what has been said, by their rest in relation to other bodies that are free to move and that yet maintain their own position and distance)—if A is pushed towards C and B towards D, the lines AC and BD being perpendicular to AB and in a single plane and the impulsions equal [and oppositely directed (Huygens's diagram shows C and D on opposite sides of the line AB)], then these bodies will move in the circumference of a circle of diameter AB, to wit with respect to the bodies among which A and B were previously at rest. [Note: the anacoluthon is in the original.] Thus A and B will have motion among themselves, that is to say in relation to one another, yet without their mutual position or distance changing.

Without one's being able to say how much the one and the other have of that motion which one commonly calls veritable, and without there being this veritable motion at all—it being nothing but a chimera, and based on a false idea.

It is the same with a single body, e.g. a wheel or globe; except that in the parts of such a body there are all sorts of different directions, not only in parallel lines as here. Now this circular motion is known either by relation to neighboring bodies that are mutually at rest and free; or by the centrifugal force that causes the tension of the thread that binds 2 bodies together—and so their circular motion would be known even if these other bodies did not exist at all. Or else, if there is only one body that rotates, the rotation causes the projection of some bodies that one might place on it; as, if it were a turning table, balls that one put on it outside the center would promptly flee and leave it. And in rotating water in a circular vessel it causes the elevation of the water toward the edges.

One knows by this that the fixed stars are mutually at rest and have received no impulsion at all to go around, because [if they had] they would separate—unless they are stuck in a solid sphere as some people used to believe. Consequently the Earth has received that [rotatory impulsion]. As one knows in another way by the clocks—that is to say, that the earth flings off more strongly toward the Equator.

Now in the circulation of 2 bodies bound by the thread AB one knows that they have received impulsion which has produced their mutual relative motion or direction; but one cannot know, by considering them alone, whether they were pushed equally, or whether only one was pushed. For if A alone had been pushed, the circular motion and the tension of the thread would have followed all the same, although the circle would then have a progressive motion with respect to the other bodies at rest.

That I have therefore shown how in circular motion just as well as in free and straight motion there is nothing but what is relative—in such a way that that is all there is to know [connoitre (H.'s orthography)—i.e., detect or recognize] about motion, and also all that one has any need to know. . . .

They say, we cannot perhaps know in what motion consists, but know only that a body which has received impulsion is moved. I reply that since we have the Idea of motion no otherwise than from change of situation of some body, or of its parts (as in circular motion), toward other bodies, therefore we are unable to imagine motion except by conceiving that change of situation to occur; because motion cannot be conceived to which the idea of motion does not conform. [Note: the last sentence is, in the original, in Latin—which lends it an aspect of enhanced formality.]

(c) No. VIII, no date determined; from the Latin:

Motion is merely relative between bodies.

It is produced by impression in either of them or in both; but, motion once effected, it cannot be discerned in which of them impression has been made. Indeed, absolutely the same effect results from either impression.

True and simple motion of any one whole body can in no way be conceived—what it is—and does not differ from rest of that body.

I long believed that a κριτηριον of true motion is to be had in circular motion, from centrifugal force. For indeed, as to other appearances, it is the same whether some disk or wheel standing next to me is rotated, or whether, that disk standing still, I am carried about its periphery; but if a stone is placed on the circumference it will be projected if the disk is turning—from which, I considered, that circumference is now to be judged to be moved and rotated truly, and not just relatively to something. But that effect manifests only this: that, impression having been made in the circumference, the parts of the wheel have been impelled in different directions by motion relative to one another. So that circular motion is relative [motion] of parts excited in contrary directions but constrained on account of a bond or connection. But can two bodies whose distance remains the same be moved relatively to one another? Indeed, in this way: if increase of distance is prevented. Contrary relative motion [then] truly obtains in the circumference.

It can be discerned whether a straight rod is moved freely and all in one direction (or is at rest, for that is the same thing), or whether its parts have received the impression of contrary motions. . . .

Most consider motion of a body true when it is carried from a determinate and fixed place in cosmic space. Wrongly. For since space is infinitely extended on all sides, what can be the definiteness or immobility of a place? Perhaps they will declare the fixed stars in the Copernican system to be at rest. They are indeed unmoved among themselves; but, taken all together, in respect of what other body will they be said to rest, or how will they differ *in re* from bodies most rapidly moved in some direction? Accordingly a body can neither be said to rest nor to be moved in infinite space, because rest and motion are merely relative.

Rightly enough Descartes, article 29 of the second part. Except that he says the same force and action is required whether AB is carried from the neighborhood of CD or the latter from the neighborhood of the former. Which is then indeed true when AB is equal to CD, but otherwise not at all. Wrongly, too, he defines motion of a body as relative to those immediately touching it. For why not likewise those farthest away?

Indistinguishable Space-Times
and the Fundamental Group

It has recently been noted (Ellis, 1971; Dautcourt, 1971; Ellis and Sciama, 1972; Glymour, 1972; Trautman, 1965) that in some general relativistic cosmologies various global features of space-time may necessarily escape determination. In contrast to classical space-time theories, the fundamental group of space-time may itself be such a feature in a relativistic space-time. A precise account of what it means for two space-times to be "indistinguishable" will permit us to prove some elementary propositions concerning the classification of indistinguishable space-times which have distinct global topologies.

The equations of familiar space-time theories are local and therefore, even assuming a complete affine connection, do not of themselves determine a unique topology for space or for space-time. The equations of Newtonian theory (as given, for example, by Trautman, 1965) permit space-time to have any topology $V \times R$, where R is the reals and V is any three-dimensional manifold admitting a complete Riemannian connection of zero curvature. There are exactly eighteen such distinct space-forms. Many topologically different Newtonian models can be distinguished empirically either by making global journeys through space or by observing systems which have made such journeys. The possibility of such journeys results not solely from the fact that Newtonian theory allows arbitrarily fast causal signals, for even very slow signals can make transits of the universe, given enough time—and if the affine connection is complete, there is always enough time.

But let us look at the case of light. If space is not simply connected—if it is a 3-torus or, let us say, the topological product of a cylinder and the reals—then there will be points p, q and spatial paths α, β such that light can leave p and reach q either by α or by β, and the path $\alpha\beta^{-1}$ will not be

NOTE: I wish to thank the National Science Foundation for support of research under grant GS 41764.

homotopic to a constant map. Thus, as long as there are sources, "ghost images" of the sources will be observable in principle, and the pattern of such images will be determined by the fundamental group of space. For example, suppose space has topology $S \times R \times R$. If an observatory and a star are located on the same cylinder, then light from the star can reach the observatory by (1) going in either of two directions from the source to the observatory, but not spiraling completely around the cylinder, or (2) spiraling completely around the cylinder (in either of two directions) any finite number of times before reaching the observatory. Light that spirals around n times will be dimmer than light that spirals around m times, $m < n$, because it will have traveled farther. Moreover, if m is a large number, the images from m and $m + 1$ spiral paths will appear closer together than will the images of m and $m + 1$ spiral paths for small m. Thus the images of the star will appear to us roughly as they do in the accompanying picture.

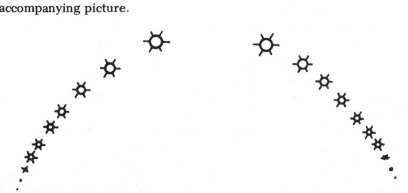

Topologies with a different fundamental group will produce other distinctive patterns. Special relativity, too, admits alternative topologies, but in this case we do not have a classification theorem. If, however, we consider only those product topologies $V \times R$, where R is timelike, then V must be a three-dimensional Euclidean space form. Now in special relativity there are not arbitrarily fast causal chains, but still we can, in principle, always determine something about the global topology of space-time. Let us write $X \ll Y$ if there is a future-directed timelike path from X to Y. Let $I^-(Y)$ be $\{X \in M: X \ll Y\}$ (and similarly, $I^+(Y) = \{X \in M: Y \ll X\}$). Consider the projection map $p: V \times R \to V$ given by $p(v, r) = v$. Then for any $Y \in V \times R$, $p(I(Y)) = V$. This means that at any time an observer in a special relativistic cosmology could in principle determine his topology in exactly the same way that an observer in a Newtonian cosmology with the same topology might do.

In Newtonian and special relativistic cosmologies we are not presented with causally inaccessible regions. In general relativity we often are, even in the simplest of cases. The Schwarzchild solution already contains a region of space-time from which, once inside it, nothing can escape. Early writers on cosmology, Weyl (1922) and Tolman (1934) for example, recognized that the determination of the curvature of space would not of itself determine the global topology of space. They assumed, however, that global topology could be determined by the appearance of ghost images and other phenomena and they did not entertain the possibility that causal inaccessibility might prevent us from making the global discriminations possible in classical cosmologies. To investigate the question we need a precise notion of what it might mean for two general relativistic cosmologies to be indistinguishable. Clearly they must be locally isometric; but as Marder (1962) has shown, that is not sufficient. Now the events of which an observer can have knowledge are exhausted by what happens at points connected to his world-line by future-directed timelike or null curves. So an intuitive requirement for indistinguishability is that there be local isometries which extend over the whole causal past of any world line. This idea can readily be made more precise.

By a space-time we shall mean a four-dimensional differentiable manifold with a smooth pseudo-Riemannian metric form of Lorentz signature. Where convenient, we identify a curve with its image on a manifold.

Definition: Two isochronous space-times, M and N, are indistinguishable if and only if for every maximal curve, σ, on M, whose tangent vector field is everywhere timelike, there is a maximal curve, τ, on N whose tangent vector field is everywhere timelike and $I^-(\sigma) = U_{x \in \sigma} I^-(x)$ is isometric to $I^-(\tau) = U_{y \in \tau} I^-(y)$, and likewise with M and N interchanged.

Remark: $I^-(\tau)$ is an open set. As long as there are no maximal timelike curves with a future-most point, we need not explicitly consider points z connected to a point x by a future-directed null curve because for any isochronous space-time, if z is connected to x by a future-directed null curve, and x is connected to y by a future-directed timelike curve, then z is connected to y by a future-directed timelike curve (see Geroch and Penrose, 1972). It should be clear that indistinguishability is an equivalence relation.

With a little strengthening, the usual cosmological assumptions amount to the requirement that space-time have a product structure, $M = V \times R$, where V is a complete three-dimensional Riemannian space of constant curvature, R the reals, and M carries the pseudo-Riemannian metric form $dt \otimes dt - R^2(t)\, dv \otimes dv$ where dt is the obvious 1-form on the reals, $dv \otimes dv$ is the Riemannian form on V, and $R(t)$ is a smooth function of the real variable t. Such a cosmological model will be called *standard*. We note that all standard space-times are assumed complete. The propositions given subsequently are stated for standard models, but all of them, save the first, apply as well if we require only that the models be products $V \times I$ (where I is an interval of the reals) and hence not necessarily complete. Many of the most popular cosmological models, e.g., the Friedman models, satisfy this weaker condition.

The following simplification is elementary:

Proposition 1: Two standard space-times, M and N, are indistinguishable if and only if for every $x \in M$ ($y \in N$) there is a $y \in N$ ($x \in M$) such that $I^-(x)$ is isometric to $I^-(y)$.

It is easy to give conditions sufficient for a standard model to have an indistinguishable but nonhomeomorphic counterpart.[1] If $M = V \times R$ is standard and (1) Δ is a nontrivial group of isometries acting freely and properly discontinuously on M; (2) for every $\delta \in \Delta$ there is an isometry g on V such that $\delta(v, t) = (g(v), t)$; and (3) for every $\delta \in \Delta$ and for every $x \in M$, if $\delta \neq I$ implies that $I^-(x) \cap I^-(\delta x)$ is empty, then M is indistinguishable from the quotient space-time M/Δ. Such conditions are not, however, sufficiently informative; we should like, in addition, quasi-local topological conditions sufficient to guarantee that a standard model has a covering from which it is indistinguishable. Ideally such conditions should represent the kind of information given by ghost images. It is evident that in general no purely topological conditions together with local isometry will be necessary and sufficient for two standard space-times to be indistinguishable.

Consider a point, x, in a standard model $M = V \times R$. Let $\alpha: [0,1] \to M$, $\beta: [0, 1] \to M$ be curves such that (1) $\alpha(0) = \beta(0) = x$; (2) $\alpha(1) = \beta(1) \in I^-(x)$; and (3) the tangent vector fields to α and β are timelike. The product curve $\alpha\beta^{-1}$ is then a closed loop through x. Let C be the set of all closed loops formed from all pairs of curves, α, β, meeting the above conditions, together with the constant map e: $[0, 1] \to \{x\}$. C is understood to contain

$\beta\alpha^{-1}$ if it contains $\alpha\beta^{-1}$. Now form the class, S, of all products of loops in C. Just as in homotopy theory, we proceed to define an equivalence relation on the curves in S and turn the set of equivalence classes into a group. We take two curves in S to be equivalent if they are homotopic by a homotopy every curve of which is itself in S. That is, curves ω, $\tau \in S$ are equivalent if and only if there is a continuous map $F: [0, 1] \mathbf{X} [0, 1] \to M$ such that

$$F(0, t) = \omega(t)$$
$$F(1, t) = \tau(t)$$
$$F(t, 0) = x$$
$$F(t, 1) = x$$

and for every $u \in [0, 1]$ the curve $\sigma_u(t) = F(u, t)$ is in S. Denote the equivalence class of $\sigma \in S$ by $[\sigma]$ and define the product $[\sigma] \cdot [\tau]$ of $[\sigma]$ and $[\tau]$ to be $[\sigma\tau]$, the equivalence class of the product curve $\sigma\tau$, and similarly define $[\sigma]^{-1}$ to be $[\sigma^{-1}]$. The standard treatments of the fundamental group readily show that the relation on S given above is an equivalence relation, that the product and inverse operations on the equivalence classes are well defined, and that the equivalence classes form a group under the operations. We denote the group thus defined by "$r(x)$."

Proposition 2: Let $M = V \mathbf{X} R$ be a standard model, $x \in M$, $p: M \to V$ the projection map. Then p induces a surjective group homomorphism $p_*: r(x) \to \pi_1(pI^-(x), px)$.

Proof: Since for α, $\beta \in S$, the projection map p generates a homotopy of $p\alpha$ and $p\beta$ from a homotopy of α and β, it is obvious that the map p_* taking the homotopy class of α to the homotopy class of $p\alpha$ is well defined and a homomorphism. To show that p_* is surjective, we argue that any generator of $\pi_1(pI^-(x), px)$ contains a curve that is the composition of two other curves which are the projections of suitable timelike future-directed curves in $I^-(x)$.

Since the space-times in question are standard, whether or not a differentiable curve α in $pI^-(x)$ ending in $p(x)$ is the projection of a timelike future-directed curve in $I^-(x)$ depends only on the length of α. If $pI^-(x)$ is bounded, then there will be a number $r > 0$ such if α has length $< r$, then α is the projection of a timelike curve. In fact, if t_x is the time coordinate of x, we may set $r = \displaystyle\int_{-\infty}^{t_x} \frac{dt}{R(t)}$. (If $I^-(x)$ is not bounded, i.e., $pI^-(x) = V$,

54

then p_* is obviously surjective.) If U is the universal Riemannian covering of V, then any arc-component of $pI^-(x)$ will be an open r-ball in U. Let $q: U \to V$ be the covering map, and let $\tilde{X} \subset U$ be an arc-component of $q^{-1}pI^-(x)$ and denote by $\tilde{x}$ the pre-image of px in $\tilde{X}$. If g is a generator in $\pi_1(pI^-(x), px)$, then there is a curve $\alpha \in g$ such that α is the composition of $q(\tilde{\alpha}_1)$, $q(\tilde{\alpha}_2)$ where $\tilde{\alpha}_1$, $\tilde{\alpha}_2$ are geodesic segments in $\tilde{X}$. $\tilde{\alpha}_1$ is an arc begin-ning at $\tilde{x}$ and ending at a point $\tilde{a} \in X$; $\tilde{\alpha}_2$ is an arc beginning at $\tilde{b} \in \tilde{X}$ and ending at $\tilde{x}$; and $q(\tilde{a}) = q(\tilde{b})$. The length of $\tilde{\alpha}_1$ and of $\tilde{\alpha}_2$ is less than r in each case, and the same must be true of $q(\tilde{\alpha}_1)$ and $q(\tilde{\alpha}_2)$. Thus there are future-directed timelike curves s_1, s_2 in $I^-(x)$ that project onto $q(\tilde{\alpha}_1)$ and $q(\tilde{\alpha}_2)$ respectively, and the element of $r(x)$ containing $s_1 \cdot s_2^{-1}$ is mapped to g by p_*. It follows that p_* is surjective.

The group $r(x)$ could, ideally, be calculated from the information avail-able from ghost images. In Newtonian cosmology $r(x)$ is necessarily isomorphic to the fundamental group of space, but in general relativistic cosmology it need not be. When it is not, the next result gives us a partial classification of indistinguishable counterparts.

Proposition 3: Let $M = V \times R$ be a standard space-time with expansion function $R(t)$. Suppose that for all $x \in M$, $i_*p_*r(x)$ is a proper subgroup of $\pi_1(V, px)$, where $i: pI^-(x) \to V$ is the inclusion map. Let G be any normal subgroup of $\pi_1(V, v)$ such that for all $z \in M$ with $pz = v$, G contains a conjugate of $i_*p_*r(z)$. Then there is a standard space-time $N = \tilde{V} \times R$ indistinguishable from M, and $\pi_1(\tilde{V})$ is isomorphic to G.

Proof: Given a subgroup G of $\pi_1(V, v)$, there is a covering $(\tilde{V}, q)$, $q: \tilde{V} \to V$, of V such that for $q(\tilde{v}) = v$, $q_*\pi_1(\tilde{V}, \tilde{v}) = G^2$ where q_* is the injective homomorphism of the fundamental group induced by q. We claim, first, that if G is a subgroup of $\pi_1(V, v)$ satisfying the hypothesis of the theorem and $(\tilde{V}, q)$ a covering of the kind just mentioned, then for any $z \in M$ such that $pz = v$, $pI^-(z)$ is an admissible set. To show this, it suffices to prove that q, when restricted to any arc-component of $pI^-(z)$, has a continuous inverse.

$pI^-(z)$ is a connected, locally arc-wise connected, open set. Let $\tilde{A}$ be an arc-component of $q^{-1}pI^-(z)$ and let $\tilde{v} \in \tilde{A}$ be such that $q(\tilde{v}) = v = pz$. Let $\tilde{b}$, $\tilde{c} \in \tilde{A}$ and suppose that $q(\tilde{b}) = q(\tilde{c}) = b \in V$. Choose arcs $\tilde{\beta}$, $\tilde{\gamma}$, in $\tilde{A}$ from $\tilde{v}$ to $\tilde{b}$ and to $\tilde{c}$ respectively. Then letting $\beta = q(\tilde{\beta})$ and $\gamma = q(\tilde{\gamma})$, $\beta\gamma^{-1}$ is a loop in $pI^-(z)$ through $pz = v$. The collection of all subgroups $q_*\pi_1(\tilde{V}, \tilde{x})$ for $\tilde{x} \in q^{-1}(v)$ is exactly a conjugacy class of subgroups of $\pi_1(V, v)$.[3] By the hypothesis of the theorem, $i_*p_*r(z)$ is conjugate to a subgroup of $G = q_*\pi_1(\tilde{V}, \tilde{v})$, so by proposition 2, $i_*\pi_1(pI^-(z), pz)$ is conjugate to a subgroup of G. Now the homotopy class $[\beta\gamma^{-1}]$ of $\beta\gamma^{-1}$ is an element of $i_*\pi_1(pI^-(z), pz)$, and therefore there must be some $\tilde{x} \in \tilde{V}$ such that $[\beta\gamma^{-1}] \in q_*\pi_1(\tilde{V}, \tilde{x})$. Since q_* is injective, the corresponding lift of $\beta\gamma^{-1}$ must be closed in $\tilde{V}$. But G is normal, and hence either every lift of $\beta\gamma^{-1}$ is closed or no lift is closed; therefore every lift of $\beta\gamma^{-1}$ is closed. This proves that $b = \tilde{c}$ and hence q restricted to any arc-component of $q^{-1}pI^-(z)$ is injective. Since q is an open continuous map, the restriction of q to any arc-component of $q^{-1}pI^-(z)$ is a homeomorphism.

V is a differentiable manifold, and there is a unique differentiable structure on $\tilde{V}$ for which q is a differentiable map of maximal rank.[4] We take $\tilde{V}$ to be endowed with this structure and, letting u be the metric form on V, define a Riemannian metric $\tilde{u}$ on $\tilde{V}$ by $\tilde{u}(X, Y) = u(q_*X, q_*Y)$ for all vectors, X, Y, in the tangent space of any point in $\tilde{V}$. Then every admissible set is isometric to any of its arc-components, and $\tilde{V}$ is a Riemannian space of the same constant curvature as V. The group D of deck transformations of $(\tilde{V}, q)$ are isometries acting freely and properly discontinuously on $\tilde{V}$; since $pI^-(z)$ is admissible for any $z \in M$, for any arc-component $\tilde{A}$ of $pI^-(z)$, and any $d \in D$, $d \neq I$, $d\tilde{A} \cap \tilde{A}$ is empty.

Now consider the space-time $N = \tilde{V} \times R$ with the same expansion function, $R(t)$, as obtains on M. We claim that N is indistinguishable from M. The group Δ of isometries of N of the form $\delta(\tilde{v}, t) = (d\tilde{v}, t)$ for $d \in D$ acts freely and properly discontinuously on N. Thus if $\phi: N \to M$ is the map defined by $\phi(\tilde{v}, t) = (q(\tilde{v}), t)$, (N, ϕ) is a pseudo-Riemannian covering of M and $M = N | \Delta$, the quotient of N by Δ. It follows from the property of the group D established in the preceding paragraph that for every $y \in N$ and $\delta = (d, I) \in \Delta$, if δ is not the identity, then $I^-(y) \cap I^-(\delta y)$ is empty and $I^-(y)$ is an arc-component of an admissible set $I^-(\phi(y))$ of M. Thus for each $y \in N$ there is an $x \in M$ such that $I^-(y)$ is isometric to $I^-(y)$, and, conversely, so by proposition 1 M and N are indistinguishable.

Proposition 4: Let $M = V \times R$ be a spatially compact standard space-time; suppose compact V has zero curvature and for all $x \in M$, $r(x)$ is trivial. Then for any (topological) space-form S admitting a metric of zero curvature, there is a space-time $N = S \times R$ indistinguishable from M.

Sketch of Proof: By proposition 3 we know that because $r(x)$ is trivial for all $x \in M$, so that $i_*p_*r(x)$ is the identity element and thus a normal subgroup of $i_*\pi_1(pI^-(x), px)$, M is indistinguishable from its universal covering $U = R^3 \mathbf{X} R$. The idea is that for every Euclidean space-form S there is a space-time that is topologically $S \mathbf{X} R$ and indistinguishable from U and hence from M. This will follow if the projection on R^3 of the chronological past of every world-line in U is suitably bounded.

Let q: $R^3 \to V$ be a Riemannian covering map. From the proof of proposition 3 we know that the map ϕ: $U \to M$ given by $\phi(a, t) = (q(a), t)$ is a covering taking the chronological past of every world-line on U isometrically onto the chronological past of some world-line in M. For every γ on M, then, $I^-(\gamma)$ must be an admissible set for the covering (U, ϕ); it follows that $pI^-(\gamma)$ must be an admissible set for the covering (R^3, q) of V, so $pI^-(\gamma)$ must be simply connected.

Let $B(b, r)$ be an open ball in R^3 with radius r containing an arc-component of $pI^-(\gamma)$. Then $B(b, r)$ must also contain the projection $pI^-(\sigma)$ for some world-line σ on U. Moreover, since the volume of $pI^-(x)$ for $x \in M$ is a function only of the time coordinate and does not depend on the location of px in space, the same value of r may be chosen for every world-line σ on M. In fact, r may be taken to be the length of any curve in a generator of the fundamental group of V. It follows that for every world-line σ on U, $pI^-(\sigma)$ is contained in an open ball of radius r.

Now let S' be a (topological) Euclidean space-form. S' is homeomorphic to a space S that is the quotient of R^3 by some group G of isometries of R^3. Every such group is described by a finite set of generators and their relations so that if two sets of generators satisfying these relations generate groups G, G', then R^3/G is diffeomorphic to R^3/G'. For each group, the generators consist of translations and possibly compositions of translations and rotations, and there are no restrictions on how large the translations may be; that is, for any positive n, we can choose translations t in the generators so that the distance from x to $t(x)$ is at least n. Thus we may choose generators so that the shortest distance a point is moved by any generator (and hence by any element of G other than the identity)[5] is at least n. In particular, for any space-form S' we may choose G so that $S = R^3/G$, with S homeomorphic to S', and the shortest distance any point in R^3 is moved by any element of G other than the identity is $2r$.

The quotient space $S = R^3/G$ admits a differentiable and metric structure such that the covering map q: $R^3 \to S$ is differentiable, of maximal rank, and a local isometry. Consider the space-time $N = S \mathbf{X} R$ with the same expansion function as U. Exactly as in the proof of proposition 3, we may define a covering of N by U in terms of the covering of S by R^3, show that for all $y \in N$, $I^-(y)$ is admissible, and thus prove that N and U are indistinguishable. It follows that N and M are indistinguishable, since U is indistinguishable from M and indistinguishability is an equivalence relation.

Proposition 5: Let $M = S^3 \times R$ be standard, and suppose that the length (in radians) of the projection on S^3 of every null geodesic on M is less than or equal to π/n, $n > 0$. Then M is indistinguishable from a space-time $N = S^3/Z_m \times R$, where Z_m is a cyclic group of isometries of order $m \leqslant n$. If $M = S^3/Z_n \times R$ and $r(x)$ is trivial everywhere, then M is indistinguishable from $N = S^3/Z_m \times R$, $m \leqslant n$.

The proof of proposition 5 is omitted, since it involves no new ideas and is immediate from the structure of the groups of isometries (given in Wolf, 1967, p. 224). It should be noted that certain global assumptions will reduce or eliminate the variety of indistinguishable space-times. If it is required that space-times be standard and satisfy the global cosmological principle—that is, that the group of global isometries of space act transitively—then any two indistinguishable space-times of constant negative space curvature are isometric. The possible topologies for standard space-times of zero curvature are reduced to $R^m \times T^{3-m} \times R$, where $m < 3$ and T^{3-m} is the $(3-m)$-dimensional torus. Calabi and Marcus (1962) have shown that if the global perfect cosmological principle—that the group of space-time isometries act transitively—is introduced, then the only complete standard space-times of constant positive curvature are the De Sitter space-times.

We note some examples. The De Sitter model is a hyper-hyperboloid in five-dimensional Minkowski space, with the metric induced therefrom. The metric can be given the standard form

$$ds^2 = dt^2 - \cosh^2(t)(dX^2 + \sin^2 X(d\Omega^2)), \; 0 \leqslant X \leqslant \pi.$$

Consider the class of models, (D, n), of this kind given by the family of expansion functions

$$\cosh^2(nt)$$

where n is a positive integer. The maximum spatial coordinate distance traveled by a light beam leaving a point on the equatorial sphere is

$$\int_0^\infty \frac{dt}{\cosh{(nt)}} = \frac{\pi}{2n} \cdot$$

Thus the maximum spatial distance traveled in all of time is just π/n. By proposition 5 we have that the models (D, n) are indistinguishable from models with topology S^3/Z_m X R, $m \leqslant n$.

Consider the family of spatially open, Euclidean models, M, with cosmological constant $\Lambda \geqslant -8\pi\rho_{00}$. The metric form can be written

$$ds^2 = dt^2 - e^{1/2g(t)}(dx_1^2 + dx_2^2 + dx_3^2)$$

and we have (see Tolman, 1934, p. 403) the differential equation

$$\frac{de^{1/2g(t)2}}{dt} = \frac{8\pi\rho_{00}}{3} e^{g(t)} + \frac{\Lambda}{3} e^{g(t)}$$

which integrates to

$$g = 1/2kt + d$$

d a constant, $k = \dfrac{8\pi\rho_{00}}{3} + \dfrac{\Lambda^{1/2}}{3}$. The radial velocity of a light ray is therefore

$$dr/dt = \pm e^{-1/4(kt + d)}$$

and hence the coordinate distance traveled in all of time by a light ray leaving its source at an arbitrary time is finite. So by the argument of proposition 4, for every time t, and for every Euclidean space-form V, there is a space-time N that is topologically V X R and is indistinguishable from the space-time M_t obtained by deleting from M all points occurring at time t or earlier. All of these space-times are incomplete, but since they are strongly causal it follows from the work of Clarke (1970) that they can be made null complete by a conformal change in the metric.

Notes

1. I shall assume that the reader is familiar with the standard terminology and facts about covering spaces. See, for example, Wolf (1967), section 1.8.

2. See Wolf (1967), p. 39.

3. See Wallace (1967), p. 155.

4. Wolf (1967), p. 41.

5. I omit the argument (for seventeen different cases) that the generators can be so chosen that if n is the shortest distance a point is moved by any generator in G, then every element of G other than the identity moves every point at least a distance d.

Clark Glymour

References

1. Ellis, G. F. R. (1971). "Topology and Cosmology," *General Relativity and Gravitation*, vol. 2, p. 7.
2. Dautcourt, G. V. (1971). "Topology and Local Physical Laws," *General Relativity and Gravitation*, vol. 2, p. 97.
3. Ellis, G. F. R. and D. W. Sciama (1972). "Global and Non-Global Problems in Cosmology." In L. O'Raifeartaigh, ed., *General Relativity*. London: Oxford University Press.
4. Glymour, C. (1972). "Topology, Cosmology and Convention," *Synthèse*, vol. 24, p. 195.
5. Trautman, A. (1965). "Foundations and Current Problems of General Relativity." In A. Trautman et. al., *Brandeis Summer Institute in Theoretical Physics: 1964*. Englewood Cliffs, N.J.: Prentice-Hall.
6. Weyl, H. (1922). *Space-Time-Matter*. London: Methuen.
7. Tolman, R. (1934). *Relativity, Thermodynamics and Cosmology*. London: Oxford University Press.
8. Marder, L. (1962). "Locally Isometric Space-Times." In *Recent Developments in Relativity*. New York: Polish Scientific Publishers.
9. Geroch, R., E. Kronheimer, and R. Penrose (1972). "Ideal Points in Space-Time," *Proceedings of the Royal Society of London A*, vol. 327, p. 545.
10. Wolf, J. (1967). *Space of Constant Curvature*. New York: McGraw-Hill.
11. Wallace, A. (1967). *Algebraic Topology: An Introduction*. New York: Harcourt.
12. Calabi, E. and Marcus, L. (1962). "Relativistic Space-Forms," *Annals of Mathematics*, vol. 75, p. 63.
13. Clarke, C. J. S. (1970). "On the Geodesic Completeness of Causal Space-Times," *Proceeding of the Cambridge Philosophical Society*, vol. 69, p. 319.

Observationally
Indistinguishable Space-times

In his paper "Indistinguishable Space-times and the Fundamental Group"[1] Clark Glymour poses a criterion for the observational indistinguishability of space-time models and presents two sets of examples from the subclass of Robertson-Walker models. The underlying idea is quite intuitive.

In some space-time models studied in relativity theory any particular observer can receive signals from, and hence directly acquire information about, only a limited region of space-time. This happens, for instance, in a rapidly expanding universe in which galaxies that might try to signal one another are actually receding from one another at velocities approaching that of light. It may turn out in these cases that the information from that limited region of space-time which any one observer can have access to is compatible with quite different overall space-time structures. Two space-times are observationally indistinguishable under Glymour's criterion if, for precisely these reasons, no observer in either space-time would have grounds for deciding which of the two, if either, was his. No observer would be able to discriminate observationally between the two even if he did nothing but sit and record signals beamed at him from all directions all day long, even if the signals themselves coded all the spatio-temporal information that the sender had to offer, and even if the observer lived eternally.

Glymour is proposing a reason why the spatio-temporal structure of the universe might be underdetermined by all observational data that we could ever, even just in principle, obtain. Some claims of underdetermination in science are of a very general sort, to the effect that no body of evidence will ever force a particular scientific hypothesis upon us

NOTE: Most of the ideas in this paper arose in conversation with Robert Geroch and Clark Glymour. I have not hesitated to incorporate their many contributions. I am grateful to both.

if only we are prepared to make sufficiently sweeping revisions in relevant theory. Along these lines one might argue, and some have argued, that we can always ascribe more than one topology to space-time if we are imaginative in the invocation of phantom effects such as the duplication of all events at a distance. Such a claim may or may not be irrefutable. But it is certainly quite different from the claim that we have a choice in the ascription of topology *even without giving up any of our trustiest theories*. This is the possibility that Glymour is suggesting, if only tentatively. He makes no argument to the effect that the space-time structure of our universe, i.e., the real one, is in fact one of a pair of observationally indistinguishable space-times. The point is rather that if it were an element of such a pair, then it would be underdetermined by all observational evidence. We would then be unable to determine its global structure even if we adamantly insisted on holding on to our best theories and exploiting them fully in the attempt.

Rather than comment on the details of Glymour's proofs, I want to try to make the geometric ideas in his paper more perspicuous by considering several very simple, easily visualized examples of observationally indistinguishable space-times in two and three dimensions. In doing so I shall establish a few simple results concerning the invariance of global properties of space-times under the relation of observational indistinguishability. I shall also discuss several other relations concerning observational indistinguishability, some weaker and some stronger. The upshot of my remarks will be that the cosmologist's predicament is even worse than one thought at first. Observational underdetermination of one sort or another is more the rule than the exception.

I

Let me first rehearse a few definitions.[2] An n-*dimensional space-time* (for $n \geq 2$) is taken to be a connected, smooth, n-dimensional differentiable manifold (without boundary), endowed with a smooth, nondegenerate pseudo-Riemannian metric of Lorentz signature $(+, -, \ldots, -)$. The metric associates with each point a light cone (in the tangent space at that point). It is assumed that space-times are *temporally oriented*, i.e., that they are further endowed with a continuous, nonvanishing vector field which assigns a timelike vector to every point. The vector field distinguishes a "future lobe" in the light cone at each point.

Given two points x and y, we say y is to the *timelike future* of x and write $x \ll y$ if there is a piecewise smooth curve from x to y whose tangent vector (or vectors) at each point lies inside the future lobe of the light cone at that point—in short, if there is a *future-directed timelike curve* from x to y. If in the definition tangent vectors are permitted to be on the boundary as well as in the interior of the light cone, y is said to be the *causal future* of x, and we write $x < y$. In this case the connecting curve is called a *future-directed causal curve*. The relation $x < y$ is usually interpreted to mean that it is possible for a signal to travel from x to y; $x \ll y$ is interpreted to mean that it is possible for a heavier than light particle to make the trip. Associated with each relation is its respective past and future sets: $I^-(z) = \{y: y \ll z\}$, $I^+(z) = \{y: z \ll y\}$, $J^-(z) = \{y: y < z\}$, and $J^+(z) = \{y: z < y\}$. The I sets are open (in the space-time manifold topology) and are for this reason somewhat easier to work with than the J sets, which are in general neither open nor closed. The set $I^-(z)$ is called the *observational past of* z; it consists of those points in space-time which can possibly send a (slower-than-light) signal to z.

We can associate with each *observer* his space-time trajectory or cosmic world-line which is itself, necessarily, a future-directed timelike curve. If σ is such a world-line, the *observational past of* σ is just the union: $I^-[\sigma] = \cup \{I^-(x): x \in \sigma\}$. The idea that an observer live "eternally" is captured in the condition that his associated world-line be *future-inextendible*, i.e., that as a curve in the space-time manifold, it be extended as far as possible into the future. Such curves, by definition, have no "future end point."[3] That an observer *have* lived eternally could be captured, symmetrically, in the condition that his world-line be *past inextendible*. But as far as capacity for observation is concerned, no advantage comes through this kind of longevity. If x and y are successive points on a world-line, then $x \ll y$ and $I^-(x) \subset I^-(y)$ by the transitivity of the relation $\ll$.

We now have all the components for Glymour's definition of observational indistinguishability:

Definition: Two space-times M and M' are *observationally indistinguishable* if for every future-directed, future-inextendible, timelike curve σ in M there is a curve σ' of the same type in M' such that $I^-[\sigma]$ and $I^-[\sigma']$ are isometric; *and*, correspondingly, with the roles of M and M' interchanged.

The condition that $I^-[\sigma]$ be isometric to $I^-[\sigma']$ formalizes the condition that the portion of M which σ can possibly see over the course of his eternal lifetime is, "space-time-wise," identical with that portion of M' which σ' can possibly see over the course of his lifetime. The definition is mathematically well formed since both $I^-[\sigma]$ and $I^-[\sigma']$ are pseudo-Riemannian manifolds (without boundary) in their own right.

The simplest example of a space-time which admits no observationally indistinguishable counterpart is Minkowski space-time. Here the observational past of every future-inextendible timelike curve σ is the entire manifold; equivalently, $\text{Bnd}(I^-[\sigma]) = \phi$. The set $\text{Bnd}(I^-[\sigma])$, the boundary of $I^-[\sigma]$, may be termed the *observational horizon of* σ (the expression "event horizon" is more common).

One need not look far to find a space-time in which *all* observers have observational horizons. The light cones in Minkowski space-time are all fixed at 45°. Consider the two-dimensional plane in standard t, x coordinates with a metric whose associated light cones, while situated at 45° for t = 0, rapidly narrow to the vertical as t increases in absolute value. For example, although others would serve just as well, let the metric be $ds^2 = dt^2 - (\cosh^2 t)\, dx^2$. (Recall that $\cosh t = \frac{1}{2}(e^t + e^{-t})$). Because the cones collapse, null geodesics (trajectories of light rays) will be confined to a region of space-time of bounded x-width (see Figure 1). For the particular metric cited, they will be confined to a region of x-width π. Correspondingly, the observational past of every future-inextendible timelike curve will be confined to a region of x-width 2π.

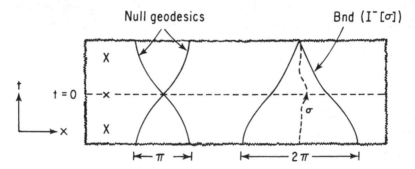

Null geodesics   Bnd $(I^-[\sigma])$

Figure 1. The covering space of two-dimensional De Sitter space-time, i.e., the t, x plane with metric $ds^2 = dt^2 - (\cosh^2 t)\, dx^2$. Light cones narrow rapidly to the vertical as $|t| \to \infty$. Every future-extendible timelike curve σ has an observational horizon of x-width 2π. Two-dimensional De Sitter space-time arises by identifying all points $(t, x + 2n\pi)$ for integers n. By introducing coordinates $\bar{t} = \sinh t$, $\bar{x} = (\cosh t)(\cos x)$, $\bar{y} = (\cosh t)(\sin x)$, it assumes the familiar form of a hyperboloid of one sheet $-\bar{t}^2 + \bar{x}^2 + \bar{y}^2 = 1$ in $\mathbf{R}^3$ with metric $ds^2 = -d\bar{t}^1 + d\bar{x}^2 + d\bar{y}^2$.

Now let M be the two-dimensional space-time just described. Let M' be the result of cutting a vertical strip in M of x-width 2π and identifying opposite sides. It is clear that M and M' must be observationally indistinguishable from one another. Since no observer in either M or M' can see beyond his 2π horizons, none will be able to determine which of the two, if either, is his. (Incidentally, M' is the two-dimensional version of De Sitter space-time.)

This first example exhibits the general features of Glymour's construction. He considers a subclass of space-times which he calls *standard*. These, like M and M', are manifolds topologically of form **R** X V carrying Robertson-Walker metrics $ds^2 = dt^2 - R(t)^2 d\sigma^2$, where $d\sigma^2$ is a smooth, complete Riemannian metric of constant curvature on V and is independent of t. Within the class he finds space-times **R** X V which admit nice families of isometries, and forms new space-times **R** X V' by taking quotient manifolds under them. The essential requirement on the isometries is that they move points sufficiently far so that the observational past of every point is disjoint from the observational pasts of all of its image points. In the present example V is **R**, V' is S^1, and the isometry in question is just the translation: $(t, x) \rightarrow (t, x + 2\pi)$.

All the examples of observationally indistinguishable space-time pairs which Glymour generates with this construction are such that either one is a covering space of the other or they share a common covering space. But examples can easily be given in which this is not the case. The building blocks for one are vertical slabs of two types, A and B, both cut from the plane. In standard t, x coordinates, both may be taken to be the set $\{(t, x): 0 < x < 2\pi\}$. Slabs A will carry the De Sitter metric from the first example: $ds^2 = dt^2 - (\cosh^2 t)\, dx^2$. Slabs B will carry the metric $ds^2 = dt^2 - (\cosh^2 t)(1 + x(2\pi - x))\, dx^2$. This metric shares the property that its associated light cones collapse to the vertical as t increases in absolute value. It approaches the metric of A smoothly along its borders so that when the two slabs are glued together (with an appropriate common boundary line inserted), the resulting double slab carries a smooth metric. The B metric also has an extra wiggle factor inserted which further narrows and then restores the cones in moving from $x = 0$ to $x = 2\pi$ for any fixed value of t. One could equally well use any other smooth wiggle factor. The point is simply to distinguish the two slabs metrically.

Now we form space-times M and M' by taking two nonisomorphic $-\omega$ $+ \omega$ sequences of A and B slabs and gluing them together. One such sequence might be . . . *ABABBA* . . . . Any sequences will do if at least one token of each slab type occurs in each. The observational past of any observer in either *M* or *M'* will be restricted to an *A* slab, a *B* slab, or an *AB* or *BA* double slab (see Figure 2). In each case, his observational past is compatible with both space-time structures. Hence *M* and *M'* are observationally indistinguishable. But by our initial choice of sequences, they are not isometric. As they stand, *M* and *M'* are homeomorphic (i.e., they have the same topology, that of the Euclidean plane). But with a simple variation on the slab theme we could distinguish *M* and *M'* topologically as well. We would need to distinguish only the component *A* and *B* slabs topologically.

Next I want to consider the condition of "observational indistinguishability after finite time." As Glymour's definition is formulated, space-times can be observationally distinguishable from each other without an observer in either one necessarily being able to distinguish between them *at any time* during the course of his life. It is sufficient that the composite, lifelong, integrated knowledge of one observer distinguish between them. A weaker condition of observational indistinguishability which Glymour considers insists that observational distinction between space-times be made *within* the lifetime of some observer.

Definition: Two space-times *M* and *M'* are *observationally indistinguishable after finite time* if for every point *x* in *M* there is a point *x'* in *M'* such that $I^-(x)$ and $I^-(x')$ are isometric; *and*, correspondingly, with the roles of *M* and *M'* are interchanged.

This seems the more natural way to formulate the condition.

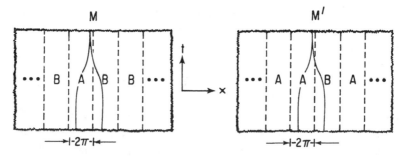

Figure 2. M and *M'* are observationally indistinguishable, although they do not share a common (metric-preserving) covering space. The observational past of a typical observer in *M* is indicated together with his counterpart in *M'*.

The two notions of observational indistinguishability are certainly not equivalent. Consider, for example, simple two-dimensional Minkowski space-time and "truncated" Minkowski space-time consisting of that portion of the former below the x-axis. The observational pasts of all points in both space-times are isometric, so the two are certainly observationally indistinguishable after finite time. But they are not observationally indistinguishable. As noted above, Minkowski space-time has no observationally indistinguishable counterpart.

There seems to be something rather unsatisfactory about truncated Minkowski space-time. Perhaps it violates our sense that the universe should satisfy what Leibniz called the "principle of plenitude." However compelling the metaphysics, a condition of *inextendibility* is often imposed on space-times. It is the condition that it not be possible to embed the space-time isometrically in another without the two being isometric. Clearly truncated Minkowski space-time *is* extendible.

If we restrict our attention to inextendible space-times, then it becomes more difficult to show that space-time pairs can be observationally indistinguishable after finite time while not observationally indistinguishable. In fact, as Glymour points out, within the class of inextendible standard space-times the two conditions are equivalent. But if standardness is not also demanded, examples showing the difference in strength are still available.

One such is found by elaborating the slab construction from the second example. Consider this time vertical "half slabs" of two types, A and B. Each is respectively that portion of its earlier counterpart falling beneath the x-axis. Let S be the set of all finite sequences of A and B slabs and consider $-\omega + \omega$ sequences in S, i.e., sequences of the form $\ldots S_{-2} S_{-1} S_0 S_{+1} S_{+2} \ldots$, which include all elements of S. Take two, in particular, which are distinct in the sense that their underlying composite sequences of A's and B's are not isomorphic. Glue all these slabs together nicely and finally glue to both of them "on top" the upper half of two-dimensional Minkowski space-time. The resulting mosaics are inextendible space-times (see Figure 3). Given any point in either M or M', its observational past intersects with only finitely many adjacent slabs and so has an isometric counterpart in the other space-time. Thus M and M' are observationally indistinguishable after finite time. But any future-inextendible timelike curve in either space-time will include in its observational past the entire manifold; hence it will include the entire finite

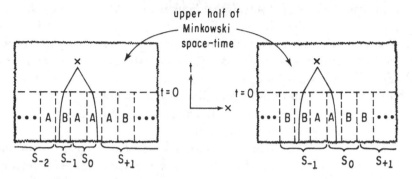

Figure 3. M and M' are observationally indistinguishable after finite time, but not observationally indistinguishable. The observational past of a typical point x in M is indicated together with its counterpart x' in M'.

sequence of A and B slabs which, by our initial choice, finds no isometric counterpart in the other space-time. Thus M and M' are not observationally indistinguishable.

In one sense, even the condition of observational indistinguishability after finite time is overly stringent. Suppose we have two space-times M and M', and suppose for every point x in M there is a corresponding point x' in M' such that $I^-(x)$ and $I^-(x')$ are isometric. Then no observer in M at any point in his life will be in a position to determine which of the two space-times, if either, is his. Yet M and M' need not necessarily be observationally indistinguishable after finite time because there might be an observer in M' who could at some time distinguish between them.

As far as the epistemological situation of the M-observer is concerned, it makes no difference what the M'-observer can or cannot determine. For this reason it is worth considering a new condition of observational indistinguishability which, unlike the first two, is not symmetric.

Definition: If M and M' are space-times, M *is weakly observationally indistinguishable from* M' if for every point x in M there is a point x' in M' such that $I^-(x)$ and $I^-(x)$ are isometric.

Quite trivially, M can be weakly observationally indistinguishable from M' without the two being observationally indistinguishable after finite time. For example, take M as in the very first example—the t, x plane with metric $ds^2 = dt^2 - (\cosh^2 t)\, dx^2$; and take M' to be either one of the two space-times in the second example, the ones built from vertical A and

68

B slabs. The observational past of every point in M, indeed the observational past of every future inextendible timelike curve in M, finds an isometric counterpart in any of the A slabs of M'. But obviously no observer in M' who ever catches a glimpse of the B portion of his space-time could think himself to be in M.

This third notion of observational indistinguishability seems a straightforward rendering of conditions under which observers could not determine the spatio-temporal structure of the universe. Yet, and this is what is most interesting, the condition of weak observational indistinguishability is so widespread in the class of space-times as to be of epidemic proportions.

There are some space-times that are not weakly indistinguishable from any other. These include space-times M in which the observational past of some point is the entire manifold, i.e., $I^-(x) = M$ for some x. The simplest such example is two-dimensional Minkowski space-time "rolled up" along the t-axis (i.e., for some $k > 0$, the points (t, x) and $(t + nk, x)$ are identified for all integers n). A more interesting example is Gödel space-time. But only these quite bizarre space-times seem to escape having counterparts from which they are weakly observationally indistinguishable. (There is a theorem lurking here.)

Let me give a geometrically intuitive argument sketch which, while falling short of a proof, suggests why this should be so. To keep things simple, let us restrict attention (much more than we have to) to space-times that are decent in their "causal structure" and have no closed or almost closed future-directed timelike curves. To be specific, let us consider only "strongly causal" space-times.[4]

Let M be one such space-time and let $\{x_i\}$ be a countable sequence of points in M, the union of whose observational pasts covers all of M, i.e., $\cup \{I^-(x_i)\} = M$.[5] Using a "clothesline construction" we can string out these $I^-(x_i)$ with appropriately chosen "space-time filler" to form a new space-time M'. (The causality assumption here disallows the possibility of the $I^-(x_i)$ folding back on themselves.) In other words, we can find a space-time M' in which all the $I^-(x_i)$ can be isometrically embedded. The space-time filler with which M' is constructed can be chosen quite arbitrarily, subject only to the constraint, of course, that it be smooth on the boundaries of the $I^-(x_i)$. Exercising this freedom we can so choose the filler as to guarantee that M' not be isometric to M. But clearly M must be weakly observationally indistinguishable from M'. Any point

$x \in M$ will be in some $I^-(x_j)$. Since $I^-(x) \subset I^-(x_i)$, $I^-(x)$ will find an isometric counterpart in the $I^-(x_j)$ portion of the clothesline.

M' as it stands suffers from being (very) extendible. But it follows from an argument of Robert Geroch[6] that every space-time has an inextendible extension. We can choose one for M', say M'', and we have enough freedom in doing so to ensure that M'' not be isometric to M. At least if M is itself inextendible, the $I^-(x_i)$ excised from it will remain unaffected by the extension from M' to M''. So for every point $x \in M$, $I^-(x)$ will still find an isometric counterpart in M''.

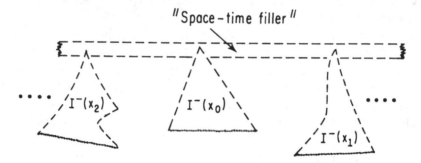

Figure 4. "Clothesline construction" by which a space-time is formed in which all the past observation sets $I^-(x_i)$ can be isometrically embedded.

This argument sketch shows that every strongly causal space-time is weakly observationally indistinguishable from some other space-time, and that if the first is inextendible, the second may be taken to be inextendible as well. If greater care is exercised in hanging the clothesline used in the construction, the argument goes through under much weaker "causality" assumptions. Indeed, it is sufficient that for some point x, $I^-(x) \neq M$.

II

To get a better feeling for the three observational indistinguishability relations it will help to consider several global properties of space-times and see whether any are preserved under them, i.e., whether it is the case that given two space-times, the first observationally indistinguishable (respectively observationally indistinguishable after finite time, weakly observationally indistinguishable) from the second, the property obtains in the first space-time only if it obtains in the second. The question is of interest because even in the presence of observational indistinguishability there is the possibility that an observer can make *some* determinations concerning the global spatio-temporal structure of the universe and so at least delimit the range of open choices.

The accompanying tabulation lists a sampling of commonly studied global properties and indicates their invariance under the three relations. The arrows indicate, of course, that if a property is not preserved under a strong sense of observational indistinguishability, it is not preserved under weaker senses (and correspondingly when it is preserved). Counterexamples for (1)–(5) and (8) are appended. The other causality properties are of some special interest because, as indicated in the table, they diverge from their respective negations when it comes to preservation under weak observational indistinguishability.

Property	OI?	OI after finite time?	Weak OI?
1. temporal orientability	No --→		
2. spatial orientability	No --→		
3. orientability	No --→		
4. inextendibility	No --→		
5. noncompactness	No --→		
6. causality	← ----------Yes		No
noncausality	← ----------------------------------Yes		
7. strong causality	← ----------Yes		No
nonstrong causality	← ----------------------------------Yes		
8. existence of a global time function	No --→		
9. existence of a Cauchy surface	← ----------Yes		No
nonexistence of a Cauchy surface	← ----------------------------------Yes		

Causality is the condition that there *not* be a closed, future-directed causal curve. If causality is violated in a space-time M, the entire violating curve will be in the observational past of some point x—any point, in fact, which lies to the future of some point on the curve. If now $I^-(x)$ is isometric to $I^-(x')$ for some point x' in a space-time M', the image of the closed curve under the isometry will itself be a closed curve in $I^-(x')$. It follows that noncausality is preserved under weak observational indistinguishability and causality is preserved under observational indistinguishability after finite time (by the symmetry of the relation).

71

On the other hand, the following simple example shows why there must be a 'No' in the third column of line 6 in the table. Take M to be two-dimensional Minkowski space-time. Construct M' by first cutting two slits in M, say $A = \{(t, x): t = 0 \text{ \& } 0 \leqslant x \leqslant 1\}$ and $B = \{(t, x): t = 1 \text{ \& } 0 \leqslant x \leqslant 1\}$, and then identifying the lower edge of slit B, excluding the corner points $(1, 0)$ and $(1, 1)$, with the upper edge of slit A, excluding $(0, 0)$ and $(0, 1)$. (See Figure 5.) M is certainly weakly observationally indistinguishable from M' since every point x in M has as an observationally indistinguishable counterpart every point x' in M' lying, say, beneath slit A. But causality is badly violated in M'; one sample closed timelike curve is indicated in the figure. M' as it stands is extendible, but it can be rendered inextendible by further identifying the upper edge of slit B with the lower edge of slit A, again excluding corner points. In a sense the points $(0, 0)$, $(0, 1)$, $(1, 0)$, and $(1, 1)$ are "missing," but there is no way they can be replaced to extend M'.

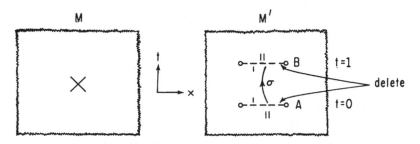

Figure 5. M is weakly observationally indistinguishable from M' although M' is not causal (σ is a sample closed causal curve). The lower edge of slit B is identified with the upper edge of slit A and the upper edge of slit B is identified with the lower edge of slit A (excluding end points). M' is inextendible.

The situation seems to be this: if causality is violated in a space-time, some observer will know about it; if on the other hand it is not violated, no observer will ever know for sure one way or the other. This does not follow from the one example, but it could be established with another clothesline construction. In addition to the other space-time segments $I^-(x_i)$, an additional causality-violating segment would have to be added to the line.

With respect to preservation under the different notions of observational indistinguishability, strong causality is quite similar to causality.

The claims made in the table are verified along the lines just indicated. The condition that there exist a Cauchy surface is a bit different.

If M is a space-time, a set $S \subset M$ is a *Cauchy surface* if S is *achronal* (i.e., for no points x and y in S is $x << y$) and if for every point z in M, every (past- and future-) inextendible timelike curve through z intersects S. For example, the surfaces $t = $ constant are Cauchy in Minkowski space-time. A simple example of a space-time that does not admit any Cauchy surface is (the covering space of) the two-dimensional version of anti-De Sitter space-time. It can be represented as the t, x plane with metric $ds^2 = dt^2 - (\cosh^{-2} x) \, dx^2$. At $x = 0$ the metric reduces to Minkowski form and the associated null cones are at $45°$. But as x increases in absolute value, the cones flatten and approach the horizontal asymptotically. (See Figure 6.) In contrast, remember that with the De Sitter metric (i.e., $ds^2 = dt^2 - (\cosh^2 t) \, dx^2$) the cones narrowed to the vertical as t increased in absolute value. No surface $t = $ constant will be Cauchy in anti-De Sitter space-time because there are inextendible curves through many points which "come in from or go out to spatial infinity" without hitting the surface. The same is in fact true of all achronal sets.

The condition that there exist a Cauchy surface is of great interest because of its usual interpretation as a condition of Laplacian determinism and the possibility of cosmic prediction.[7] If a set is Cauchy, no signal propagating causal influence can reach any point in space-time without that signal registering itself, before or after, on the set. In the absence of such a set—in anti-De Sitter space-time, for example—causal influence can "come in from infinity" without registration. For this reason even a

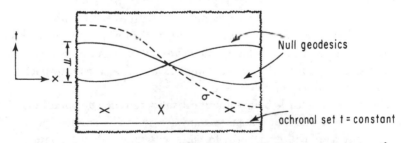

Figure 6. The covering space of two-dimensional anti-De Sitter space-time, i.e., the t, x plane with metric $ds^2 = dt^2 - (\cosh^{-2} x) \, dx^2$, which admits no Cauchy surface. Light cones rapidly flatten to the horizonal as $|x| \to \infty$. An inextendible timelike curve σ coming in from and going out to "infinity" is indicated along with a sample $t = $ constant slice which it fails to hit.

complete specification of "initial data" on a t = constant surface in anti-De Sitter space-time will not uniquely determine past and future spatio-temporal evolution.

It turns out that the existence of a Cauchy surface is equivalent to the condition of *global hyperbolicity*.[8] A space-time satisfies this condition if it is causal *and* if for all points x and y, the set $J^+(x) \cap J^-(y)$ is either empty or compact. (Note that this is not true in anti-De Sitter space-time.) Now suppose M is weakly observationally indistinguishable from M' and M fails to be globally hyperbolic. If causality fails in M then, as we know, it must fail in M' too. Suppose then that for some x, y in M, $J^+(x) \cap J^-(y)$ is neither empty nor compact. Suppose further that z is some point to the future of y, i.e., $y << z$. Then $J^+(x) \cap J^-(y)$ is contained in $I^-(z)$. (Fact: $w < y$ & $y << z \rightarrow w << z$). If z' is in M' and $\phi: I^-(z) \rightarrow I^-(z')$ is an isometry, then $J^+[\phi(x)] \cap J^-[\phi(y)] = \phi[J^+(x) \cap J^-(y)]$ will be a set in M' neither empty nor compact. Hence M' is not globally hyperbolic. Thus the nonexistence of a Cauchy surface is preserved under weak observational indistinguishability. (The same argument, of course, establishes that the existence of a Cauchy surface is preserved under observational indistinguishability after finite time.)

The following example shows, in contrast, that the existence of a Cauchy surface is not necessarily preserved under weak observational indistinguishability. Once again take M to be two-dimensional Minkowski space-time. For M' we in effect glue together the lower half of M with the upper half of two-dimensional anti-De Sitter space-time. More specifically, M' is the t, x plane with the metric $ds^2 = dt^2 - dx^2$ in the region $t \leq 0$, and the metric $ds^2 = dt^2 - (\cosh^{-2} x) \, dx^2$ in the region $t \geq 1$. For the buffer strip $0 < t < 1$ choose any metric whatsoever that smoothly connects the other two. M' is clearly an inextendible space-time without a Cauchy surface. Equally clearly, however, M is weakly observationally indistinguishable from M'. Every point in M has as a counterpart every point in M' with coordinate $t \leq 0$.

The comments made before about causality can now be paraphrased. In particular, it seems that if a space-time has a Cauchy surface, none of its native observers will ever know for sure whether it does or not!

III

The predicament of the cosmologist attempting to determine the global space-time structure of his universe has been cast as a serious one. However, an objection could be made. It is important and should be considered. The notion of weak observational indistinguishability was introduced on the suggestion that Glymour's two conditions are unnecessarily stringent. According to the objection they are not stringent enough.

The different conditions of observational indistinguishability are posed solely in terms of spatio-temporal structure. They do not mention the things and processes which populate space-time. But, the objection runs, two cosmological models are only truly observationally indistinguishable if neither underlying space-time structure *nor* its contents (stars, galaxies, background radiation, or whatever) distinguish them to any observer.

In response, the several definitions of observational indistinguishability can be extended in a straightforward way to include the physical goings-on within space-time. Let us suppose that the fundamental furniture of the universe consists of a number of *matter-fields* (e.g., an electromagnetic field) which are mathematically represented by tensor fields defined on the underlying space-time manifold, and whose dynamical histories are constrained by (partial differential) field equations. This is no more than the framework within which relativity is in fact studied. In this context, a *cosmological model* may be construed as an ordered $(n+1)$ tuple $(M, F_1, \ldots, F_n)$ whose first element is a space-time and whose remaining elements are tensor fields of the appropriate type on M, satisfying appropriate field equations. We say that two *cosmological models* $(M, F_1, \ldots, F_n)$ and $(M', F'_1, \ldots, F'_n)$ *are observationally indistinguishable* in any of the three senses defined if M and M' are observationally indistinguishable in that sense *and* if the isometries between past observational sets called for in the definitions also preserve the values of respective matter fields, i.e., for any such isometry ϕ, $\phi_*(F_i) = F'_i$ for $i = 1, \ldots, n$.

In reply to the present objection it can now be argued that observational indistinguishability between cosmological models, if not in the narrower sense between space-times, really *is* a sufficient condition for the empirical underdetermination of space-time structure. The reply seems a strong one. If one accepts the idealization of a cosmological model in the first place, then specification of the values of the various matter fields in a region of space-time completely specifies what there is to be observed—background radiation, quasars, or whatever. If, for example, the cosmological model $(M, F_1, \ldots, F_n)$ is weakly observationally indistinguishable from $(M'_1, F'_1, \ldots, F'_n)$, then nothing any observer in the

75

first model could "see" at any time in his life, no matter how discerning his instruments, would ever distinguish between the two models. It seems clear, too, that no physical theory such as any observer could ever extrapolate from the matter-fields as he sees them could possibly cut the ice between them. To the extent that he is entitled to adopt the would-be ice-cutting theory, so is his observational counterpart. By the very definition of weak observational indistinguishability it would seem that any theory supported by the observational evidence available to any one observer in $(M, F_1, \ldots, F_n)$ would have to be neutral, as between two cosmological models.

Suppose we grant now that observational indistinguishability between cosmological models (in any of the three senses) is a sufficient condition for the empirical underdetermination of space-time structure. There remains the question of existence. Suppose $(M, F_1, \ldots, F_n)$ is a cosmological model and that M as a space-time is observationally indistinguishable (in one of the senses) from some other space-time. We can ask whether there necessarily exists a cosmological model $(M', F'_1, \ldots, F'_n)$ which is observationally indistinguishable (in the same sense) from $(M, F_1, \ldots, F_n)$. At least with respect to the sense of weak observational indistinguishability the answer seems to be clearly yes! The same clothesline construction that served to generate space-times can be used to generate cosmological models as well. Instead of linking the space-times $I^-(x_i)$ we link the cosmological models

$$(I^-(x_i), F_1\big|_{I^-(x_i)}, \ldots, F_n\big|_{I^-(x_i)}),$$

connecting them with space-time and matter-fields filler.

The cosmologist's epistemological predicament, it thus appears, is not at all relieved by bringing into the picture the matter-fields that populate space-time.

Appendix

Counterexamples (see tabulation page 71)

(1) Spatio-temporal orientability conditions are not preserved under observational indistinguishability.

Definition: A space-time is:

(a) *temporally orientable* if it admits a continuous, nonvanishing timelike vector field;

(b) *spatially orientable* if it admits three (or in general $n - 1$) continuous spacelike vector fields whose vectors are at every point linearly independent;

(c) *orientable* if it is both temporally and spatially orientable, or neither.

For (a), take M to be the covering space of two-dimensional De Sitter space-time (Figure 1) and form M' by first cutting a vertical strip of width 2π in M, twisting, and then identifying opposite sides. M' is topologically a Möbius strip. M and M' are observationally indistinguishable, but only M is temporally orientable (see figure).[9]

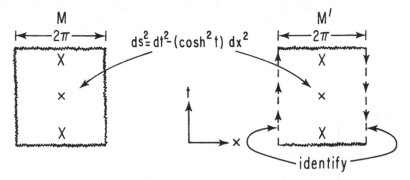

For (b) and (c), take M to be a three-dimensional version of the previous M. The observational past of every future-inextendible curve will be confined to a vertical square cylinder with sides of width 2π. Form M' by cutting out such a cylinder from M and then cross-identifying the $t =$ constant surfaces, turning them into Möbius strips. M and M' are observationally indistinguishable; but while M is spatially orientable and orientable, M' is neither (see figure).

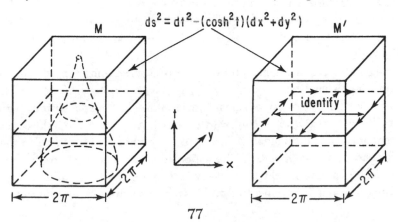

(2) The condition of inextendibility (see page 67) is not preserved under observational indistinguishability. The construction resembles that in Figure 5. Start with two copies of two-dimensional Minkowski space-time and excise from each the slit $I = \{(t, x): t = 0 \;\&\; 0 \leq x \leq 1\}$. Now identify the upper edge of each slit with the lower edge of the other (excluding end points). This will be M. It is inextendible. For M', start again with two copies of two-dimensional Minkowski space-time. From one cut the same slit $I = \{(t, x): t = 0 \;\&\; 0 \leq x \leq 1\}$. From the other cut away the closed upper half, leaving the set $\{(t, x): t < 0\}$. Now identify the upper edge of I with a corresponding section of unit width from the edge of the second space-time (see figure).

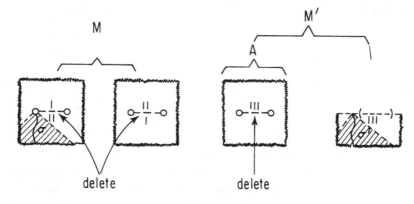

M, but not M', is inextendible. But they are observationally indistinguishable. To see this it suffices to check the few possibilities. Any future-inextendible timelike curve in M will find a counterpart in the "A section" of M'. But any future-inextendible curve on M' will also find a counterpart in M. In particular the curve σ' which runs off the edge of M' in its truncated portion finds a counterpart σ in M which runs to the "hole" $(0, 0)$.

(3) The condition of noncompactness is not preserved under observational indistinguishability.

For M we take a two-dimensional "horizontal cylinder space-time" with metric $ds^2 = (\cos x)\, dx dt + (\sin^2 x)\,(dt^2 - dx^2)$. At $x = n\pi$ the light cones are tangent to the horizontal, pointing to the right for even n and to the left for odd n. At $x = (n + 1/2)\pi$ the cones are at 45° to the horizontal (see figure). The observational past of every future-inextendible curve in M is confined to a region of x-width 2π. M' is formed from M by identifying

78

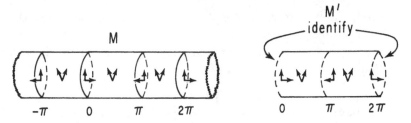

points $x = 2n\pi$. Clearly M, but not M', is noncompact. But M and M' are observationally indistinguishable.

(4) The condition that there exist a global time function is not preserved under observational indistinguishability.

> *Definition:* A smooth map t: $M \rightarrow$ **R** is a *global time function* on the space-time M if for all x, y in M, $x < y$ & $x \neq y \rightarrow t(x) < t(y)$.

The condition that there exist a global time function is equivalent to the condition of "stable causality." [10]

For M start with two-dimensional rolled up Minkowski space-time and then make excisions (as in the figure) which just prevent null geodesics, which are aligned at 45°, from circumnavigating the manifold. This space-time is causal (and strongly causal), but does not admit a global time function. Any real valued function on M which increases along causal curves will be discontinuous somewhere.

(As it stands the space-time is extendible (we can replace the excisions). But without changing its cone structure or causal properties, we can render it inextendible by multiplying its metric by a conformal factor ϕ^2 which appropriately goes to ∞ as the slits are approached [see figure].) M'

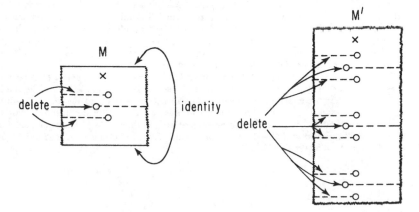

is taken by "unrolling" M. M and M' are observationally indistinguishable since no observational past of any future-inextendible curve in either extends beyond the excision barriers. But only M' admits a global time function.

Notes

1. See also Clark Glymour, "Topology, Cosmology and Convention," *Synthèse* 24 (1972): 195–218.

2. A comprehensive treatment of work on the global structure of (relativistic) space-times is given in: S. W. Hawking and G. F. R. Ellis, *The Large Scale Structure of Space-Time* (Cambridge: Cambridge University Press, 1973). See also Roger Penrose, *Techniques of Differential Topology in Relativity* (Philadelphia: Society for Industrial and Applied Mathematics, 1972). More accessible than either is Robert Geroch, "Space-Time Structure from a Global Viewpoint," in B. K. Sachs, ed., *General Relativity and Cosmology* (New York: Academic Press, 1971).

3. A future end point need not be a point on the curve. The definition is this: If M is a space-time, I a connected subset of R, and $\sigma: I \to M$ a future-directed causal curve, a point x is the *future end point* of σ if for every neighborhood O of x there is a $t_0 \in I$ such that $\sigma(t) \in O$ for all $t \in I$ where $t > t_0$, i.e., σ enters and remains in every neighborhood of x.

4. A space-time is *strongly causal* if, given any point x and any neighborhood O of x, there is always a subneighborhood $O' \subset O$ of x such that no future-directed timelike curve which leaves O' ever returns to it.

5. A countable cover of this form can be found in *any* space-time M, strongly causal or not. Since M is without boundary, for every y in M there is an x in M such that $y << x$, i.e., $y \in I^-(x)$. So the set $\{I^-(x): x \in M\}$ is an open cover of M. But M has a countable basis for its topology (Robert Geroch, "Spinor Structure of Space-Times in General Relativity I," *Journal of Mathematical Physics* 9 (1968): 1739–1744.) So by the Lindelöf Theorem there is a countable subset of $\{I^-(x): x \in M\}$ which covers M.

6. Robert Geroch, "Limits of Spacetimes," *Communications in Mathematical Physics* 13 (1969): 180–193.

7. See John Earman, "Laplacian Determinism in Classical Physics" (to appear) and Robert Geroch's paper in this volume.

8. Robert Geroch, "Domain of Dependence," *Journal of Mathematical Physics* 11 (1970): 437–449. (A somewhat different but equivalent definition of global hyperbolicity is used.)

9. There is a problem of how to define observational indistinguishability in a nontemporally orientable space-time (the definition given presupposed temporal orientation). But under any plausible candidate, M and M' in the example would come out observationally indistinguishable. One could associate with every inextendible timelike curve σ all the points that are connected with some point on the curve by another timelike curve. (In a temporally oriented space-time this would be the union $I^+[\sigma] \cup I^-[\sigma]$.) *Even these* sets in M and M' would find isometric counterparts in the other.

10. A space-time is *stably causal* if there are no closed causal curves *and* if there are no closed causal curves with respect to any metric close to the original. (This can be made precise by putting an appropriate topology on the set of all metrics on the space-time manifold.) Note that in the space-time M of the following example the slightest flattening of the light cones would allow timelike curves to scoot around the barriers. The equivalence is proven in S. W. Hawking, "The Existence of Cosmic Time Functions," *Proceedings of the Royal Society* A, 308 (1968): 433–435.

——————ROBERT GEROCH——————

Prediction in General Relativity

1. Introduction

There are at least two contexts within which one might place a discussion of the possibilities for making predictions in physics. In the first, one is concerned only with the actual physical world: one imagines that he has somehow learned what some physical system is like now, and one wishes to determine what that system will be like in the future. In the second, one is concerned only with the internal structure of some particular physical theory: one wishes to state and prove, within the mathematical formalism of the theory, theorems that can be interpreted physically in terms of possibilities for making predictions.

Of the two, the second context certainly seems to be the simpler and the more direct. Indeed, it is perhaps not even clear what the first context means. One's only guide in making a prediction in the physical world is one's past experiences in the relationship between the present and the future. But it is precisely the collection of these experiences, systematized and formalized, which makes up what is called a physical theory. That is to say, one seems to be led naturally from the first context to the second. One would perhaps even be tempted to conclude that the two contexts are essentially the same thing, were it not for the fact that it seems never to be the case in practice that one's past experiences lead in any sense uniquely to a physical theory; one must, at some point, make a choice from among several competing theories in order to discuss prediction. Thus one might divide a discussion of prediction in physics into two parts: (1) the choice of a physical theory and (2) the establishment and interpretation of certain theorems within the mathematical formalism of that theory.

Consider, as an example, Newtonian mechanics. Suppose that we wish to describe within this theory our solar system, which we idealize as

NOTE: Supported in part by the National Science Foundation, Contract No. GP-34721X1, and by the Sloan Foundation.

follows: the sun and planets are represented by ten mass points, subject to Newtonian gravitational forces. Because of the structure of the differential equations of the theory, one can determine, given the positions and velocities of these points at any one instant of time, their positions and velocities at all later times. Such predictions are of course made routinely in the case of the solar system and are later confirmed, with remarkable agreement, observationally. Let us now attempt to express this activity in terms of some theorem in Newtonian mechanics. We take, as the statement of our theory, the following: "The world is described by points in Euclidean space, each of which is assigned a mass, and which move with time according to a given law of force between them." One might conjecture, within this theory, a mathematical result of the following general form: "Given the positions and velocities of some collection of mass points at some particular time in some region of Euclidean space, there is one and only one solution of the equations of motion, in that region, for later times." But this particular conjecture, at least, is false, for one has the option of having additional mass points, initially outside the given region (and hence not included in the initial data), which subsequently move into the region and influence the motion of our original mass points. In fact, our conjecture is not even true if we further demand that the fixed region be all of Euclidean space. One can, within Newtonian mechanics, construct a solution representing two rocket ships which bounce between them a mass point with ever increasing speed. The result is that the rocket ships accelerate in opposite directions; if the speeds are adjusted correctly, the ships can be made to escape to the "edge" of our Euclidean space in finite time, leaving nothing behind. The time-reverse of this situation, then, is also a solution of the equations, a solution which allows objects to "rush in from infinity," influencing the later development of our system without ever having been included in the initial data.

In fact, there seems to be no theorem in ordinary Newtonian mechanics that suggests possibilities for prediction. Our conjecture above would, presumably, be true if we required in the conjecture that the fixed region be all of Euclidean space and, furthermore, that no information come into the system from infinity. But this result would not, at least to me, suggest prediction, for it constrains both the initial state of the system and its future behavior. One might, instead, consider an alternative theory, e.g., that above, but with the additional proviso that the only admissible solutions are those in which the total number of mass points remains the same with time. (In fact, there are apparently some technical difficulties with such a theory. For example, one must restrict the class of allowed force laws to guarantee existence of admissible solutions, and the passage to a more realistic version in which mass points are replaced by a continuous mass distribution may be tricky.) We emphasize that the new theory is identical with the old as far as observational evidence in our World is concerned, for exotic systems such as that described above have not been observed. Nonetheless one can easily imagine that the two theories will differ markedly in terms of what theorems, suggestive of the possibilities of prediction, they will admit.

The purpose of this paper is to introduce and discuss a few issues relating to the question of prediction in the general theory of relativity.[1] The remarks above are intended to justify the rather narrow framework in which we shall operate. Our theory is standard general relativity. We have a smooth, connected, four-dimensional manifold M, whose points represent "events" (occurrences in the physical world having extension in neither space nor time). There is on this manifold a smooth metric g_{ab} of Lorentz signature, which describes certain results of measuring spatial distances and elapsed times between pairs of nearby events. To simplify the discussion, we shall suppose also that our space-time M, g is strongly causal, i.e., that every point has a small neighborhood through which no timelike curve passes more than once. Observers are described by timelike curves in space-time, light rays by null geodesics, etc. Other physical phenomena are described by tensor fields on space-time, subject to differential equations (e.g., electromagnetic phenomena by the Maxwell field, subject to Maxwell's equations). Our goal is to formulate definitions and theorems within this mathematical framework.

2. Domain of Dependence

It is clear that the difficulties associated with Newtonian mechanics arise from the feature of that theory that it does not restrict the speeds of particles. There is, however, such a restriction in relativity, in which the limiting speed is that of light. One might guess, therefore, that it will actually be easier to discuss prediction in relativity than in Newtonian mechanics. This turns out to be the case, a fact which finds expression in the notion of the domain of dependence.

Let M, g be a space-time. Let S be a three-dimensional, achronal (i.e., no two points of S may be joined by a timelike curve) surface in M. The (future) domain of dependence[2] of S, $D^+(S)$, is the collection of all points p of M such that every future-directed timelike curve in M, having future endpoint p and no past endpoint, meets S. For example, if S is a three-dimensional, spacelike disk in Minkowski space (Figure 1), then $D^+(S)$ is the "cone-shaped region" shown. The point q is not in $D^+(S)$, for the future-directed timelike curve γ in the figure fails to meet S.

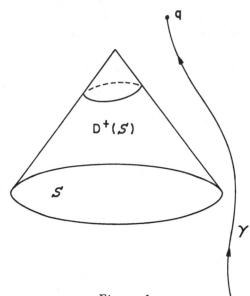

Figure 1

The physical meaning of this definition is the following. The surface S represents "a region of space at some instant of time." Signals in general relativity travel along timelike curves. For q in $D^+(S)$, every such curve to q must have met S, i.e., in physical terms, every signal which could possibly influence the state of affairs at q must have been registered, in some sense, on S. For q not in $D^+(S)$, signals could reach and hence influence the physics at q without having been registered on S. In short, one expects that a sufficiently detailed knowledge of what is happening on S (i.e., at the "initial time") should determine completely what is happening at each point of $D^+(S)$. This physical picture is in fact supported by a collection of theorems in general relativity. The detailed statement depends on the type of matter or fields considered; as an example, we take electromagnetic fields. Electromagnetism is represented by an antisymmetric, second-rank tensor field on space-time, subject to Maxwell's equations (say, without sources). The theorem, in this case, reads as follows: Given the electromagnetic field on S, there is at most one extension of that field to $D^+(S)$, subject to Maxwell's equations. That is to say, the physical situation (in this example, the electromagnetic field) is uniquely determined at any point q of $D^+(S)$, given the situation on S.

We emphasize that the domain of dependence is essentially a relativistic concept. For example, for S of "spatial size" one light-year, $D^+(S)$ will "extend into the future" for about one year in time, i.e., only until signals from outside S have time to move into our region. That there is no analogous notion in Newtonian mechanics is the source of the examples in the previous section.

It is tempting to conclude that this definition essentially exhausts what can be said within our theory: what can be determined from initial data (on S) is precisely what is in $D^+(S)$, and so all that remains is to work out the properties of this $D^+(S)$, its dependence on S, etc. That the situation is not so simple can be seen in the following example. Let M, g be Minkowski space-time, and let S be a spacelike, three-dimensional plane in M. Then $D^+(S)$ is the entire region to the future of S, as shown in Figure 2. We next consider a second space-time, M', g', which is Minkowski space-time with a small, closed, spherical "hole" removed, and a similar surface S' in this space-time. Then $D^+(S')$ is as shown in the figure. The point is that these two space-times, both legitimate within our theory, look identical in the immediate vicinities of their respective surfaces, although they are of course quite different in the large. For example, the only solution of Maxwell's equations in M, g that vanishes on S is the solution that also vanishes to the future of S, while there are solutions of Maxwell's equations in M', g' that vanish on S' and yet do not vanish to the future of S'. (Such a solution must, as already noted, vanish in $D^+(S')$, but it need not vanish in the region indicated in the figure because, physically, "electromagnetic radiation can emerge from the hole.") Suppose, then, that one has decided that our universe, at some time, looks like a neighborhood of S in M, g and that there are no electromagnetic fields present. Could one conclude that no electromagnetic fields will later be seen? Clearly, from this example, one could not. Similar, but more elaborate, examples can be constructed for other situations. In what sense, then, can one make any physical predictions within the general theory of relativity?

It is clear that the mechanism of the example above is the fact that,

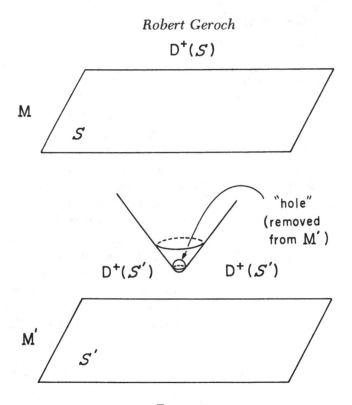

Figure 2

although S determines what happens in $D^+(S)$, what this $D^+(S)$ will be, and in particular how "large" it is, requires knowledge not only of S, but also of the space-time M, g in which S is embedded. Even the imposition of Einstein's equation (which we ignored in the example above) permits only the determination of the geometry in $D^+(S)$, and hence does not prohibit the construction of similar examples by "cutting holes in space-time." Apparently, the situation is that, although the notion of the domain of dependence expresses well what there is of the relationship "present determines future" in general relativity, it is nonetheless difficult to find therein a totally satisfactory formulation, from the physical viewpoint, of this relationship.

Thus general relativity, which seemed at first as though it would admit a natural and powerful statement at prediction, apparently does not. It seems to me that the only cure is to attempt to do for general relativity what we discussed earlier for Newtonian mechanics—change the theory.

We here describe, as an example of the possibilities available along these lines, one such.

Call a space-time M, g *hole-free* if it has the following property: given any achronal, three-dimensional surface S in M, and any metric-preserving embedding Ψ of $D^+(S)$ into some other space-time M', g', then $\Psi(D^+(S)) = D^+(\Psi(S))$. That is to say, we require that the domain of dependence, in M', of the surface $\Psi(S)$ in M' be the same as the image by Ψ of the domain of dependence of S in M. Minkowski space-time, for example, is hole-free (as, indeed, are the standard exact solutions in general relativity). On the other hand, Minkowski space-time, with a hole as in Figure 2, is not hole-free. (Let Ψ be a metric-preserving mapping from $D^+(S')$ in that example to Minkowski space-time.) This definition, then, provides an intrinsic characterization of space-times that have been constructed by cutting holes (although an imperfect one: Minkowski space-time to the past of a null plane is hole-free by this definition). Note that one could not accomplish the same objective by simply insisting that space-times not be constructed by cutting holes in given space-times, for this characterization involves not only the space-time itself but also its mode of presentation. Similarly, "maximally extended" is no substitution for "hole-free," for there are space-times that satisfy the former and not the latter.

One might now modify general relativity as follows: the new theory is to be general relativity, but with the additional condition that only hole-free space-times are permitted. As far as observational consequences in our world are concerned, the two theories are identical, since non-hole-free space-times never arise in any practical applications. The new theory, however, admits a simple and natural theorem which suggests prediction: if S and S' are achronal, three-dimensional surfaces in hole-free space-times M, g and M', g', respectively, and if there is a mapping from S to S' which preserves all fields, then there is such a mapping from $D^+(S)$ to $D^+(S')$. This result is in fact practically a restatement of the definition.

It might be of interest to understand better the strength and role of this definition, as well as the scope of other possibilities.

3. Prediction

In the previous section, we were concerned with the relationship between what is happening in one region of space-time (the present) and what is happening in some other region (the future). The word "predict," however, suggests not only the existence of such relationships but also the existence of some agent who gathers the initial data and actually makes a claim about the future. In the present section, we describe within the theory such agents.

Consider first the following example. Let S be a small, three-dimensional, spacelike disk in Minkowski space-time (Figure 3). Then, as we have remarked, initial data on S determine the physical fields in $D^+(S)$, in particular at point p. Let us now introduce an observer, represented by timelike curve γ, who is to actually make this prediction regarding p. In order to make his prediction, our observer must first collect the data from S, a task he carries out as follows. At point r, our observer sends out a swarm of other observers, who fan out, experience, and record

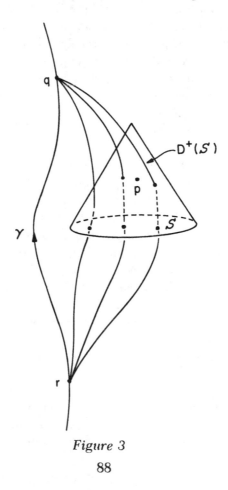

Figure 3

every part of S. They then return to the original observer with this infor-mation, meeting him at point q. Thus by point q our observer has assem-bled all the relevant information and is prepared to make his prediction regarding point p. But note that p lies to the past and not to the future of q. In physical terms, by the time our observer gets around to making his prediction regarding p, p has already happened; he makes a retrodiction rather than a prediction.

It is clear that the problem in the example above arises because the other observers cannot exceed the speed of light in returning to the original observer, whence they arrive too late for a genuine prediction. It is also clear why in Newtonian mechanics, with no limit on the speeds of signals, no distinction need be made between "determination" and "pre-diction."

Other choices of S and q in Minkowski space-time lead to the same result: retrodiction. Indeed, it is perhaps not immediately clear whether or not one can construct any examples in which genuine prediction is possible in the theory. It turns out that there are such examples. Let M, g be the space-time obtained by removing from Minkowski space-time two small, spacelike, three-dimensional disks, as shown in Figure 4, and iden-tifying [3] the lower edge of disk A with the upper edge of disk B. Thus, for example, a timelike curve entering A from below will re-emerge from the top of B. Let the surface S, the point p, and the timelike curve γ, repre-senting our observer, be those shown. Then, since every future-directed timelike curve to p meets S, p is in $D^+(S)$. Our observer, however, can now gather his initial data by point q, where p is not in the past of q. At the still later point v, our observer can finally learn of point p and so can then check his prediction observationally. In this space-time, then, pre-dictions are possible.

It is interesting to note that it is an essential feature of the example above that the observer verifies his prediction only indirectly—by reach-ing a point v to the future of p—rather than directly by passing through p. That direct verification is also possible is shown by the following example. Let the space-time M, g be the Einstein universe, so the spatial sections are three-spheres and time is the real line (Figure 5). Our observer has, at q, collected the data from S, while $D^+(S)$ is the entire future of S. Thus all the experiences of the observer beyond q could have been predicted at q.

It is convenient to isolate the essential features of these examples by means of a definition. Let M, g be a space-time. For x any point of M,

89

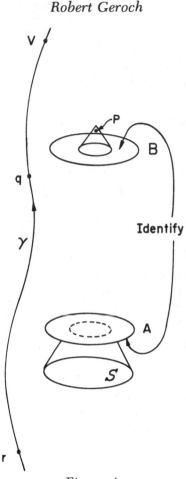

Figure 4

denote by $I^-(x)$, the past of x, the set of points that can be reached from x by past-directed timelike curves. Now fix any point q of M, and denote by $P(q)$ the set of all points x such that *every* past-directed timelike curve from x, without past end point, enters $I^-(q)$, but such that $I^-(x)$ is *not* a subset of $I^-(q)$. We shall call this set $P(q)$ the domain of prediction of q. For example, for q any point of Minkowski space-time, every point x either has the property that $I^-(x) \subset I^-(q)$ or has the property that some past-directed timelike curve from x fails to meet $I^-(q)$. Hence the domain of prediction of each point q in Minkowski space is empty.

The physical meaning of this definition is as follows. The point q repre-

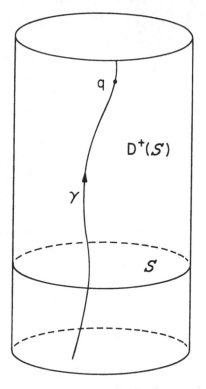

Figure 5

sents the point (of our predicting observer) at which all the information has been collected. Then the set $I^-(q)$ represents that region of space-time from which information could reach q. The first condition for membership of x in $P(q)$ requires, physically, that every signal that could affect x must have come from $I^-(q)$, i.e., that every such signal could have been recorded and carried to q. The second condition requires essentially that x not be in $I^-(q)$, i.e., that we have a prediction at x rather than a retrodiction. This interpretation is supported by the following, easily proved, result: point x is in $P(q)$ if and only if $I^-(x) \not\subset I^-(q)$, and, in addition, there is a three-dimensional, achronal surface S in $I^-(q)$ with x in $D^+(S)$. It follows immediately, for example, that, in the examples of Figures 4 and 5, p is a point of $P(q)$. Thus we interpret the domain of prediction of q as "the region of space-time that can be predicted from q."

Assuming that the definition above accurately reflects the physical notion of "making predictions in general relativity," what remains is to study its consequences. We give one example. We saw in the example of Figure 4 that $P(q)$ is nonempty but contains no points to the future of q ("predictions could not be verified directly"). In the example of Figure 5, on the other hand, $P(q)$ includes points to the future of q, and, in that example, we have a "closed universe." In fact these observations are a special case of a more general result, namely: Given a space-time M, g, and a point q of M such that $P(q)$ contains a point to the future of q, then M, g is a closed universe, in the sense that it admits a compact spacelike surface. (This result is essentially a corollary of a theorem [4] of Earman's that a Cauchy surface to the past of a point must be compact.) In physical terms, "predictions which can be verified directly arise only in closed universes." Why should this strange result follow from just the basic principles of general relativity? Are there any other similar theorems about the domain of prediction?

4. Conclusion

We can conveniently summarize by comparing general relativity and Newtonian mechanics. For our purposes, there are apparently two essential differences between the two theories: (1) signal speeds are unlimited in Newtonian mechanics but limited in general relativity; and (2) the space-time framework is fixed once and for all in Newtonian mechanics (Euclidean space plus time) but not in general relativity. The notion "future from present" seems to arise far more simply and naturally in general relativity than in Newtonian mechanics because of the limitation on signal speeds in the former. On the other hand, the freedom in the space-time model in general relativity leads to new difficulties not present in Newtonian mechanics. Finally, the question of the collection of initial data, while irrelevant in Newtonian mechanics, leads, because of limitations on signal speeds, to additional complications in general relativity.

The notion of observational indistinguishability [5] leads to a classification of properties of space-times according to their interaction with this notion. In a similar way, one could classify properties as deterministic and nondeterministic, and as predictive and nonpredictive.

Notes

1. For a careful and thorough discussion of this issue, see J. Earman, "Laplacian Determinism in Classical Physics," preprint.

2. See, for example, R. Geroch, "Domain of Dependence," *Journal of Mathematical Physics* 11 (1970): 437; S. W. Hawking, G. Ellis, *The Large-Scale Structure of Space-time* (Cambridge: Cambridge University Press, 1974).

3. For a discussion of this construction, see, for example, R. Geroch, "Space-Time Structure from a Global Viewpoint," in R. Sachs, ed., *Relativity and Cosmology* (New York: Academic Press, 1971), p. 71.

4. J. Earman, private communication.

5. C. Glymour, this volume; D. Malament, this volume.

C. J. S. CLARKE

Time in General Relativity

Introduction

In this essay I shall discuss two intimately related properties of time, or of the temporal aspects of phenomena, which, having at first been developed through philosophical arguments, have now come to play central roles in general relativistic physics.

The first property is the *directionality* (or *anisotropy*) of time, a phrase which I use loosely to refer to the fact that the peculiarities of the events around us on Earth can consistently be used to distinguish between two possible ways of ordering time. In other words, no one is for long in doubt about which way to run a film through a projector. This is one of the most basic physical facts about the world we live in.

If this directionality is assumed to exist at all points of a general relativistic model of the universe, then the possible structure of the space-time is accordingly restricted: it is called *time-orientable*, meaning that its metric structure is such that at each point one can designate one temporal direction as "positive" and the other as "negative," with the designations at neighboring points agreeing. The assignment of positive and negative can be made arbitrarily at one point, but it is then fixed at all other points by this requirement of neighboring agreement (continuity). If the physicist goes on to say that one direction is *future* and the other *past*, then through the wide connotations of these words he tacitly assumes that by far the most proper form of physical argument is the prediction of the future from the past.

The second property I shall discuss is again drawn from experience and then generalized to a mathematical principle: time is *strictly linear*. My history, and, as far as I know, the history of any object, is describable as a (finite or infinite) linear extension and is not like a circle. Nor, in our normal experience, can two distinct instants[1] of a person's or an object's history be physically contemporaneous; indeed, there is much in our experience that weighs against the mere possibility of, for example, my

94

use of a time-machine to make me contemporary with an event I have already experienced. This is reflected in our language, which so assumes the linearity of time that any contrary statement (such as that in the previous sentence) is liable to be internally inconsistent unless one restricts the usual implications of the words used.

The mathematical correlative of such a time-trip, or of a circular history, is a curve in space-time which is closed,[2] in that it returns to its starting point, and yet is timelike, locally describing a possible history for an object. It is almost invariably assumed in general relativity that space-time is *causal*, meaning that such closed timelike curves do not occur. While time-orientability asserted that an absolute distinction between past and future could be made locally near every point, causality implies that this distinction is meaningful on the entire history of any object, in which an event once past can never be regained.

The discussion of these properties in general relativity differs from the parallel arguments in philosophy. For instance, many philosophical writers claim that the "normal" properties of time are logically necessary if temporal language is to have anything like its everyday meaning, or if we are speaking of a world in which human discourse and action as we know it is possible. But relativity uses a language where words have new technical meanings to discuss cosmological models in parts of which human action, or even existence, is undoubtedly not possible! Yet the distinction is all too often ignored, and the conclusions of philosophical arguments are uncritically used, mainly by physicists, to justify various mathematical restrictions on space-time. I hope to exemplify here how such restrictions should be sought only through a discussion which is consistently within the general relativistic context.

Directionality

While few would deny that on Earth time is directional in the sense described above, controversy centers on whether the fact is a physical law in its own right or a consequence of laws and contingent circumstances together. I shall first examine the standard example of work in support of the latter case, showing not only the greater explanatory power of this approach but also the deep consequences of this controversy for general relativity, extending far beyond the classification of types of scientific explanation. In the example to be described, the laws used are the laws of electrodynamics in the form of partial differential equations that are indifferent with respect to any distinction between past and future; they are combined with the contingent fact that the universe (or at least a very large part of it surrounding us) is in that kinetic state which, with our usual assignment of time direction, we call expansion.

The effect of this circumstance was examined by Sciama[3] in the context of the usual idealized cosmological model (the homogeneous isotropic Robertson-Walker solutions of the Einstein equations for a perfect fluid). In models like this [4] it is possible to express the electromagnetic field at any point p in one of two extreme forms, or as any mixture of the two: (1) a sum of contributions from all the particles lying in a region to the past of p, each radiating in the usual way so that the radiation recedes from the particle as one progresses to the future, together with a contribution from radiation already present at the past boundary of the region; and (2) a sum of contributions from particles in a future region, each radiating "backward" (the radiation receding from the particle as one progresses toward the past), together with a contribution from radiation at the future boundary.

Both descriptions are mathematically admissible for any solution of Maxwell's equations. Why, then, is it customary always to use (1)? The answer is clear in the idealized model used by Sciama, in which only (1) has the property that, as the contributing region used is extended progressively further into the past, the contribution from radiation entering the region can be taken to become less and less, while the contribution from the radiating particles tends to a finite value. In the limit, the field at p is then represented as due simply to the sum of contributions from all the particles to its past. If, however, we try to perform a similar limiting procedure with (2), we find that both the contribution from the particles and the contribution from radiation at the boundary increase without limit (on the simplest analysis)[5] as the region is extended to the future. Thus the only representation of the electromagnetic field which could describe it as being produced solely by particles is that in which the particles radiate in the usual time-sense, relative to the time-sense[6] defined by the kinetics of the universe. (One might note in passing that this example illustrates the process of transition from time-symmetric *laws*, expressing only neutral connections between temporally neighboring events, to a system in which the motions of particles *cause* the field. On the present analysis, the effect must follow its cause because 'causation' is linked to the kinetic structure of the universe.)

This argument is not merely a specimen instance of one way in which contingent facts can impart one sort of directionality to temporal phenomena. As is well known, there are good reasons for believing that it describes in simplified form the central process that determines the directionality of time in any reasonable model of the universe, all the various possible arrows of time being dependent on the electromagnetic arrow. A full proof of this belief has yet to be given, but the lines which it would take are fairly clear. The kinetic state of the universe, through electromagnetic phenomena, determines a direction of time with respect to which matter loses heat by radiating it away into space. This condition of thermal disequilibrium then gives a thermodynamic directionality to physical processes; from this the anisotropy of time as expressed through recording processes and the second law of thermodynamics could plausibly arise from Reichenbach's "branch system" argument.[7]

My aim in this section is to explore the implications for general relativity, if it be accepted that the viewpoint I have just described is generally valid—the view, that is, that all processes characteristic of the directionality of time have their sense determined by the large-scale kinetic structure of the universe. I shall suggest that if such a determination takes place, then the direction of time is not some metaphysical absolute that must be related to a relativistic model by an interpretative convention: rather, it is grounded in the kinetic structure of the model itself.

This proposition has drastic consequences. For if time is thus kinetically determined, then there is no reason to expect it to have all the properties which it would possess as a primary absolute. For instance, even if a time-coordinate can be defined in the model, the interpretation of the sense of this coordinate—whether it measures time "forward" or "backward"—must be determined by the intrinsic physical properties of the model; in particular it may vary from place to place[8] and be in some places undefinable.

An example illustrates the physical importance of this and reveals the difficulties that arise. Consider a homogeneous and spherically symmetric star collapsing into a black hole. The appropriate solution[9] is the Schwarzchild metric outside the star, joined onto the "interior Schwarzchild metric" inside. Now this latter is identical with the Robertson-Walker cosmological solution discussed by Sciama *with the time direction reversed*, which suggests that the time-sense as determined intrinsically might be anomalous. And, indeed, when Sciama's argument is applied to points inside the black hole, it is found that one can realize the representation (2) of the field as being caused by future motions of charges. Whether or not (1) is also allowed depends on the state of motion of the matter in the universe before the black hole forms.

In this situation one reaches quite different physical predictions according to the attitude adopted toward time. Usually a certain direction of time, with respect to which the star is collapsing, is taken as an a priori datum for physical reasoning. Causation and explanation are strictly unidirectional; the task of physics is to explain or predict later stages of the system in terms of the earlier stages that give rise to them. Typically, these earlier stages might be regarded as "initial conditions" from which the system evolves. In the case of the black hole one can find a spacelike hypersurface (a Cauchy surface) whose physical condition determines uniquely the conditions everywhere in the space-time. Then one can, for instance, argue that a small departure from exact symmetry on this "initial" hypersurface does not hinder the formation of a black hole.[10]

But suppose that we apply to the region inside the black hole arguments based on the "reversed" time direction kinetically determined there. Such arguments will be qualitatively like those usually developed in Robertson-Walker cosmologies, but time-reversed. In cosmology, for example, galaxies are usually regarded as having been formed by the gravitational amplification of small fluctuations of density present at a very early time in an otherwise homogeneous universe. The origin of these fluctuations is often sought in quantum processes which, very early on, introduce a random element into a cosmos which initially was quite homogeneous. In this way it is hoped to provide an explanation of the occurrence of the galaxies which, if successful, should account for their observed distribution of sizes and angular momenta. The direction of time enters twice: once in designating the conditions at the $t = 0$ boundary of space-time as *initial* conditions which can be postulated a priori; then again in giving a directionality to the *growth* of quantum fluctuations, which are regarded as being statistically independent in accordance with Penrose and Percival's[11] analysis of the directionality of time.

If the time-sense in the black hole were determined intrinsically, then we could postulate a *time-reversed "growth" of perturbations*, in the same way as in the cosmological case. The physical consequences are then dramatic: it turns out that such perturbations grow indefinitely large near the boundary (horizon) of the black hole. In this analysis the conventional black hole model is an unstable configuration which cannot physically exist!

Such is the confidence placed in reasoning from an a priori time-sense that virtually no physicist would draw the conclusion that black holes do not exist. They would rather say either that the model in question is grossly unrealistic in its description of the final singularity (which may be equally true in the cosmological case) or that one must restrict one's reasoning to the domain in which everything is normal, outside the black hole, trusting that this domain will remain unaffected by processes in the interior, however extreme. In either case, they are then able to fall back on an a priori time-sense, which is the "real" direction of time—with respect to which an anomalous part may perhaps appear to be running backward. But if an *intrinsically* determined time-sense is so rejected, then we must recognize and justify the alternative: a mode of explanation which is unsymmetric with respect to time and which applies the time-sense determined by processes near the earth to the entirety of the universe, irrespective of the nature of the processes elsewhere; or which seeks to establish an absolute time *extrinsic* to the physical universe to govern its evolution. This is the dilemma to which I shall return in the final section, after examining the second conventional property of time.

Cyclic versus linear histories

Perhaps the most important result in modern relativity theory has been the prediction by Hawking and Penrose of the necessary occurrence of singularities in general relativistic models of the universe. The theorems they prove use various assumptions about the reasonableness of matter, together with a condition (strong causality) intermediate between the nonexistence of closed timelike curves (causality) and the existence of a global time coordinate (stable causality).[12] These theorems are applied to regions at the center of a collapsing star or to the early stages of the universe—regimes which are totally unlike any of which we have experience. Yet the philosophical arguments on which the assumption of causality rests[13] collapse if only the slightest departures from everyday experience are contemplated. The danger of transplanting the conclusions of philosophical arguments to an alien relativistic context could hardly be better illustrated.

A selection of some recent arguments[14] against "cyclic time" will bear the point out.

(a) Cyclic time is postulated only when one first considers the possibility of events recurring periodically, and then joins time up on itself to form a circle. (Thus, if event *B*, a recurrence of *A*, is in all respects similar to *A*, then it should be regarded as being numerically the same event.) But, the argument runs, this is fallacious because it overlooks the essential role of time: the provision of a framework which enables us to speak of qualitatively similar but numerically distinct events. Time determines the identity of events, not vice versa.

(b) In any case, the reason for postulating cyclic time in (a) is not even self consistent; for, in saying that an event recurs we are implying that there must be *two* events, one of which is a *recurrence* of the other.

(c) One of time's definitive features, part of the essence of the concept, is its linearity: an extension without linearity could therefore not be called 'time' without gross abuse of language. This arises because 'time,' through its basic definition, is linked to our consciousness and the ideas of before and after, which require linearity.

(d) Physical processes are essentially directional (enabling us to think of time as directional). But in a cyclic time all processes are periodic and so cannot have any unidirectional trend.

(e) Cyclic time is inconsistent with our undoubted participation in the world as agents. For, suppose I travel backward in time and meet my former self at an earlier age. As a free agent, what is to prevent me from drawing a gun and shooting my former self, which is a logical impossibility?

The first four of these can be dealt with summarily. The objections (a) and (b) are irrelevant to relativity because the construction of cyclic time which is countered in (a) is *not* the reason for postulating closed timelike curves in general relativity. These curves are postulated only when they are forced by the dynamics of the universe; they do not arise from making identifications in a periodic universe and, in general, no periodic universe exists from which such a causality-violating universe could be derived. For example, in the Taub-NUT or Kerr solutions, the progressive development of the universe causes the null-cones to tip so that closed curves which in one region are spacelike become timelike in another region. In neither of these models is it possible to "unwind" the time so as to remove the anomaly.

Argument (c) may be quite proper, but it is directed against our use of the word 'timelike' to describe these curves, not against their existence.

The case (d) is interesting, as it rests on a confusion between the directionality of *time* and the directionality of *systems in time,* a distinction which I have discussed at length in the previous section. We can in fact quite well have the physical processes in all relevant systems directional while time as a whole is cyclic. An example of this is provided by the steady-state model with the time coordinate "rolled up" to become periodic; this is possible because the *metric* is static, although the physical processes within that metric are directional (the universe expands). But in any case we are not interested in situations in which the entire universe has a cyclic time coordinate, but in those where there may be just one curve [15] which violates causality.

Thus we are left with (e), seemingly the most powerful argument as it rests on a clear logical contradiction. One could point out that in the realms of astrophysics under consideration (the very earliest phases of the universe or the final stages of a collapse) one cannot conceive of the presence of human beings, whether free agents or not. But this consideration alone will defeat (e) only if one is prepared to accept the position that the existence or non-existence of closed timelike curves is to be determined by the ability of human beings to withstand the climate. A factual and physical matter such as causality should not depend on such a criterion, which is not only physically arbitrary, but is also dependent on the level of technology at our disposal. Therefore I shall argue against (e) directly, showing that closed timelike curves can occur even in regions of the universe occupied by normally functioning human beings. I shall include the idea of free will, not only because I hold it to be an important fact of our experience that cannot yet be discussed satisfactorily in other terms, but also because free will produces the most powerful form of (e): my arguments will hold a fortiori if free will is not referred to.

While a will which could never be exercised would be nonsense, it is likewise unreasonable to demand that will should always achieve its ends: possession of free will does not imply omnipotence in its execution. Herein lies the solution to the apparent logical paradoxes of acausality: it turns out, from purely physical reasoning, that in a universe with closed timelike curves the laws of physics manifest themselves in an abnormal manner ("normality" being established by the behavior of physics in a universe without such curves). This abnormality is precisely such as to frustrate the execution of any wish whose accomplishment would create a logical antinomy. One's acts of will still count in the world as partial causes of what occurs; the totality of events which transpire if a certain will is exercised is different from that which would obtain were that will not exercised. The only consequence of the acausality is that the result of an act of will is not always what would be expected on a naïve analysis based on "normal" experience.

To see how this is so, consider the case already cited of a person who meets his former self in circumstances in which, if physics were normal, he would be able to shoot him. Then, as a preliminary step in the analysis, let us replace the complex human being by a simple automaton which nonetheless exhibits the abnormal physics referred to. This apparatus [16] is to consist of a gun, a target, and a shutter so arranged that the impact of a bullet on the target will trigger the shutter so as to move in front of the gun. It pursues a causality-violating curve in space-time in such a way that two points on the object's world line A and B, with B later in the object's history than A, are physically contemporaneous and disposed as in Figure 1 so that the gun at B is aimed at the target at A and the shutter is initially up at A.

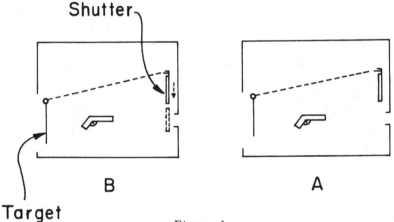

Figure 1

Suppose now that the machine "shoots its former self": the gun at B is fired, either by an automatic timing mechanism or by the intervention of a human being making a conscious decision. If the shutter in B were still up, the bullet would strike the target at A, which would cause the shutter in B to be down, a contradiction. But if the shutter were down in B, then the bullet would be stopped, the target in A would not be hit, and the shutter in B should still be up: the shutter is up if, and only if, it is down; the situation is logically impossible.

First I shall make a classical analysis, allowing for quantum effects later. Classically, as Feynman and Wheeler[17] pointed out, all the equations involved are *continuous*, and the argument to a contradiction just given is fallacious because it assumes that it is possible to set up a discontinuous situation in which the shutter is either up or down, with no intermediate state. The position of the shutter at B, x, say, is a continuous variable on which depends continuously the angle by which the bullet is deflected; this is in turn continuously related to the force of impact on the target and to the speed y with which the shutter in A is triggered to start descending. Thus $y = f(x)$, where f is a continuous function which is large for $x = 0$ (shutter right up) and zero for $x = 1$ (shutter right down). Suppose that the proper time in the apparatus between A and B is T, so that $x = Ty$; then the physical processes we have described, each a normal classical process, give rise to the equation $x = Tf(x)$ for x. This will always have at least one solution corresponding to the shutter just grazing the bullet so that it is deflected and gives the target a glancing impact, marginally triggering the shutter. Paradox is thereby avoided.

The general features of this situation are applicable to all such paradoxical arrangements. At a local level the ordinary equations of physics can be written down. They must then be solved in a global context which is abnormal. Consequently, the solution is abnormal, in that it corresponds to a type of behavior which in a causal universe would have only an infinitesimal chance of occurring. In an acausal universe miracles can occur quite often, and one must set aside one's normal judgment as to what is likely and what is unlikely.

It might seem that quantum processes are peculiarly discrete and so might produce a real discontinuity. (Actually this is open to doubt: Schrödinger's cat[18] is indeed either alive or dead; but is this a property of atomic decays, or of Geiger counters, cyanide capsules, and cats?) But in any case, if we work in the quantum domain, then we must recognize that such discreteness, if it occurs, is accompanied by indeterminacy. The continuous deterministic evolution that characterizes both the classical equations that we have just examined and the Schroedinger equation becomes (in a way still highly disputed) a *probabilistic* evolution of discrete possibilities when translated into observed outcomes. Hence it is of no avail to replace the mechanical gun and target by a quantum mechanical decaying atom and Geiger counter, for example. What one might gain in discreteness one loses in indeterminacy: even if the shutter were to close fully, there would still be a finite probability of the emitted particle tunelling through it and so triggering the counter.

When a closed timelike curve brings about a coincidence of events—such as the shutter just grazing the bullet—which would be grotesquely implausible under normal circumstances, then, and only then, normal concepts of causation and likelihood are completely disrupted. But if we are concerned only with a few timelike curves, and not with a completely cyclic time, then such coincidences will be seen as the exceptions to the normal behavior in which the concepts of causality, free will, and so on are grounded. In particular, the occasional closed timelike curve [19] will not alter the psychology of taking a free decision: it will merely alter the consequences of that decision.

It is now not difficult to imagine a way in which these factors might operate in the fully human example with which I started. The gun-toting protagonist is free to choose whether or not to shoot. If he decides against it, then there is no paradox. But if he decides to shoot, then his hand will waver and he will only graze his former self, his unexpected weakness being caused, not by divine intervention but by the flesh wound which he thereby inflicts on his former self!

Causes in general relativity

I have argued first that a general relativistic model of the universe may have no globally valid way of assigning a local sense of time, so that it may not even be time-orientable; and second, that even where a time sense can be defined, there is no reason to suppose that space-time need be causal. The most cogent objection to these ideas, in my estimation, has yet to be examined: my argument has been within the context of a physical theory whose aim is to explain the structure of the universe; but, if there is no conventional pattern of cause and effect, can any account be offered which is in any sense an explanation, and not a mere description?

This objection is usually based on too narrow an attitude to explanation—the attitude of assuming that the only possible explanation is of the "Cauchy problem" type in which a system is explained entirely in terms of physical laws and its *initial* conditions. In these last few paragraphs I shall briefly give my reasons for believing that this approach is neither necessary nor sufficient for achieving a reasonable explanatory scheme.

There are certainly some cases in which the Cauchy prediction approach might be reasonable. For instance, if we were concerned with a system of which we ourselves were external observers, then predictive arguments could be part of the prediction-test-hypothesis cycle of Popperian methodology. The scientist could gather information about, or experimentally create, the initial situation of the system. Then its state would be examined after a few seconds, days, or years to see if the theory had been supported or refuted. Clearly in this case the time which enters the theory when a prediction is made is totally linked to the laboratory time in which the physicist operates, and so must be directionally uniform and topologically linear. Yet even here, in a laboratory system, the scheme of Cauchy data may not be relevant. In studying gas contained in a cylinder in which a piston is moving, any explanation must include the externally imposed motion of the piston as part of the data.[20] Such an explanation, though not Cauchy, would be regarded as proper and scientifically illuminating.

In general relativity we are ourselves within the system, and any predictive arguments we may use about the universe are not in "real time": the mathematical process of explaining the present state in terms of an earlier state is separate from the historical process of testing the theory (a situation which is in practice almost always the case). Thus, if we wish, we can be Popperian in our methodology without using strict prediction in the mathematical models of the universe. And it may well be that we cannot use Cauchy data arguments in these models, since the work of Yodzis, Müller zum Hagen, and Seifert[21] has recently shown that "naked singularities" can occur in the universe. Data have to be given on these, just as on the surface of the piston in the cylinder. If this is so, then we must abandon any hope of being able to explain the universe in terms of initial conditions only.

If we allow types of explanation other than those based on Cauchy data, then to me there seems to be no pressing reason for retaining the sort of temporal causation that is required for the "initial condition" sort of explanation. Instead we may have to try to understand the universe in terms of laws of physics acting within a context which may be highly noncausal, having as data the conditions on all the boundaries[22] of the space-time which are not future relative to a locally determined time-sense. Such data can then be assigned according to the same criteria as are at present used in cosmology: they may be, for example, "simple" or "chaotic." As in normal predictive cosmology the data may not be given arbitrarily, but are subject to constraints which in some cases can be thought of as arising from interactions between different datum-points through the space-time.

According to the conventional view the universe can be seen as evolving from its initial condition like a watch wound up at the moment of creation.[23] This is not the case with the view I am proposing. According to my view it is not even correct to say that the data can be "prescribed," as in the piston example, since the structure of the boundaries on which the data reside is itself determined by those data. There is a web of interconnecting causation, proceeding in all temporal directions, which explains the universe as a spatio-temporal whole. By conventional standards the explanation thus achieved may appear post hoc, in that one cannot directly state acceptable data and then decide what universe results; rather, one must examine any proposed model as a whole to decide whether its data are acceptable at an explanatory level.

If this novel position is forced on us by, for example, the observation of naked singularities, or of the time-reversed instability of a black hole, then one can imagine two possible lines of development. In the first, it may prove possible to accommodate all our observations in a model in which the boundaries and the data on them are very simple: a special case of this is the standard homogeneous cosmological model usually used at present. Then all the complexity which we observe in the universe is a consequence of the interplay of physical laws within space-time. If this proves workable then a real explanation of this complexity could indeed be achieved, irrespective of causality.

The second possibility would be that the complexity of the universe could only be swept onto the boundary of space-time, not disposed of. This could happen with conventional cosmology as well as with an unconventional causality structure: adopting the wound-up-watch model does not in itself guarantee explanatory power. If this were to happen, then, in the absence of any wider theory which in turn explained the boundary data, we should have to admit that cosmology was more descriptive and less explanatory than its recent practitioners have hoped.

Notes

1. By an "instant of a history" I refer to what is usually idealized as a point on a world-line: a part of the world-tube of a person or object which has a small enough temporal extent to be regarded as being within the "now" of some relevant observer. Because this instant is a localized concept, the statement here is not tautologous.

2. 'Closed' is used in the sense of 'compact without boundary' (as in 'closed universe,' etc.).

3. There is a full presentation of this argument, followed by a critical discussion, in D. W. Sciama's chapter "Retarded Potentials and the Expansion of the Universe," in T. Gold and D. L. Schumacher, eds., *Symposium on the Nature of Time* (Ithaca: Cornell University Press, 1967), pp. 55–67.

4. This is possible in any globally hyperbolic universe. See, for example, F. Friedlander's *The Wave Equation in Curved Space-time*, Cambridge Monographs on Mathematical Physics 2 (Cambridge: Cambridge University Press, 1975).

5. In fact a high limit is eventually reached because of the onset of correlations between the movements of charges at different places. See note 3 above. This does not affect the validity of the conclusion, however.

6. I use 'sense' to mean one of the two possible orientations of a line or curve.

7. H. Reichenbach, *The Direction of Time* (Berkeley: University of California Press, 1971 reprinting). From the physicist's point of view the weakest link in his argument (and in later versions of it by other authors) is the lack of a general account of *how* branch systems in a relatively low entropy state should be formed actively in such a disequilibrium situation. The work of Prigogine and his collaborators has now clarified this to some extent.

8. This might suggest the possibility of two intercommunicating worlds whose time senses were opposite, a situation whose possibility has often been opposed by philosophers. In fact this cannot happen since it is a consequence of Sciama's argument that two worlds in mutual intercommunication must have the same time-sense, if they have any at all.

9. See, for example, B. K. Harrison, K. S. Thorne, M. Wakano, and J. A. Wheeler, *Gravitation Theory and Gravitational Collapse* (Chicago: University of Chicago Press, 1965).

10. R. H. Price, "Nonspherical Perturbations of Relativistic Gravitational Collapse: I Scalar and Gravitational Perturbations" and "II Integer-Spin, Zero-Rest-Mass Fields", *Physical Review* D5 (1972): 2419–2438 and 2438–2454.

11. O. Penrose and I. C. Percival, "The Direction of Time", *Proceedings of the Physical Society* 79 (1962): 605–616.

12. A space-time M with metric g is called stably causal if the space-times (M, g') are causal for all g' sufficiently near g (in the fine topology on metrics). This is the case if and only if the space-time has a global time-coordinate. See S. W. Hawking and G. F. R. Ellis, *The Large Scale Structure of Space-time* (Cambridge: Cambridge University Press, 1973).

13. The dependence of singularity theorems on causality assumptions has been greatly lessened by the work of F. J. Tipler, *Causality Violation in General Relativity*, University of Maryland Ph.D. thesis, (1976).

14. These arguments are derived, with heavy paraphrase, from R. Lucas, *A Treatise on Time and Space* (London: Methuen, 1973); but I am responsible for the form which they take here.

15. Metaphorically speaking! If there is one closed timelike curve, then there is an infinity of them, but they may still occupy only a small volume of space-time.

16. This example, and its resolution, comes from J. A. Wheeler and R. P. Feynman, "Classical Electrodynamics in Terms of Direct Interparticle Action," *Reviews of Modern Physics* 21 (1949): 425–434.

17. See note 16, and also A. Peres and L. S. Schulman, "Signals from the Future," *International Journal of Theoretical Physics* 6 (1972): 377–382.

18. For a critical discussion of Schrödinger's cat from a modern viewpoint see B. S. De Witt, "Quantum Mechanics and Reality," *Physics Today* (September 1970): 30–35.

C. J. S. Clarke

19. See note 15.

20. I am indebted to Professor A. Taub for this point.

21. H. Müller zum Hagen, P. Yodzis, H.-J. Seifert, "On the Occurrence of Naked Singularities in General Relativity," *Communications in Mathematical Physics* 34 (1973): 135–148; 37 (1974): 29–40.

22. By the "boundary of space-time", I mean the *b*-boundary (B. G. Schmidt, "A new Definition of Singular Points in General Relativity," *General Relativity and Gravitation* 1 (1971): 269–280). Data which can be expressed as scalars on the frame bundle sometimes have limiting values on this boundary (C. J. S. Clarke, "The Classification of Singularities," *General Relativity and Gravitation* 6 (1975): 35–40), and it might be hoped that this would generalise to genuinely singular situations.

23. See, for example, J. C. Graves, *The Conceptual Foundations of Contemporary Relativity Theory* (Cambridge, Mass.: M. I. T. Press, 1971).

—————————JOHN EARMAN—————————

Till the End of Time

1. Introduction

What could it mean to say that time has a beginning or an end? Is it possible that time has a beginning or an end? In this paper I shall not be concerned with these questions in their full generality, for I shall be concerned only with physically interesting possibilities. I cannot specify at the outset what is to count as a physically interesting possibility in the present context—substantial discussion will be needed to uncover the factors relevant to such a specification. In the sense in which I am using it, the notion of a physically interesting possibility is broader than that of a physical possibility; any actual physical possibility is a physically interesting possibility, but not conversely, although every physically interesting possibility must be intimately related to actual physical possibilities. It would seem good strategy to discuss physical possibilities first, before proceeding to the murkier concept of physically interesting possibilities. This would indeed be sound strategy, except for the fact that we do not know what counts as a physical possibility in the present context. Thus it is necessary to plunge right into murkier waters.

The particular approach that I shall explore is certainly not the only one, nor do I claim it is the best. However, it does have a virtue, albeit a negative one; it reveals that we are not now in a position to give meaningful answers to the questions posed above, and that in order to arrive at such a position it is necessary to settle a number of other questions first, some of which belong to mathematics, some to physics, and some to metaphysics.[1] Since the recognition of ignorance is often the first step toward wisdom, it is to be hoped that the way will be paved for more positive results.

2. Aristotle and Leibniz on the Beginning and End of Time

Initially, Aristotle's theory of time seems to allow for the possibility of a beginning or an end for time. According to Aristotle, time is the measure or numerable aspect of motion. Hence, if motion has a beginning or an end, time would have a beginning or an end. But Aristotle forecloses this possibility, for he says that time cannot have a beginning or an end, and that, therefore, motion must also be eternal. His argument here is that a first or a last instant of time is impossible since in conceiving of an instant of time we must conceive of it as being preceded and succeeded by other instants.[2]

Today this argument does not seem very compelling. We have learned to be suspicious of any argument that purports to prove the impossibility of Y by seeking to demonstrate the inconceivability of Y and by utilizing the premise that if Y is inconceivable, Y is impossible. I shall not stop to give a detailed analysis of Aristotle's argument, for my main concern is to examine the implications of modern science for the questions at issue, and from the point of view of modern science, Aristotle's theory of time has only a curiosity value. For from the modern point of view, time must be seen as the temporal aspect of the more fundamental entity, space-time, and modern science countenances space-time structures which, in a precise sense, do not harbor any physical change and whose temporal aspects are, in a precise sense, infinite in both past and future.[3]

However, Aristotle's views do raise some points that help to focus the issues under discussion. Aristotle seems to take the statement, "Time comes to an end," to mean, "There is a last instant for time." He does not seem to have conceived of the possibility that time could *come to an end* without coming to *an* end—that time could, so to speak, "run out" in the future direction without there being a last instant. Suppose, for example, that time can be represented by the metric space (I, d_e) where I is the open interval $(-\infty, +1)$ of the real line $\mathbb{R}$ and d_e is the usual Euclidean metric (it is understood that the positive direction of $\mathbb{R}$ corresponds to the future direction of time). In this case there is no last instant for time, but time is finite in the future in the sense that there is a finite upper bound on how far one can go in the future direction from any given point of time, i.e., for any $x \in I$, there is a finite N_x such that for any $y \in I$ where $y > x$, $d_e(x, y) < N_x$.[4] I suppose Aristotle might have responded that such a possibility is not a real one, and that to conceive of an instant of time, we must conceive of other instants which precede and succeed the first by as great an interval as we like. But such a response should be taken, I think, as indicating that its author has limited powers of conception.

Conversely, the existence of a future end point for time is not sufficient by itself to guarantee that time comes to an end in the sense intended, at least not if we are willing to extend the concept of a metric space (X, d) so that d may be mapping from $X \times X$ to $[0, + \infty]$. For example, let X be the extended real line $\widetilde{\mathbb{R}}$ which is obtained from $\mathbb{R}$ by adjoining the points $\pm \infty$ to $\mathbb{R}$, defining $- \infty < x < + \infty$ for any $x \in \mathbb{R}$. Define $d(x, y) = d_e(x, y)$ and $d(\pm \infty, x) = d(x, \pm \infty) = + \infty$ for $x \in \mathbb{R}$; $d(+ \infty, + \infty) = d(- \infty, - \infty) = 0$, and $d(+ \infty, - \infty) = d(- \infty, + \infty) = + \infty$. Time as represented by (X, d) has first and last instants, but they are infinitely far in the past and future so that time never runs out in the past or future directions.

Thus in what I shall refer to as the Aristotelian conception of the beginning (end) of time, two elements are involved: (1) time is finite in the past (future) in some appropriate sense, and (2) there is a past (future) end point for time. I shall begin by investigating condition (1); only in the latter part of the paper shall we be in a position to deal with (2) in a meaningful way. It might be thought that if we have a model which satisfies condition (1), we can always make it into a model which illustrates the Aristotelian conception of the beginning or the end of time by adjoining end points. We shall see, however, that things are not so simple; the illusion of simplicity is fostered by two pernicious tendencies: first, the tendency to think of time as an autonomous entity rather than as an aspect of space-time, and second, the tendency to think of time as being represented by an interval of the real line. In these respects, the discussion of this section and, indeed, most of the discussion of these matters in the philosophical literature has been misleading.

I turn now to Leibniz's views. Leibniz believed that there are possible worlds which have a beginning or an end. But when combined with the following argument, his Principle of Sufficient Reason poses a problem for the actuality of any such world: "Time can be continued to infinity. For since a whole of time is similar to a part, it will be related to another whole of time as its part is to it. Thus it must always be understood as being capable of being continued into another greater time."[5] So for any possible world W which has, say, a beginning, there is another world W' which extends W to past infinity. What sufficient reason could God have for actualizing W rather than W'? And does not the Principle of Plenitude suggest that He would choose W' over W?

Leibniz addresses himself to these questions in a letter to Bourguet.[6] If nature is "always equally perfect, though in variable ways, it is more probable that it had no beginning" (because, presumably, W' would then be more perfect than W). On the other hand, if nature decreases in perfection as we go backward in time, and if the perfection decreases at such a rate that, within a finite period, we reach zero and negative perfection, then it is more probable that time has a beginning (since, presumably, W would then be more perfect than W'). Leibniz simply left the matter hanging since he was unwilling to commit himself on whether the perfection of nature changes at such a rate as to make a beginning for time probable, and even on whether it is changing at all.

As we shall see below, relativity theorists have had to confront issues similar to those with which Leibniz struggled. But we shall also see that certain relativistic space-time models circumvent the main problem.

3. The Meaning of Temporal Finiteness within a Relativistic Space-time Framework

If we take seriously the notion that time must be thought of as the temporal aspect of space-time, we are led to ask what sort of space-time structure would illustrate the notions of the beginning and the end of time. In what follows, I shall work with relativistic space-times, and this for two reasons. First, and most obviously, current evidence indicates that actual space-time is relativistic. The fact that the possibilities to be discussed below can be constructed within this framework makes for some initial confidence that they will lead to physically interesting possibilities. Secondly, and less obviously, certain of the possibilities to be discussed cannot be realized within the orthodox Newtonian framework.

Definition 1. A *relativistic space-time* S is a triple $\langle M, g, \nabla \rangle$ where M is a connected, four-dimensional differentiable manifold, g is a Lorentzian metric for M, and ∇ is the unique symmetric linear connection compatible with g.[7]

According to general relativity theory, which will be taken as our guide to physically interesting possibilities, space-time structure and the distribution of matter-energy are not independent. Thus we must consider cosmological models.

Definition 2. A *cosmological model* $\mathfrak{M}$ is a pair $\langle S, \mathcal{E} \rangle$ where S is a space-time and $\mathcal{E}$ is an energy-momentum tensor.[8]

Conditions for a physically interesting example of how time can have a beginning or an end will be conditions on cosmological models. However,

I shall concentrate primarily on conditions which the space-time $\mathcal{S}$ must satisfy, for once we have settled on these conditions, it will prove easy to find a cosmological model which incorporates $\mathcal{S}$ and which satisfies all the conditions that one might plausibly impose on $\mathcal{S}$ and on $\mathcal{S}$ and $\mathcal{E}$ together.

It will be assumed that $\mathcal{S}$ is temporally orientable so that it can be assigned a consistent time sense or direction, and that one of the two possible time orientations has been singled out so that one can meaningfully speak of the future direction of time.[9] The restriction imposed by this requirement is not a strong one; for if $\mathcal{S}$ is not temporally orientable, there always exists a covering space-time that is. And in any case, the questions at issue do not even arise unless $\mathcal{S}$ possesses the assumed feature. We shall shortly see that other features as well need to be assumed for the same reason.

It is easy to display relativistic space-times that are finite in their temporal aspects. For example, start with Minkowski space-time $\mathcal{S}_{Min}$ and obtain the hypertorodial space-time $\mathcal{S}_{roll}$ by rolling up M_{Min} in the spatial sense and then rolling up the resulting space in the temporal sense.[10] The temporal aspect of $\mathcal{S}_{roll}$ is finite in a straightforward sense, but this sense of finiteness is not appropriate for our present concerns; for time in $\mathcal{S}_{roll}$ is not bounded in the past or future in any appropriate way, and, indeed, there is no past or future in the usual sense, since every point of M_{roll} lies to the past and future of itself and every other point.

The moral is that for $\mathcal{S}$ to illustrate how time can be finite in the past or future, something like time in the usual sense must be present to start with. Exactly which temporal features $\mathcal{S}$ must display is open to debate, but certainly $\mathcal{S}_{roll}$ does not qualify. Some of the features that might plausibly be required are given in the following definitions.

Definition 3. The (time-oriented) space-time $\mathcal{S}$ has *global time order* if and only if there do not exist any nontrivial, closed, future-directed, timelike curves.[11]

If $\mathcal{S}$ has a temporal order according to definition 3, then the relation $\prec$ on $M \times M$ where $x \prec y$ (read "x is chronologically earlier than y") is defined to hold just in case there is a differentiable map $\sigma \colon [0, 1] \to M$ such that $\sigma(0) = x$, $\sigma(1) = y$, and $\dot{\sigma}(\lambda)$ is future-pointing for all $\lambda \in [0, 1]$, is transitive and asymmetric. Hence M is partially ordered by $\preccurlyeq$, where $x \preccurlyeq y$ holds just in case $x \prec y$ or $x = y$.

113

John Earman

Definition 4. $\mathcal{S}$ possesses a *global time function* if and only if there is a smooth map $t: M \rightarrow \text{IR}$ such that $t(x) < t(y)$ whenever $x < y$.

The existence of a time order does not necessarily imply the existence of a global time function. If $\mathcal{S}$ does have a global time function, we can project out a one-dimensional time T: an "instant" (a point of T) corresponds to a time slice of M, i.e., a set $S_\lambda \underset{\text{def}}{=} \{x: x \in M$ and $t(x) = \lambda, \lambda =$ constant$\}$. The points of T are totally ordered by the relation of temporal precedence induced on T by the natural order of the slices S_λ.

Definition 5. Suppose now that $\mathcal{S}$ has a global time function t. Consider the field of future-pointing unit normals to the S_λ. If this field is sufficiently smooth, its integral curves will form a timelike congruence. t is *metric* if and only if each pair of the S_λ are a constant space-time distance apart as measured along the curves of the congruence.

If t is metric by definition 5 and T is the one-dimensional time associated with t, we can find a metric space (X, d), where X is an interval of IR, and a one-one correspondence $f: T \rightarrow X$ such that for $a, b \in T$, $d(f(a), f(b))$ is the space-time distance between the slices corresponding to a and b. However, this representation of T as a metric space can be misleading for our purposes. *If* there is a timelike curve of maximal proper length from some point $x \in M$ to S_λ, then this curve must be a geodesic normal to S_λ; and the curves of the congruence in definition 5 are geodesics which are normal to S_λ. But it is *not* always the case that there is a timelike curve of maximal length from some arbitrary $x \in M$ to S_λ. If the S_λ have the Cauchy property defined below, then such a maximal curve will always exist.

Definition 6. A time-slice S of the space-time $\mathcal{S}$ (i.e., a properly embedded spacelike submanifold without boundaries) is a *Cauchy surface* of $\mathcal{S}$ if and only if every future-directed timelike curve without end point intersects S once and only once.

Suppose that $\mathcal{S}$ possesses a global time function. No future-directed timelike curve can intersect any of the corresponding S_λ more than once. We ordinarily assume that if a process "goes on forever" (has no past or future end point), then there must be for each instant of T a stage of the process dated by that instant. This assumption will not be satisfied unless the S_λ have the Cauchy property.

Since the existence of a global time function and the existence of Cauchy surfaces are strong requirements, it would seem desirable to have

114

a definition of temporal finiteness that does not rely on these requirements. Towards this end, one might try

Definition 7. Let $\mathcal{S}$ be a time ordered space-time. A *half-curve* of $\mathcal{S}$ is
a curve which has one end point and which has been extended as far as
possible in some direction from that point. A future- (past-) directed
timelike half curve is *complete in the future (past)* if and only if proper
length as measured along the curve from its end point assumes arbitrarily
large values; otherwise the curve is *incomplete in the future (past)*. $\mathcal{S}$ is
finite in the future (past) or *future- (past-) bounded* if and only if every
future- (past-) directed timelike half-curve is incomplete in the future
(past).

Note that *any* relativistic space-time contains some incomplete timelike
half-curves. However, in the more commonly known cases, all timelike
half-curves of bounded acceleration are complete; in particular, all
timelike half-geodesics are complete, geodesics being curves of zero acceleration.

Unfortunately, definition 7 is unsatisfactory. Start with Minkowski
space-time $\mathcal{S}_{Min}$ (which for sake of illustration is taken to be two-
dimensional), and remove all those points on or above some chosen null
hypersurface. (Note: if $\mathcal{S} = \langle M, g, \nabla \rangle$ is a space-time, then the result of
removing a closed set of points from M and restricting g and ∇ to the
remainder is again a space-time.) The resulting space-time $\mathcal{S}_{null}$ (see
Figure 1) will be finite in the future according to definition 7. But should
$\mathcal{S}_{null}$ be counted as finite in the future in the sense being sought? It is

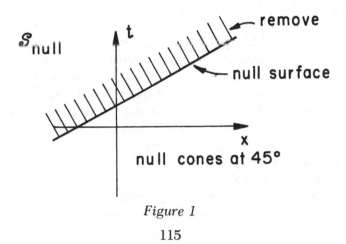

Figure 1

true that in $\mathcal{S}_{null}$ every possible observer will "run out of time." But time itself can be said to go on forever, since for every metric time function t, the metric of the associated one-dimensional time T is not future-bounded.

A somewhat more complicated example gives an even clearer indication of the shortcomings of definition 7. Start again with $\mathcal{S}_{Min}$ and remove spacelike hypersurfaces at regular intervals as indicated in Figure 2. The resulting space-time $\mathcal{S}_{strip}$ is finite in both the past and future according to definition 7; but clearly its temporal aspect is not finite in the intended sense.

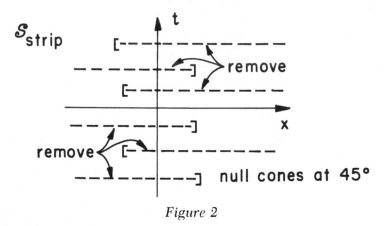

Figure 2

The moral is that the collection of individual timelike curves may give a misimpression of temporal finitude; global time functions may have to be consulted in order to gain an accurate impression. But the exact form the consultation should take is not easy to specify. Start yet again with $\mathcal{S}_{Min}$, and let (x, t) be a pseudo-Cartesian coordinate system for M_{Min}. Obtain the truncated space-time $\mathcal{S}_{trun}$ by deleting all those points whose temporal coordinates satisfy $t \geq 1$ (see Figure 3). As restricted to M_{trun}, t is a global metric time function for $\mathcal{S}_{trun}$, and the associated temporal metric is future-bounded. Consider, however, another pseudo-Cartesian coordinate system (x', t') for $\mathcal{S}_{Min}$. As restricted to M_{trun}, t' is also a metric time function for $\mathcal{S}_{trun}$, but the metric of the associated time is *not* in general future-bounded.

My own reaction is that, in the sense intended, $\mathcal{S}_{trun}$ should be regarded as finite in the future. The existence of a time function t' that

116

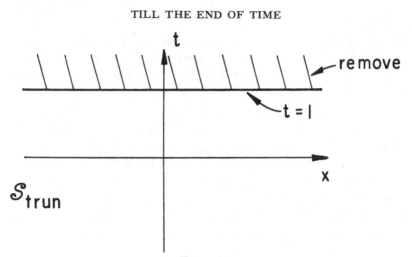

Figure 3

makes $\mathcal{S}_{trun}$ look infinite in the future does not detract from this conclusion since the original time function t has a preferred status with respect to the present problem: the t = constant time slices are Cauchy surfaces for $\mathcal{S}_{trun}$ whereas the t' = constant hypersurfaces are not. If my view is correct, then definition 7 is satisfactory when applied to space-times with a Cauchy surface. In such a case there exists a global time function all of whose time slices are Cauchy surfaces. And past (future) boundedness in the sense of definition 7 is equivalent to the past (future) boundedness of the metric associated with such a time function.

However, as the examples of $\mathcal{S}_{null}$ and $\mathcal{S}_{strip}$ reveal, definition 7 cannot be applied with confidence beyond this restricted class of cases. It is difficult to think of a general criterion that can be so applied. Suppose, for example, that $\mathcal{S}$ has a global metric time function but that no global metric time function for $\mathcal{S}$ has time slices with the Cauchy property. Shall we then decree a democracy among metric time functions and say that $\mathcal{S}$ is temporally finite if and only if all the temporal metrics of the one-dimensional times associated with the metric time functions are appropriately bounded? This would be too strong a requirement. If we remove one additional point from $\mathcal{S}_{trun}$ then there are no Cauchy surfaces in the resulting space-time, and, consequently, it would not be counted as temporally finite in the future according to the criterion under consideration. This is an unacceptable result.

At this juncture it must be strongly emphasized that the above exam-

117

ples are of more than mere academic interest; for physical space-time singularities, e.g., "curvature singularities," can play the role of the deleted regions. And such singularities have been shown to be a pervasive feature of the solutions to Einstein's field equations (see below).

In summary, unless a space-time satisfies some very strong causality conditions, it may not make sense to ask, "Is time finite or infinite?" Even when such conditions are satisfied, it is difficult to formulate an adequate general criterion for temporal finiteness, if indeed one exists. That the finite-infinite distinction becomes blurred in the relativistic context is an interesting point, and it deserves more attention in the philosophical literature. But having made it, I propose to avoid it in what follows by concentrating on cases in which we can agree that an appropriate type of finiteness obtains.

Before closing this section I want to comment on the meaning of the Aristotelian conceptions of the beginning and the end of time. It might be thought that the way to illustrate these conceptions within the relativistic context is to modify the picture of space-time so that a space-time can be a manifold with boundary, the boundary being the disjoint union of two spacelike three-manifolds, one of which can be interpreted as the "first instant" and the other as the "last instant." Prima facie, the existence of such boundaries is problematic. Their existence prompts one to ask such questions as "What explains the appearance of these boundaries?" and "What is beyond these boundaries?" Perhaps such questions can be dismissed as not legitimate. Or perhaps there are quite straightforward answers, e.g., the answer to the second question may be simply "Nothing!" What is problematic is how in general to attach a "future (past) boundary" to a space-time that is future- (past-) bounded. In some cases there is an obvious and natural procedure, but such cases turn out not to provide physically interesting illustrations of how time can have an end or a beginning, or so I shall argue below. For the moment, let us set aside the questions of boundaries and of how to implement condition (2) of the Aristotelian conceptions of the beginning and the end of time.

4. Truncated Space-times

Although the example of the truncated Minkowski space-time S_{trun} presented above is a trivial example, it poses a nontrivial problem for the philosophy of time: are there good reasons for believing that actual space-time cannot be truncated in the way S_{trun} is?

If one holds that there are no good reasons, then one would be wise to receive extreme unction as soon as possible, since time may run out any second now. Moreover, such a position implies a profound skepticism with respect to our knowledge of the past and future; it implies that we do not in fact know the great bulk of the things we ordinarily claim to know about the past and future. For if a person has no reason to believe p, then he does not know p; and if p is a presupposition of q and he does not know that p, then he does not know that q.[12] And, needless to say, most of our ordinary knowledge claims about the past and future presuppose that there is a past and future. Thus my claim to know that I existed five seconds ago (as measured along my world-line), or that barring certain catastrophes, the universe will still exist five seconds from now, presupposes respectively that actual space-time is not truncated in the past (future) in such a way that proper length along my world-line as measured backwards (forwards) in time from now never reaches five seconds; hence, if I have no good reason to believe that space-time is not truncated in the manner described, I do not know that I existed five seconds ago or that the universe will exist five seconds from now.

One's initial reaction to such skepticism is apt to be that it is too absurd to be taken seriously. Whether or not this reaction can be sustained remains to be seen.

On the other hand, if one holds that there are good reasons to reject $\mathcal{S}_{trun}$ and its like as a model for actual space-time, then these reasons must be supplied. We shall see that this order is less simple to fill than one might think at first glance. But before going into details, it will be helpful to contrast the above examples with another sort.

5. Another Example—Temporally Finite but Untruncated Space-time

Definition 8. A space-time $\mathcal{S}' = \langle M', g', \nabla' \rangle$ *properly extends* the space-time $\mathcal{S} = \langle M, g, \nabla \rangle$, if and only if M is isometrically embeddable as a proper subset of M. $\mathcal{S}$ is *inextendible* if and only if there is no $\mathcal{S}'$ that properly extends $\mathcal{S}$.

Can there be a space-time that is inextendible and yet is finite in the past or future? If so, the finiteness cannot be the result of truncation surgery.[13] The answer is affirmative; specific examples will be studied below (see section 8).

119

We can now see one of the virtues of the four-dimensional point of view, from which time is seen as an aspect of space-time. For from the one-dimensional point of view, there is no distinction between the temporal finiteness of $\mathcal{S}_{trun}$ and the inextendible space-times referred to above. And from the one-dimensional point of view, it is hard to fault Leibniz's assertion that time is always capable of being extended into a greater time. But not so in the four-dimensional view! Another virtue of the four-dimensional view will emerge below; namely, that general relativity provides a mechanism for realizing temporal finiteness in space-time.

6. Arguments against Truncation

There are physical considerations, both observational and theoretical, that can be brought to bear against certain kinds of geodesic incompleteness. For example, we never observe particles in inertial motion simply to pop out of existence, and conservation principles weigh against our ever observing this. But such considerations do not seem to operate against $\mathcal{S}_{trun}$; for here it is the whole matrix of existence that, so to speak, pops out.

(It is worth noting that in the present context, the problems attending the notion of space coming to an end are not wholly different from those attending the notion of time coming to an end. If we truncate Minkowski space-time in the spatial sense (e.g., delete all those points whose pseudo-Cartesian coordinates (x, t) satisfy $|x| \geq A$, where A is a positive constant), all the ancient problems arise about what happens when we poke a spear at the "edges" of space; but these problems arise here in a form not contemplated by the ancients, and, in fact, in the present context they are part and parcel of the problems involved in interpreting incomplete timelike curves. On the other hand, there are important differences between space and time. For example, adding the time slice consisting of all those points of M_{Min} whose temporal coordinates satisfy $t = 1$ as a future boundary for M_{trun} does not help to resolve any problems. But it is possible to deal with the timelike incompleteness arising from truncating in the spatial sense by adding a spatial boundary (in the above example, the set of all points satisfying $|x| = A$) and by treating this boundary as a rigid reflecting barrier; now, of course, the problems raised by the ancients do arise in the form they contemplated.)

Nor do the basic laws of relativistic physics help to rule out truncated space-times. Thus, for example, it follows from the local nature of Einstein's field equations that if the cosmological model $\langle \mathcal{S}, \mathcal{G} \rangle$ is a solution, so is $\langle \mathcal{S}_{trun}, \mathcal{G}_{trun} \rangle$.

At this point, common sense cries out: "If my world line cannot be extended backwards in time from now more than five seconds, then most of my memory impressions are mistaken. But this is absurd." But exactly where does the absurdity lie? There is no logical contradiction, nor—apparently—any inconsistency with the basic laws of physics. Of course, it does seem that the likelihood that the universe switched itself on, so to speak, a few seconds ago in such a way that people are endowed with the memory impressions they in fact have, is very low. But this estimate of likelihood is surely not an estimate of probability based on either observed relative frequencies or on the implications of the laws of physics. The point becomes clearer when we contrast the present case with a case in which we assume that space-time is not truncated in the past, and ask for the probability of the spontaneous appearance of fossillike objects. Here laws and observed relative frequencies can be brought to bear, e.g., statistical mechanics tells us that the probability of a spontaneous fluctuation producing the 'fossils' is vanishingly small. No such argument is available in the case in which we ask for the likelihood of the universe being truncated in the past. The estimate of low likelihood in this case seems simply an expression of our unwillingness to accept a certain kind of explanation. But whether or not this unwillingness can be backed by philosophically respectable reasons is the question at issue.

Moreover, even if memory impressions, 'fossils' and the like did provide reasons for rejecting truncation in the past, there still remains the case of truncation in the future. Here I am inclined to think that the best we can hope for is a pragmatic vindication of the rejection of future-truncated space-times as models of the actual universe. Long and frustrating experience has revealed the fruitlessness of attempting to refute radical skepticism with respect to our knowledge of the external world. However, one can hope to show that such skepticism is inquiry-limiting and in this way justify proceeding as if the skepticism were false. And so I believe it is with future truncation.

Intuition would have it otherwise—the actual space-time world cannot be truncated in the future. Nor is this intuition raw and untutored. Indeed, it is not much of an exaggeration to say that this intuition is the starting point of recent research on gravitational collapse. Consider the original Schwarzschild vacuum solution to the Einstein field equations. (The reader need not be familiar with the details in order to follow the argument.) Focus on an observer whose world-line approaches the Schwarzschild radius. It is found that the observer would have to use up an infinite amount of Schwarzschild coordinate-time in order to reach this radius. But another calculation shows that only a finite amount of proper time elapses. Therefore, either the observer encounters some sort of violent agency at the Schwarzschild radius and is snuffed out of existence, or he simply runs out of time and ceases to exist, or he crosses over the Schwarzschild radius into a region of space-time not covered by the Schwarzschild coordinates. The first possibility can be ruled out, since it can be shown that the space-time metric is perfectly regular as one approaches the Schwarzschild radius (the so-called Schwarzschild singularity is a coordinate singularity, not a real singularity). The second possibility is ruled out on the grounds that "it would be unreasonable to suppose that the observer's experience could simply cease after some finite time, without his encountering some violent agency." [14] We are left then with the last alternative. If we admit this, we are driven to ask what happens to a massive object with exterior Schwarzschild field once it collapses within its Schwarzschild radius. This is the beginning of the story of gravitational collapse, a story that cannot be told here. What should be told here is that the argument just considered is essentially an application of Leibniz's Principle of Sufficient Reason. Had Leibniz lived to read it, he would no doubt have claimed that it provides an illustration of his doctrine that physics rests on metaphysical principles.

Without taking a stand on this doctrine, I am simply going to assume in what follows that certain kinds of truncated space-time do not provide physically interesting examples of how time can be finite. The final judgment about the justifiability of this assumption must await further investigation.

7. Conditions for Physically Interesting Finite Pasts and Futures

The above discussion leads to the following conditions (or more properly, to condition schema, since the details are left open) on a cosmological model $\langle S, \mathcal{E} \rangle$ if it is to serve as a physically interesting example of how time can be finite in the past (future):

(1) S possesses certain properties, among which are the property of

having a global time order and, possibly, the properties of definitions 4–6.

(2) S is past- (future-) bounded.

(3) S has certain features, among which is, perhaps, the feature that there are no negative energy densities (see below).

(4) S and S together satisfy certain conditions, among which is, perhaps, that of being a solution of Einstein's field equations.

(5) $\langle S, S \rangle$ is maximal with respect to (1), (3), and (4), i.e., there is no $\langle S', S' \rangle$ which properly extends $\langle S, S \rangle$ and which satisfies (1), (3), and (4).

The condition most open to question is (5). Finiteness in the past or future might seem so objectionable that we would expect nature to go on building even if she could do so only by employing building blocks that do not satisfy (1), (3), and (4). This suggests that we strengthen (5) to

(5') S is inextendible.

On the other hand, (5) and (5') may prove to be too strong. If nature starts by building a cosmological model $\langle S, S \rangle$, we cannot require her to go on building until a model that is maximal with respect to properties P is reached if there is no such maximal space-time. And unless the set of properties P is chosen with care, it may not be provable that every space-time having properties P is contained in a space-time that is maximal with respect to P. If $S = \langle M, g, \nabla \rangle$ has a Cauchy surface C and if we consider only those extensions $S' = \langle M', g', \nabla' \rangle$ of S such that the image $C' = \phi(C)$ of C under the embedding map is a Cauchy surface of S', then it can be shown that there is a unique maximal extension $\tilde{S}$ of S.[15] But although $\tilde{S}$ may be maximal in this sense, it may not be a good model for illustrating how time can be finite. For by deleting portions of Minkowski space-time S_{Min}, we can obtain a space-time S_{del} which is maximal in this sense and which is past and future-bounded; but S_{del} is just as objectionable as the original truncated space-time S_{trun} discussed in section 4.[16]

Another approach is to use the concept of a framed space-time, a space-time $S = \langle M, g, \nabla \rangle$ together with an orthonormal tetrad at some point of M. If we restrict attention to framed space-times and extensions in which the embedding map carries the preferred frame of the one onto the preferred frame of the other, it can be proven that every space-time is contained in a maximal space-time.[17] However, the use of preferred frames is somewhat artificial. And it is not obvious that any framed space-time with properties *P* can be extended to a framed space-time which is maximal with respect to *P*.

All this suggests that (5) may have to be weakened to

(5″) $\langle \mathcal{S}, \mathcal{E} \rangle$ is not extendible to an $\langle \mathcal{S}', \mathcal{E}' \rangle$ which satisfies (1), (3), and (4) and which itself is not past- (future-) bounded.

However, (5″) may be too weak. For there are truncated cosmological models that cannot be extended as a solution to Einstein's field equations to a model that is not past- or future-bounded (consider, for example, truncated versions of the Friedmann models—see the following section); such truncated models will again be as objectionable as the original one.

Fortunately we can carry forward the discussion without having to decide precisely what form of the inextendibility condition to impose; for we shall see that there are cosmological models which satisfy the strongest form of the inextendibility condition and which have the other properties we desire.

8. Some Physically Interesting Examples of Temporally Finite Space-times

The considerations of the preceding section are closely related to the definition of "singular space-time" adopted by a number of physicists: $\mathcal{S}$ is said to be singular if no extension of $\mathcal{S}$ is timelike geodesically complete (i.e., no extension is such that every timelike half-geodesic is complete). Several theorems have been proved about the existence of singularities in the cosmological models of the general theory of relativity. These theorems will not in general be relevant to our present concerns; for we are not interested in timelike incompleteness in general—we are interested only in cases in which the incompleteness is of such a global nature that time runs out for all possible observers and in which $\mathcal{S}$ satisfies the other conditions discussed in section 7. There is one general class of cases in which the singularity theorems will be of general interest; namely, those in which $\mathcal{S}$ contains a Cauchy surface. For, in the first place, such an $\mathcal{S}$ will possess a global time function. Second, if $\mathcal{S}$ is timelike geodesically incomplete, we may be able to prove that the incompleteness is of a global nature. Moreover, the existence of a Cauchy surface adds a bonus—it makes possible the implementation of Laplacian

Determinism; thus it may be possible to predict (retrodict) from the state of the universe at a given instant, the end (beginning) of time.

In fact, general relativity predicts that in many seemingly physically interesting cases, time is finite in the past or future. More precisely, let $\langle \mathcal{S}, \mathcal{S} \rangle$ be a cosmological model such that Einstein field equations (without cosmological constant) [18] are satisfied and $\mathcal{S}$ possesses a Cauchy surface C. Then if the convergence of the future-pointing unit normals to C is everywhere greater than some positive constant C_0 (this means that at the instant corresponding to C, the universe is everywhere contracting at a rate at least as great as that given by C_0), and if a condition on the energy-momentum tensor is satisfied (in typical cases, this condition can be violated only by having negative energy densities or large negative pressures), then as measured from C, no future-directed timelike curve of $\mathcal{S}$ has a proper length greater than $3/C_0$.[19]

The Friedmann cosmological models satisfy the hypotheses of this theorem and/or the temporal converse of the convergence condition; thus they are finite in the past and/or in the future. Moreover, the space-times involved are inextendible—none of them can be embedded as a proper subset of any space-time, much less a solution to Einstein's field equations, and they are therefore maximal in the strongest sense. Thus they would seem to qualify as physically interesting examples of how time can be infinite in the past or future.[20]

9. Examples Reconsidered

How physically realistic are such models? Two opposing views on this question in particular and on the singularity theorems in general have emerged. According to one view, no space-time can be regarded as being physically realistic unless it is timelike geodesically complete. The fact that timelike geodesic incompleteness is a pervasive feature of the cosmological models of general relativity is taken as an indication that something is drastically wrong with the theory.[21] The opposing attitude is that nothing is wrong with the theory and that, therefore, we must learn to live with singularities. Thus C. Misner[22] argues that since observational evidence together with Einstein's theory suggests that time in our universe has a beginning, we had better get used to the notion that time has an "absolute zero."

However, Misner goes on to say that *"the universe is meaningfully infinitely old because infinitely many things have happened since its be-ginning.*"[23] Viewed as a way of making the notion of the beginning of time more palatable, this statement is unexceptionable; but viewed in another light, it threatens to undermine the analysis given above. For in support of this statement, Misner introduces a time scale Ω which is related logarithmically to the proper time scale and which is "attractive as a primary standard" in that "significant epochs (e.g., galaxy formation, nucleo-synthesis hardon era, etc.) are spaced at reasonable intervals of Ω."[24] But on the Ω scale, the universe is infinitely old. Therefore, if the Ω scale were accepted as the "primary standard," time would have no beginning on the above analysis since it would not be past-bounded on the Ω scale. And, it might be asked, what can justify the use of the proper time scale to the exclusion of all other time scales? The answer is that it may be useful to employ the Ω scale or some other scale in discussing some phenomenon, e.g., galaxy formation; but if one is to believe the theory in the context of which the discussion is taking place, space-time is equipped with a metric that gives the measure of space-time distances, and it is this metric that we must use in answering the question, "Does time have a beginning or an end?"

10. Singularities and the End Points of Time

Misner uses the term "singularity" to cover not only the case of timelike geodesic incompleteness but also the case in which there is some "in-finity," e.g., infinite curvature or infinite mass density. In suggesting that a singularity may well have occurred in our universe at some finite proper time in the past, he presumes that the singularity may well involve some such infinity. But this sort of talk is not consistent with the point of view we have adopted so far, for such talk assumes there are "singular points" at which the curvature or mass density becomes infinite; but we have been assuming that space-time is a Lorentzian manifold and that, in particular, the metric of M is everywhere non-singular (i.e., defined and differentiable at every point of M).[25] Thus, in speaking of a Friedmann universe, we mean Friedmann space-time $\mathcal{S}_F$ with singular points omitted, and in the statement that $\mathcal{S}_F$ is not extendible to a larger space-time $\mathcal{S}_F'$, $\mathcal{S}_F'$ must be taken to be without singular points. (For example, the proof of the inextendibility takes the following form for the spatially closed Friedmann space-time which is finite in the past and future. Assume for purposes of contradiction that M_F can be isometrically embedded as a proper subset of M_F' and let $\phi\ (M_F)$ denote the image of M_F under the embedding map. Then we can find a point p ϵ $(M_F' - \phi(M_F))$ and a sequence of points x_i ϵ $\phi(M_F)$ such that the sequence converges to p and the scalar curvature $\mathcal{R}(x_i)$ diverges as $x_i \to$ p. But this yields a contradiction, since the curvature invariants are differentiable functions of space-time position.)

There is good reason not to admit singular points as part of the space-time manifold M: unless some limitation is put on the type of singularity we admit, we shall not have any theory at all, and there does not seem to be any good motivation for picking out the admissible singularities or for prescribing how many and in what configuration they will be placed on the manifold.[26]

Still, it is possible that singular points can be treated by joining them to M so as to form a "boundary" for M. Some such move must be made if our picture of space-time is to accommodate the notion that the scalar curvature "becomes infinite." (It is at this point that I part company with R. Swinburne,[27] who argues that if the best confirmed cosmological theory implies that there is a time t_0 in the past at which matter was infinitely dense, then we can conclude that the universe (which Swinburne takes to mean the collection of all physical objects that are spatially related to the earth) came into existence after t_0, since an infinitely dense state of matter is physically impossible. But if it really is physically impossible, then the conclusion we must draw from Swinburne's argument is that the theory in question is false. Moreover, on my interpretation of relativity theory, models like those of Friedmann do not imply that there was or will be a time at which matter was or will be infinitely dense; rather, they entail the more radical consequence that time itself is finite in the past or future. On the other hand, such models do imply that singular states do "occur"—not at any point of time, but at ideal points which are attached to space-time by some procedure yet to be described.)

Also we would like to have a means of illustrating the Aristotelian conception of the beginning or end of time, i.e., a means of representing first or last instants of time.

Mathematically, what we want is a prescription for associating with any future- (past-) bounded space-time $\mathcal{S}$ a "future (past) boundary" for $\mathcal{S}$. One immediately runs into a problem if this task is interpreted in the following literal way: for any future- or past-bounded space-time $\mathcal{S} = \langle M, g, \nabla \rangle$, find a manifold with boundary $\overline{M}$ such that $\overline{M} = M \cup \partial(M)$

and the boundary $\partial(M)$ is the disjoint union of the "future boundary" and the "past boundary." In the case of the truncated Minkowski space-time $\mathcal{S}_{trun}$, the obvious and natural way to obtain $\overline{M}_{trun}$ is to take the "future boundary" to be the set of all points whose temporal coordinates satisfy $t = 1$. But the very reason why there is an obvious procedure in this case—the fact that there is an extension $\mathcal{S}_{Min}$ of $\mathcal{S}_{trun}$ so that the future boundary of $\mathcal{S}_{trun}$ can be derived by taking the closure of M_{trun} in M_{Min}—means that condition (5′) of section 7 is violated. And it can be shown that for any time-orientable space-time $\mathcal{S} = \langle M, g, \nabla \rangle$, if there exists a manifold with boundary $\overline{M} = M \cup \partial(M)$ such that the boundary $\partial(M)$ is a spacelike three-manifold, then $\mathcal{S}$ is extendible to an $\mathcal{S}' = \langle M', g', \nabla' \rangle$ so that $\partial(M) = Cl(M) \cup Cl(M' - M)$ where Cl denotes the operation of taking the closure in M'. Thus we have a dilemma. If $\mathcal{S}$ is, say, a future-bounded and inextendible space-time, then the "future boundary" for $\mathcal{S}$ cannot be represented as a boundary of the space-time manifold of $\mathcal{S}$, at least not in the way the time slice $t = 1$ of Minkowski space-time does for $\mathcal{S}_{trun}$. On the other hand, if $\mathcal{S}$ is extendible to a larger space-time, it violates condition (5′) and thus may fail to qualify as an interesting example of how time can come to an end. To sum up, the straightforward way of trying to illustrate the Aristotelian conceptions of the beginning and end of time seems to be blocked.

A more sophisticated approach is needed. In order to illustrate the possibilities, I shall briefly describe the g-boundary approach.[28] For any given space-time $\mathcal{S} = \langle M, g, \nabla \rangle$, this approach associates with each incomplete timelike geodesic an ideal end point. An equivalence relation is defined on the timelike geodesics, and two ideal end points are identified just in case their corresponding geodesics are equivalent modulo this relation. The resulting set of ideal points forms the g-boundary $\partial_g(M)$; a topology is defined for $\partial_g(M)$, and a prescription is given for attaching $\partial_g(M)$ to M to form a manifold $\overline{M}_g$ with g-boundary. In some cases a differential and metric structure can also be defined for $\partial_g(M)$. $\overline{M}_g$ will *not* in general be a manifold with a boundary. However, if $\overline{M}$ is a manifold with boundary $\partial(M)$ such that every incomplete timelike geodesic of M strikes $\partial(M)$, then given the space-time $M \equiv \overline{M} - \partial(M)$, the g-boundary approach can be used to extend M to $\overline{M}$.

In the case of the spatially closed Friedmann universe, the g-boundary does not consist, as one might expect, of two points corresponding to the initial and final singular states; rather, it consists of two three-spheres S^3, and topologically, Friedmann space-time-plus-g-boundary is the product of S^3 and a closed and bounded interval of IR. Similarly, if the g-boundary approach is applied to a conical space-time with vertex removed, it does not give back the vertex point and only the vertex point. However, a modification of the equivalence relation on the timelike geodesics will lead to this result.[29] In addition to alternate schemes for equating timelike geodesics, there are alternative topologies and metric structures for $\partial_g(M)$. There doesn't seem to be any sense in choosing one alternative once and for all to the exclusion of all others; different alternatives may be better for illuminating different aspects of "singularities." Consequently there is no one right way to represent the "first" or "last moment of time."

11. Before the Beginning and After the End

The truncated Minkowski space-time of section 4 illustrates how time can be finite in the future; but the success of this example is the success of stipulation. We stipulate in effect that time is finite in the future by erasing an infinite portion of Minkowski space-time; there seems to be no reason to believe—and some reason not to believe—that this particular stipulation could be physically realized. In contrast, the Friedmann cosmological models provide a more interesting illustration in that they provide us with a mechanism for realizing temporal finiteness. Still, it could be claimed that these examples succeed only by virtue of stipulation; this time the stipulation specifies what we are to count as a space-time. Why, it might be asked, could we not change our picture of space-time so that other regular regions of space-time are joined onto the "initial" and "final" singularities of the Friedmann models, making them passing episodes in a longer history?

I shall approach this question somewhat obliquely by considering first another question. We have assumed that space-time is a connected manifold. Why, it might be asked, could we not change this picture to allow for the existence of other regions of space-time totally disconnected from ours? Well, clearly we could. But the question remains as to what the new picture of space-time amounts to, and in particular, what is meant by the "existence" of these other regions. Does every possible connected space-time "exist" (in this new sense of existence) as a region (of the new extended space-time) that is disconnected from ours? If so, the new picture does not really contain any innovation; the "existence" of a disconnected space-time region such that _____ means no more than it is possible that _____. Disconnected regions of space-time are only a device for picturing other possible worlds. On the other hand, if not every possible space-time "exists" as a region disconnected from ours, how do we tell which of them "exists"? For the "existence" of the other regions to make any empirical difference, these other regions must interact in some fashion with our region. This interaction cannot be described in anything like the usual spatio-temporal terms we use to describe causal interactions since, by hypothesis, these other regions do not enjoy any of the usual spatio-temporal relations with our region. What sort of interaction, then, can it be?

Does talk about what happens "before" and "after" the "initial" and "final" singularities in the Friedmann universes make any more sense than talk about events that happen "out there," where "out there" indicates a region of space-time disconnected from ours? If the singularities are true space-time singularities and not just regions where matter is very dense—and we have assumed that they are true space-time singularities—then they seem to separate us from the "other" regions that "join on" "before" and "after" just as effectively as disconnectedness separates us from the other regions "out there." The "before" and "after" regions might just as well be other possible worlds.

This interpretation is opposed to the more usual picture of an "oscillating universe." In the case of the spatially closed Friedmann universe, there are formal solutions to the differential equations governing the temporal behavior of the radius R of the universe which, if taken literally, would allow one to picture the universe as oscillating between the singular points where $R = 0$. (There are other solutions in which r does not oscillate.) But here as elsewhere, one can be misled by taking a picture too literally, for from the point of view presented above, these mathematical solutions are purely formal. As we have seen above, the "singular points" at which $R = 0$ can be represented only by sophisticated mathematical techniques, and on some representations they are not even points but rather three-spheres. Second, although continuity considerations can be used to help characterize certain aspects of singularities, they do not seem to provide a means of carrying us "through" the singularities and into "other" space-time regions on the "other side" of the singularities. Compare the situation here with that in Newtonian gravitational theory of point mass particles. When the particles collide, Newton's $1/r^2$ law blows up. Under certain conditions, however, solutions to the equations of motion can be extended through the singularities by means of analytic continuation. This procedure is possible only because of the constant background of Newtonian space-time, which gives the means of defining relevant senses of continuity of solutions. But in the relativistic case, the space-time background itself becomes singular, and hence no means of defining a "continuous" extension through the singularity is at hand.

I am not claiming that we could never come to possess any empirically well-grounded principle that would "carry us through" space-time singularities. What I do claim is that we do not now possess such a principle and that it is difficult to see how such a principle could be constructed and confirmed within present relativity physics.[30]

Two final points. First, the arguments given in this section take for granted certain elements of the currently accepted picture of space-time. This picture may well be dropped in the future in favor of some radically different picture; but this is a matter which at the present time must be left to writers of science fiction. Second, if the sentiments of this section are rejected, then one must conclude that no physically interesting example of how time can be finite in the future or past can be constructed within the current framework of physics. Such a negative conclusion would be interesting in its own right.

12. Concluding Remarks

Many of the above considerations rely heavily on the space-time metric. This is no accident, for most of the crucial distinctions I have drawn cannot be made in terms of topological, or affine, or even conformal structure. Some philosophers hold that the metric element is "nonintrinsic" and "conventional" in a way that, say, the topological structure is not. I do not share this view, but I wish to point out that if it is correct, then the answers to many of the questions which philosophers have asked about the beginning and end of time are matters of convention.

Unless the line of analysis I have pursued in this paper is very misleading, the answers to the questions posed at the outset lie somewhere in a thicket of problems growing out of the intersection of mathematics, physics, and metaphysics. This paper has only located the thicket and engaged in a little initial bush-beating. This is not much progress, but knowing which bushes to beat is a necessary first step.

Some philosophers will be disappointed that the thicket is populated by so many problems of a technical and scientific nature. On the contrary, I

John Earman

am encouraged by this result because it shows that a long-standing philosophical problem has a nontrivial and, indeed, a surprisingly large content. Moreover, this result is a good illustration of the artificiality and danger of trying to separate philosophy from science. If I am right, some of the best philosophy of time is being done today by physicists or, as I would prefer to say, by natural philosophers. Conversely, this work inevitably brings a confrontation with traditional philosophical problems.

Notes

1. The history of philosophy seems to me to be littered with disputes which could not possibly have been resolved in a satisfactory manner by the disputants, simply because they were not in a position to settle the relevant questions.

2. See *Physics*, Bk. VIII, 251b.

3. See my paper "Space-Time, or How to Solve Philosophical Problems and Dissolve Philosophical Muddles Without Really Trying," *Journal of Philosophy*, 67 (1970): 259–277. See also S. Shoemaker, "Time Without Change," *ibid.*, 66 (1969): 363–382.

4. It might be objected here that although time as represented by the metric space $((-1, +1), d_e)$ is finite in both the past and the future according to this criterion, time is not "really" finite since in the topology induced by the metric, $(-1, +1)$ is noncompact. But all this shows is that compactness does not always capture the relevant sense of finiteness.

5. L. E. Loemker, ed., *Leibniz: Philosophical Papers and Letters* (New York: D. Reidel, 1970), p. 669.

6. *Ibid.*, p. 664.

7. Technically g is a symmetric tensor field of type $(0, 2)$ which is defined on all of M, which is nondegenerate, and which has index 1 or 3. The differentiability classes of M and g are left open since nothing important in what follows turns on this question. The technical definition of ∇ is not needed here; for present purposes, the important fact about ∇ is that it allows us to define a notion of parallel transport on M and hence a notion of space-time geodesic (i.e., auto-parallel curve).

8. Technically, $\mathcal{E}$ is a symmetric tensor field of type $(2, 0)$.

9. For more details, see the first reference of note 3 and my paper, "An Attempt to Add Some Direction to 'The Problem of Director of Time,' " *Philosophy of Science*, 41 (1974): 15–47.

10. Let (x, t) be a pseudo-Euclidean coordinate system for M_{Min}. First identify those points with coordinates (x_1, t_1) and (x_2, t_2) just in case $t_1 = t_2$ and $x_1 = x_2$ modulo m; then identify two points of the resulting space just in case their spatial coordinates are the same and their temporal coordinates are equal modulo n (m, n positive integers).

11. I will use "curve" ambiguously to denote a map $\sigma : [0, 1] \rightarrow M$ and the image $\sigma([0, 1])$.

12. Although the notion of presupposition may be problematical in general, it seems transparent enough in the following examples.

13. Actually, things are more complicated than this, for a space-time $\langle M, g, \nabla \rangle$ may be inextendible but still locally extendible in the sense that there is an open subset $U \subset M$ such that: (1) the closure of U in M is noncompact, but (2) $\langle U, g|_U, \nabla|_U \rangle$ can be isometrically embedded in a larger space-time so that the closure of the image of U is compact. Since there are enough complications already, I shall not discuss this one here. For some implications of local extendibility for determinism, see my paper "Laplacian Determinism in Classical and Relativistic Physics," forthcoming in the University of Pittsburgh series in the Philosophy of Science.

14. R. Penrose, "Gravitational Collapse," *Revisita del Nuovo Cimento*, 1 (1969): 252–275.

15. For a precise statement and proof of this result, see Y. Choquet-Bruhat and R.

Geroch, "Global Aspects of the Cauchy Problem in General Relativity," *Communications in Mathematical Physics*, 14 (1969): 329–335.

16. For example, let M_{del} consist of all those points of M_{Min} in the intersection of the future lobe of the null cone at $(0, 0, 0, 0)$ and the past lobe of the null cone at $(0, 0, 0, 1)$, (x, y, z, t) being a pseudo-Euclidean coordinate system. $C = \{(x, y, z, t): t = 1/2 \text{ and } x^2 + y^2 + z^2 < 1/2\}$ is a Cauchy surface for S_{del}. S_{del} cannot be extended in such a way that C remains a Cauchy surface. This example is due to Choquet-Bruhat, *ibid.*

17. See R. Geroch, "Singularities," in M. Carmeli, ed., *Relativity* (New York: Plenum Press, 1970).

18. In many cases the addition of a cosmological term will not block the singularity results.

19. See S. W. Hawking, "The occurence of singularities in cosmology," *Proceedings of the Royal Society of London A*, 294 (1966): 511–521.

20. The Friedmann models were chosen as illustrations because of their simplicity; but precisely because of their simplicity—in particular, their high degree of symmetry and the absence of any pressure term in their equation of state—they are somewhat unrealistic. The singularity theorem cited above applies to models that are much more realistic in these respects.

21. See R. Penrose, "Structure of Space-Time," in C. M. DeWitt and J. A. Wheeler, eds., *Battelle Rencontres* (New York: W. A. Benjamin, 1968).

22. "Absolute Zero of Time," *Physical Review*, 186 (1969): 1328–1333.

23. *Ibid.*, p. 1331.

24. *Ibid.*

25. When I speak of curvature, I am referring to the rational scalar invariants formed from the Riemann Tensor and its derivatives. The so-called irrational scalar curvature invariants can become infinite in regions where the metric is nonsingular.

26. See J. A. Wheeler, "Geometrodynamics and the Problem of Motion," *Reviews of Modern Physics*, 33 (1961): 63–78.

27. *Space and Time* (New York: St. Martin's Press, 1968), chapter 15.

28. See R. Geroch, "Local Characterization of Singularities in General Relativity," *Journal of Mathematical Physics*, 9 (1968): 450–465.

29. See D. A. Feinblum, "Global Singularities and the Taub-NUT Metric," *Journal of Mathematical Physics*, 11 (1970): 2713–2720. Recently, a seemingly more satisfactory definition of space-time singularities has been given by B. G. Schmidt, "A New Definition of Singular Points in General Relativity," *General Relativity and Gravitation*, 1 (1971): 269–280. But this definition is too complicated to be reviewed here.

30. One possible exception has been mentioned in C. J. S. Clarke's contribution to this volume; namely, that the laws of gravitation might be put in a distributional form such that singularities could be countenanced as part of the space-time manifold. However, one would guess that there are solutions to Einstein's field equations which (1) are temporally finite and (2) cannot be extended in even a distributional sense.

The Causal Theory of Space-time

> "It is not sufficient to say that Einstein's clocks and measuring rods are *ideal* ones: for, before we are in a position to speak of them as being ideal, it is necessary to have some clear conception as to how one could, at least theoretically, recognize ideal clocks or measuring rods in case one were ever sufficiently fortunate as to come across such things; and in case we have this clear conception, it is quite unnecessary, in our theoretical investigations, to introduce clocks or measuring rods at all."

> A. A. Robb

1. Introduction

The special theory of relativity forced a radical revision of classical views about the causal structure of the world. One of the most obvious changes involved the possible rates of causal propagation. According to the special theory there is a finite upper limit to the speed of causal chains, whereas classical causality allowed arbitrarily fast signals. Foundational studies, such as that of Reichenbach (1969), soon revealed that this departure from classical causality in the special theory is intimately related to its most dramatic consequences: the relativity of simultaneity, time dilation, and length contraction. By now it has become clear that these kinematical effects are best seen as consequences of the geometrical structure of Minkowski space-time, which in turn incorporates a nonclas-

NOTE: I am greatly indebted to Adolf Grünbaum, Allen Janis, David Malament, Howard Stein, and John Stachel for many helpful remarks and discussions. Geoffrey Matthews (Indiana University) is responsible for many of the details of the construction of section III. Above all, I wish to thank my colleague at Indiana University, J. Alberto Coffa, for many invaluable discussions, decisive criticisms, and an extremely careful and critical reading. As a result of his efforts, this essay certainly contains less error, and, one hopes, more of the truth. To be sure, any errors that remain are entirely my own responsibility (unless they were overlooked by Coffa). Finally, I would like to thank Loretta Moses for her patience and skill in typing this manuscript.

This research was supported, in part, by National Science Foundation grant SOC73-05753.

sical theory of causal structure. However, it has not been widely recognized that the converse of this proposition is also true: *the causal structure of Minkowski space-time contains within itself the entire geometry (topological and metrical structure) of Minkowski space-time.*

Most likely, this remarkable result has been overlooked for a variety of reasons. First of all, it is by no means obviously correct; its demonstration requires a good deal of logical and mathematical effort, and until recently, the existing proofs—in the works of Robb (1914, 1936)—have lacked economy and elegance. Second, the causal definability of the geometry of flat space-time is in direct conflict with the view that the determination of geometry *necessarily* involves the introduction of metrical conventions. Finally, the inability of *classical* causality to provide more than merely an account of temporal *order* provides plausible grounds for regarding this as a fundamental limitation of *any* causal theory; thus while special relativity has been cited by defenders of a causal theory of time as providing fresh support for the causal theory of *temporal order*, these same causal theorists have not seen that the causal resources of this theory are so powerful as to yield in addition a purely causal account of temporal metric, spatial geometry, and the topology and metric geometry of Minkowski space-time. Even in general relativity, causal structure determines the topological, differentiable, and conformal structures of a wide class of physically "reasonable" space-times; the metric of these space-times, however, is not causally determinate (see section VIII below).

In the following section, the classical causal theory is outlined and its ability to provide a basis for the construction of time and space is investigated. It is shown that the classical theory of causal structure permits at most the construction of an (absolute) temporal *order*; the source of this limitation of classical causality is exhibited. Section III introduces the causality-related notions to be used throughout, developing these from the single symmetrical relation of causal connectibility. In section IV, the causal theory of the *topology* of Minkowski space-time is then outlined, and its adequacy as a basis for the topology of the manifolds of general relativity is discussed briefly. In section V the linear structure of Minkowski space-time is constructed out of causality; Robblike methods of developing its metrical structure are sampled. These constructions illustrate the way in which the causal theory of Minkowski space-time bridges the gap between "qualitative" and metrical properties of space-time. Adequacy proofs for the causal construction of the linear structure of Minkow-

135

ski space-time are provided. In section VI it is shown how the causal cone, together with the affine structure of Minkowski space-time, may be used to define its standard metrical structure: congruence, orthogonality, and the space-time interval. In section VII, the causal definability of congruence in Minkowski space is again demonstrated using group-theoretical methods and a theorem of E. C. Zeeman (1964). The construction of the linear structure of Minkowski space on this basis is more or less immediate, and is indicated in outline. Finally, in section VIII, some objections to the causal theory are considered, and the extent to which the causal theory may be extended to general relativity is discussed.

II. Classical Causality and Its Limits

Leibniz seems to have been the first to conjecture that the structure of time and space might be reducible to causality (1956), and the basic ideas of his causal theory are both simple and ingenious. The *simultaneity* of two distant events is defined as their lack of causal interaction. Thus if we choose a given event and consider all those events in the universe neither affecting nor affected by the chosen event, we hope to obtain an *acausal* "slice" of the world: a set of events no two of which causally interact. Identify such a slice with an instant of time, and define time itself as the set of all such slices or instants. Now order the slices or instants using the causality relation between events chosen from within each slice. Instant *A* (slice *A*) is then prior in time to instant *B* (slice *B*) just in case every event in *A* causally influences every event in *B*. In this way the temporal order of the world is reduced to causal relations between its events.

So far, only the order *between* acausal slices has been mentioned. What of the order *within* each slice? This, according to Leibniz, is where space appears in the construction. The events within each acausal slice have an order, presumably the same for all slices, and this order is just the *spatial* order of the world.

Even this rough sketch of the causal theory shows that the success of its constructions will require considerable cooperation from the world's causal structure. For example, consider the definition of simultaneity as absence of causal interaction. The definition alone cannot guarantee that the resulting slices are acausal and mutually exclusive; yet if two slices were to share an event, the account of temporal order fails to be unique. Consider the unfortunate situation depicted in Figure 2.0. The instants $I(e_1)$ and $I(e_3)$ intersect, yet we may suppose that, nevertheless, they are both acausal world slices. But now consider the instant $I(e_2)$, not depicted above. Clearly both e_1 and e_3 are in $I(e_2)$, yet e_1 and e_3 are causally connectible. Thus $I(e_2)$ is not an acausal slice, contrary to the intention of the causal theory.

Leibniz was aware of this problem and explicitly set down a postulate which handled the difficulty. However, this and similar questions are better dealt with after the theory has been formulated more precisely.

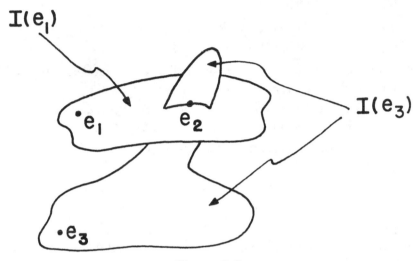

Figure 2.0

For working purposes let us use the relation of causal precedence, and read "e_1 CP e_2" as "event e_1 causally precedes event e_2." Since there is no intention here to build into the relation CP a *direction* of causal order, "CP" could just as well be read "causally succeeds" throughout. This awkwardness might be avoided in a "purer" approach by beginning with a three-place relation of 'causal betweenness,' and then choosing an arbitrary pair of causally connected events to orient the event-manifold, and define CP accordingly.[1]

To begin with, let us assume that the relation of causal precedence is irreflexive, asymmetrical, and transitive, i.e.,

Postulate I. For all events e_1, e_2, e_3:
  (1) not e_1 CP e_1,
  (2) if e_1 CP e_2, then not e_2 CP e_1, and
  (3) if e_1 CP e_2, and e_2 CP e_3, then e_1 CP e_3.

We may now define the relation of causal interaction or *causal connectibility*[2] ($e_1 \; \gamma \; e_2$) in the natural way:

Definition 2.0. $e_1 \; \gamma \; e_2$ iff $e_1 \; CP \; e_2$ or $e_2 \; CP \; e_1$.

The following proposition now follows easily:

Proposition 2.1. For all events e_1, e_2, and e_3:
   (1) not ($e_1 \; \gamma \; e_1$)
   (2) if $e_1 \; \gamma \; e_2$, then $e_2 \; \gamma \; e_1$.

The Leibnizian definition of simultaneity is now given by:

Definition 2.2. e_1 Simul e_2 iff not ($e_1 \; \gamma \; e_2$).

Instants (of Time) are now event slices all of whose members are simultaneous with some event or other.

Definition 2.3. Let U be the set of all events. Then A *is an instant* iff $A \subseteq U$, and for some event e_0 in U, $A = \{e \mid e \text{ Simul } e_0\}$.

Assuming the success of these definitions, the set of all instants may be ordered as follows:

Definition 2.4. Let I_1 and I_2 be instants. Then I_1 *temporally precedes* I_2 ($I_1 < T \, I_2$) iff there are events $e_0 \in I_1$ and $e'_0 \in I_2$, such that $e_0 \; CP \; e'_0$.[3]

For this account to succeed, simultaneity must be an equivalence relation on the class (U) of all events. Proposition 2.1 and definition 2.2 yield reflexivity and symmetry for simultaneity at once; i.e., we have

Proposition 2.5. For all events e_1 and e_2,
   (1) e_1 Simul e_1,
   (2) if e_1 Simul e_2, then e_2 Simul e_1.

However, the transitivity of simultaneity does not yet follow, for we need assurance that if e_1 and e_2 are simultaneous, there is no other event e_3, causally connected to e_1 but *not* causally connected to e_2. As Figure 2.0 has shown, if this were to occur, the instant containing e_2 would also contain causally connected events (e_1 and e_3). Instants would not then be acausal sets, as intended. Thus we need to postulate that such a situation never arises:

Postulate II. (Leibniz postulate). If e_1 and e_2 are simultaneous and e_1 is causally connected with e_3, then e_2 is also causally connected with e_3.[4]

It now follows easily that Simul is an equivalence relation on U, i.e., we have:

Proposition 2.6. If e_1 Simul e_2 and e_2 Simul e_3, then e_1 Simul e_3. Simul is an equivalence relation on U.

Proof: Suppose e_1 Simul e_2, e_2 Simul e_3, yet not e_1 Simul e_3. Then $e_1 \gamma e_3$, and by postulate II, $e_2 \gamma e_3$, thus contradicting e_2 Simul e_3. Hence we must have that e_1 Simul e_3. That Simul is an equivalence relation now follows from proposition 2.5. Done.

As a consequence, we also have justification for the above definition (2.4) of temporal precedence.

Proposition 2.7. Let I_1 and I_2 be nonempty instants. Then $I_1 < T\, I_2$ iff for all $e_1 \,\epsilon\, I_1$, $e_2 \,\epsilon\, I_2$, $e_1\, CP\, e_2$.

Hence instants are acausal sets (i.e., no two events in an instant are causally connectible), and the order of instants has been defined consistently.

Clearly the success of the theory presupposes the correctness of the Leibniz postulate; here some further clarification is needed. To begin with, the postulate seems to be too strong. Suppose that events e_1 and e_2 are distant yet simultaneous. Will there not be, in some cases at least, an event e_3 causally connected with e_2, say $e_2\, CP\, e_3$, yet completely uninfluenced by the remote happening e_1? Not that some causal chain from e_1 *could not* terminate at e_3; but *do* such chains always so terminate, as a matter of actual fact?

Such a commitment may be avoided by construing our primitive causal relations modally, i.e., 'causally connected' becomes 'causally connectible,' etc.; we might even interpret the relation of 'causal connectibility' as a relation between events that is physically on a par with 'causally connected,' only more general.[5] In any case, any such weaker interpretation should be construed so as to have causal connectedness entail causal connectibility, but not conversely. Postulates I and II, under such an interpretation, would, a priori, stand a better chance of being true.[6]

Of course, the most serious objection to the *truth* of the Leibniz postulate is that (even taken modally) it fails in the space-times of special and general relativity. In special relativity we have instead:

Postulate II* (Einstein postulate). For any distinct events e_1 and e_2 such that e_1 Simul e_2, there exists an event e_3 such that e_3 is causally connectible with e_2, yet e_1 Simul e_3.

At first sight, it seems that if this postulate were true, the result would be the immediate demise of the causal theory. For while the Leibniz postulate (together with postulate I) guarantees that simultaneity will be an equivalence relation, the Einstein postulate virtually guarantees that it will not, since, according to this postulate, the only acausal instants are unit sets. Hence if we go on to postulate (as we must in any case) the existence of at least two simultaneous events, the Einstein postulate guarantees that an instant exists which is not acausal. However, it will soon be clear that the overthrow of the Leibniz postulate, rather than jeopardizing the causal theory, is a *necessary* condition for its *complete* success. In order to see this clearly, let us take up the account of the classical causal theory from where we left off.

So far, instants and a temporal order have been defined. However, as it stands, the theory of temporal order is incomplete. Are the causal ordering and its derivative temporal ordering discrete, dense, continuous, or of some other order type? We have not yet laid down any postulates to settle this question. However, their formulation in causal terms alone presents no difficulty. Thus we need add only the following postulate:

Postulate III. (Causal continuity).

(1) For any event e there are events e_1 and e_2 such that $e_1 \, CP \, e \, CP \, e_2$. (No first or last event.)

(2) For any events e_1, e_2 such that $e_1 \, CP \, e_2$, there is an event e such that $e_1 \, CP \, e \, CP \, e_2$. (Density.)

(3) For any nonempty sets of events A and B which are such that for any $e_1 \in A$, $e_2 \in B$, we have $e_1 \, CP \, e_2$, there exists an event e_0 such that if $e_1 \in A$, $e_2 \in B$, $e_1 \neq e_0 \neq e_2$, then $e_1 \, CP \, e_0 \, CP \, e_2$. (Continuity; cut Axiom, cf. Tarski, 1961.) Intuitively, this axiom states that if A and B are any sets of events such that the elements in A causally precede every element in B, then there is some event (e_0, say) which separates A and B. The axiom fails for the rational numbers, as $A = \{x \mid x^1 < 2\}$ and $B = \{x \mid x^2 > 2\}$ show, since the required separator e_0 would be $\sqrt{2}$—not a rational.

(4) There is a denumerable set K, $K \subseteq U$, such that for any two events e_1, e_2, where $e_1 \, CP \, e_2$, there is an event e in K such that $e_1 \, CP \, e \, CP \, e_2$. (Existence of a denumerable dense subset.)

It is now easy to show that the temporal order relation $<T$ on the set (Time) of all instants yields a structure (⟨Time, $<T$⟩) which is isomorphic to the real numbers in their standard ordering.[7] In other words, instants,

140

ordered by temporal precedence, have the same type of order as the reals in their standard ordering.

Still, the construction of a causal theory of space and time has barely begun. We must now go on to define: (1) space; (2) spatial topology and order; (3) the geometry or metric of space; and (4) the metric of time. First, the problem of space.

It was mentioned earlier that Leibniz spoke of space as the order of coexisting, i.e., simultaneous, events. It would be more correct to say that Leibniz is here characterizing spatial *order*, rather than space itself. An ordered *instant* is not (a) space; a space is a set of points that persist *through* the set of instants, and so is best construed as just the set of such persisting "points." Once space has been constructed, then we may turn to the question of the order of these points, and indeed might well attempt to base this order upon some relation between events within instants. The problem then comes down to defining points, and this just amounts to specifying for any given event e_0, exactly what other event in any chosen instant occupies the *same* spatial point or place. Thus in Figure 2.1, let $I(e_0)$ be the instant that contains e_0. Let I' be some other instant. What event e_0' in I' is at the same place as e_0? Once we know this, we may define a (spatial) point as the largest class of events all of which are at the same place. What we need then is a function that yields a different "fibre" (see Figure 2.1) through each event in a given acausal slice, with no two fibres intersecting, and each fibre meeting each instant in exactly one event. Each fibre is a spatial point; a space is the set of all fibres. Thus the problem of space, within the causal theory, is that of defining either a unique fibration of instants, or at least a distinguished class of such fibrations.

The Leibniz postulate precludes a nonarbitrary solution to this problem. Consider again the slices $I(e_0)$ and I'. Suppose that we assign e_0' and e_0 to the same spatial point or fibre as in the figure, rather than assigning, say, event e_1 in I' to the same spatial location as e_0. But I' is a causally definable equivalence class (by the Leibniz postulate), so e_1 must sustain the *same* causal relations to e_0 as does e_0'. Thus a nonarbitrary association of e_0' and e_0 cannot be made on causal grounds alone; a fortiori, the same is true for any fibration of U. Hence the space problem cannot be solved utilizing only the resources of classical causality.

Typically, extra-causal aid is sought from point-particles and their dynamics. If it were physically possible to distinguish a privileged set of

141

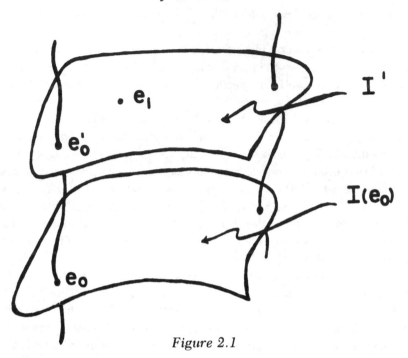

Figure 2.1

noncolliding particles, their world-lines might then be taken as our fibres or points, thereby defining a unique (absolute) space. A weaker approach is to distinguish a *family* of such sets of particles dynamically (inertial systems), with the result that we have a number of spaces, one for each system. Some of the awkwardness that results from having so many spaces on hand would then be reduced should it turn out that these spaces share the same (say Euclidean) geometry. But, once again, how are we to construct the geometry of space out of causality alone? If the metric is of the right sort (namely, Riemannian; cf. Hicks, 1971, pp. 70–71), we may use it to define the spatial topology; so let us consider only the spatial metric.

For simplicity's sake, consider the case in which we have constructed a unique space using a single dynamically distinguished set of point-particles (particles "at rest in absolute space"). The causal situation is completely characterized by the simultaneity slices, their temporal order, and the fact that any two events within a single particle (or fibre) are causally connectible. Once again, in order to make distance determinations possible, additional ontic resources are necessary. Typically, ex-

142

tended physical objects are now introduced, with a distinguished subset of these deemed "rigid." It is commonly believed, correctly, I think, that within the context of a classical causal theory and classical dynamics, there is an ineradicable element of physical convention which is involved in distinguishing rigid from nonrigid extended bodies; but this question is not at issue here.[8]

Similar considerations apply to the temporal metric. The continuity of the temporal order, as Riemann (1973) pointed out, precludes basing the temporal metric on the cardinality of the intervals between instants, for the simple reason that all intervals have the same cardinality. Once again, causality requires external aid, and typically, this is provided by *periodic* material processes (clocks). Just as in the case of rigid bodies, the conventions involved in distinguishing isochronous from irregular periodic process naturally and rightly emerge as crucial within such a construction.

The above arguments may be summarized as follows. The unaided classical causal theory provides no more than a theory of temporal order. To go beyond this, extra-causal considerations are necessary. Typically, dynamically free point-particles, rigid bodies, or periodic material processes are now introduced to construct space, its geometry, and the temporal metric. This weakness of classical causal theory derives in large part from the Leibniz postulate with its unique acausal slices of simultaneity. As a result, *any map of the universe U onto itself which preserves the temporal order relation will also be a mapping which preserves the total causal structure of U*, i.e., such a map will be a causal automorphism of *U*. This means, roughly, that we may stretch or shrink the classical time axis and "distort" the acausal slices at will, while preserving all of the causal relations of *U*. It thus follows (see theorem 2.10 below) that space, spatial geometry, and temporal metric cannot be constructed from classical causality alone. While the preceding arguments have appealed to this feature of classical causal structures in a more or less intuitive way, it is now time to proceed to demonstrate this rigorously.

Let us call any model of postulates I and II (the Leibniz postulate) a *Leibniz space-time*. In other words:

Definition 2.8. (U, CP) is a *Leibniz space-time* iff U is a nonempty set, CP is a two-place relation on U, and for all events $e_1, e_2, e_3 \in U$,

(1) $<e_1, e_1> \notin CP$ (irreflexivity),

(2) If $<e_1, e_2> \in CP$, then $<e_2, e_1> \notin CP$ (asymmetry),

(3) If $<e_1, e_2> \in CP$ and $<e_2, e_3> \in CP$, then $<e_1, e_3> \in CP$ (transitivity),

(4) If $<e_1, e_2>$, $<e_2, e_1> \notin CP$ and $<e_2, e_3> \in CP$ or $<e_3, e_2> \in CP$, then either $<e_1, e_3> \in CP$ or $<e_3, e_1> \in CP$ (Leibniz postulate).

A causal automorphism of a Leibniz space-time is just a one-one map of its universe U onto itself which preserves the causality relation CP; that is:

Definition 2.9. Let (U, CP) be a Leibniz space-time. Then h *is a causal automorphism of* (U, CP) iff h is a bijection: $U \to U$ such that for all e_1, $e_2 \in U$, $<e_1, e_2> \in CP$ iff $<h(e_1), h(e_2)> \in CP$.

The causal automorphisms are the symmetries of a causal space; i.e., they are the ways of transforming it that do not alter its causality relation.[9] What we want to show now is that any mapping that merely preserves the *order of instants* in a Leibniz space-time is a causal automorphism (the converse is trivial). This then is the root of the classical causal theory's inability to go beyond an account of temporal order *no matter how that theory is extended.* Here is the theorem and its simple proof.

Theorem 2.10. Let (U, CP) be a Leibniz space-time and $<T$ the temporal precedence relation defined as in definition 2.4 above. Then h is a causal automorphism of (U, CP) iff h is a bijection of U onto itself that preserves $<T$, i.e., for any instants I_1, I_2 (as defined in definition 2.3 above), $I_1 <T I_2$ iff $h[I_1] <T h[I_2]$.

Proof: The "only if" part is trivial. So suppose $I_1 <T I_2$ iff $h[I_1] <T h[I_2]$, where $h: U \to U$ is a bijection.

(1) Suppose $e_1 CP e_2$. Let $I(e_1)$ and $I(e_2)$ be instants containing e_1 and e_2, respectively. By definition 2.4, $I(e_1) <T I(e_2)$, so $h[I(e_1)] <T h[I(e_2)]$. But $h(e_1) \in h[I(e_1)]$ and $h(e_2) \in h[I(e_2)]$. Hence by proposition 2.7, $h(e_1) CP h(e_2)$.

(2) Suppose $h(e_1) CP h(e_2)$. Let $I(e_1)$ and $I(e_2)$ be as in (1). Now $h(e_1) \in h[I(e_1)]$ and $h(e_2) \in h[I(e_2)]$. Since (by proposition 2.6) instants are equivalence classes, we must have $h[I(e_1)] = I[h(e_1)]$ and $h[I(e_2)] = I[h(e_2)]$. By definition 2.4, $I(h(e_1)) <T I(h(e_2))$, so we must have $h[I(e_1)] <T h[I(e_2)]$. But then $I(e_1) <T I(e_2)$, so by proposition 2.7, $e_1 CP e_2$. Thus $e_1 CP e_2$ iff $h(e_1) CP h(e_2)$, so h is a causal automorphism of $<U, CP>$. Done.

One example of an application of this result will serve for our purposes. Let R^n be an n-place relation holding only between events within an instant and defined solely in terms of the causal relation CP. (For the sake of fixing ideas, consider R^n to be a four-place candidate for spatial congruence, the equidistance of e_1, e_2 and e_3, e_4.) Since R^n is assumed definable in terms of CP, any mapping that preserves CP must also preserve R^n. Let h_0 be any causal automorphism which preserves instants. Consider h_0 restricted to the instant $I_j(h_0/I_j)$. Tamper with h_0/I_j in any way you like as long as you leave the result $h_0^*/I_j: I_j \to I_j$. Clearly h_0^* preserves temporal precedence ($<T$) if h_0 does. Hence *any* h_0^* preserves R^n as well. So if $R^n(a_1, \ldots, a_n)$ holds in I_j for distinct a_i, $R^n(b_1, \ldots, b_n)$ must hold for *all* distinct b_i in I_j. Defining a spatial congruence relation is clearly impossible: a Leibniz space-time has far too many automorphisms.

The theorem upon which these results are based does not involve the assumption of temporal continuity (postulate III above) in its proof; in other words, the definition of a Leibniz space-time does not demand the temporal continuity of such spaces. Were we to go on now to postulate a *discrete* temporal order, a temporal metric would be obtainable by counting instants. Obviously, the (time) order-preserving maps would preserve this counting metric as well. (Clearly, a causally defined spatial metric is still impossible.)

However, the addition of the temporal continuity postulate leads immediately to the causal undefinability of a temporal metric. For the betweenness-preserving maps of the real line are just its topological homeomorphisms; so too then for instants and their order. Thus in Leibniz space-times with temporal continuity, causal structure is preserved by *any (topological) homeomorphism* of the set of instants. As in the case of space, no nontrivial temporal metric exists which is invariant under such a large set of transformations.

These results show that, for *Leibniz* space-times, temporal continuity precludes a causally definable temporal metric. But it should not be thought that continuity and homogeneity *alone* are the source of this metrical impotence. Rather, it is the continuity of time *taken together with the Leibniz postulate* that permits the proof of these results. As will become clear presently, continuity and homogeneity are properties of Minkowski causal spaces as well, but by virtue of the Leibniz postulate failing in such spaces and the Einstein postulate (postulate II* above) holding in its stead, the above result (theorem 2.10) fails to obtain. Indeed, according to a theorem of Zeeman (1964), any causal automorphism of Minkowski space-time maps a pair of congruent space-time intervals onto another pair of congruent intervals, although all space-time intervals, considered as sets of point-events, are continuous and have the same cardinality. [10]

John A. Winnie

III. Foundations of the Causal Structure of Minkowski Space-time[11]

The construction of Leibniz space-time began with the two-place relation of causal precedence. In this section we shall see that the causal structure of Minkowski space-time may be derived from the *symmetrical* two-placed relation of causal connectibility.[12] Causal betweenness may then be defined in terms of causal connectibility, the space-time may be oriented, and the standard asymmetrical relations of causal and chronological precedence may be constructed easily. Along the way, it needs to be shown that the notions so defined do coincide with their intended counterparts within a standard interval formulation of the geometry of Minkowski space-time. Hence we shall need some such formulation to use as a reference throughout. The following definition will suffice for the present.

Let M be a nonempty set, and I^2 be a function from $M \times M$ into the real numbers. The pair (M, I^2) is said to be a *Minkowski space-time* if and only if there is a one-one function ϕ from M onto R^4 in which the interval I^2 is given by:

$$I^2(e_1, e_2) = -(\Delta x^0(e_1, e_2))^2 + \Sigma_k (\Delta x^k(e_1, e_2))^2, \ k = 1, 2, 3$$

where $\Delta x^i(e_1, e_2)$ is $(x^i(e_1) - x^i(e_2))$, and the functions x^i are the coordinate functions of ϕ (i.e., $x^i \equiv u^i \circ \phi$, where $u^i(a_1, \ldots, a_n) \equiv a_i$). The coordinate functions $\{x^i\}$ derived from ϕ will be called a *Lorentz coordinate system*; x^0 is its time function. Frequently the function x^0 will be denoted by "t," while "x," "y," and "z" will be used for x^1, x^2, and x^3, respectively. (A signature of $(-, +, +, +)$ is used throughout.) The topology of M is that of R^4 and is induced by the function ϕ; in other words, if $A \subseteq M$, then A is open in Minkowski space-time just in case its image, $\phi[A]$, is open in R^4 with its standard topology.

As our single causal relation we now take the relation (γ) of causal connectibility. Intuitively, event e_1 is causally connectible to event e_2 just in case a signal (massive or massless) can be sent from e_1 to e_2, or conversely. In terms of (M, I^2), $e_1 \ \gamma \ e_2$ just in case $I^2(e_1, e_2) \leqslant 0$. Note that, under this interpretation, the relation γ is both symmetrical and reflexive. However, γ is *not* transitive, since for any two events e_1 and e_3 which are simultaneous (not $e_1 \ \gamma \ e_3$), there will always be an event e_2 in their common causal futures.

146

It is important to realize that the above characterization of causal connectibility is not intended to be a *definition* of this relation. The present version of the causal theory takes γ as its only descriptive primitive relation. An axiomatic account, such as Robb's (1914), would now proceed to lay down a number of postulates for γ, each of which would be provably true under the above interpretation. However, an axiomatization of γ is not needed to show the adequacy of the causal theory of Minkowski space-time. For this purpose, we need only show that all geometrical notions in Minkowski space-time may be built up by explicit definitions from the causal connectibility relation. In order to do this we proceed on parallel "tracks." Having paired γ with its intended Minkowski interpretation, we may now consider another standard relation R on Minkowski space-time and propose (---γ---) as a construction of R. The proof of this construction's adequacy consists in showing that if γ has its intended interpretation, then this is true of (---γ---) as well. We continue in this way until we have provided constructions for a set of standard geometrical relations on Minkowski space-time which we know determine its geometry.

The following construction provides an example of this procedure. The intention is to construct the 'causal betweenness' relation from causal connectibility. The intended interpretation of "$CB\ e_1e_2e_3$" is that the relation holds just in case there is a nonspacelike parameterized curve in M through e_1, e_2, and e_3 such that the parameter λ_2 of e_2 (i.e., $f(\lambda_2) = e_2$) is between that of e_1 and e_3 (i.e., $\lambda_1 \leq \lambda_2 \leq \lambda_3$ or $\lambda_3 \leq \lambda_2 \leq \lambda_1$). An equivalent condition (in Minkowski space-time) is that e_1 and e_3 are causally connectible and e_2 is in the closure of the Alexandroff interval[13] of e_1 and e_3. Obviously, other equivalent characterizations of causal betweenness are possible, and we may use our general knowledge of Minkowski space-time to establish their equivalence to either of the above conditions. From the standpoint of the causal theory, the interesting fact is that causal betweenness may be characterized solely in terms of causal connectibility, as the following definitions and the justifying theorem show.

First of all, consider the set of all events $J(e)$ causally connectible with a given event e. This is just the set of all events e_1 such that $I^2(e_1, e) \leq 0$ and so is definable as

Definition 3.0. $J(e) = \mathrm{df.}\ \{e_1 \mid e_1\ \gamma\ e\}$.

Clearly this definition is accurate, so a justifying theorem will not be given. However, a proof of the adequacy of the following definition of *causal betweenness* is provided below.

Definition 3.1. CB $e_1e_2e_3$ = df. $e_1 \, \gamma \, e_3$ and $J(e_2) \subseteq J(e_1) \cup J(e_3)$.

Proposition 3.2. CB $e_1e_2e_3$ if and only if e_2 is causally between e_1 and e_3.

Proof: (A). Suppose e_2 is causally between e_1 and e_3. Then e_1 is causally connectible to e_3, and if $e_1 = e_3$, $e_2 = e_1 = e_3$. Hence, trivially, $J(e_2) \subseteq J(e_1) \cup J(e_3)$. Suppose $e_1 \neq e_3$, with e_2 causally between e_1 and e_3. Let e^* be causally connectible with e_2 (i.e., $e^* \in J(e_2)$). Then there is a causal curve from e_2 to e^*, and a causal curve from either e_1 or e_3 to e_2.[14] Hence there is a causal curve from either e_1 or e_3 to e, so $e \in J(e_1) \cup J(e_2)$.

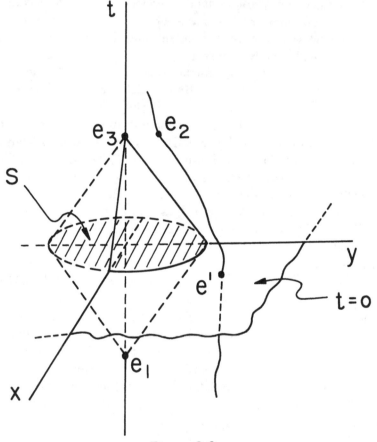

Figure 3.0

148

(B). Suppose e_2 is not causally between e_1 and e_3, with $I^2(e_1, e_3) <$ 0. Let $\{x^i\}$ be a Lorentz coordinate system in which $t(e_1) = -1$, $t(e_3) = +1$ and $x^k(e_1) = x^k(e_3) = 0$ $(k = 1, 2, 3)$. Then the intersection of the $t = 0$ hypersurface with the set of all events causally between e_1 and e_2 is the set of all events S with coordinates $t = 0$ and $\Sigma (x^k)^2 \leqslant 1$ (see Figure 3.0). Those events causally between e_1 and e_3 are just the domain of dependence of S, i.e., for all and only such events e^* will it be the case that *every* endless timelike curve through e^* meets S (Geroch, 1970). Hence if e_2 is not causally between e_1 and e_3, there is an endless timelike curve through e_2 which does not meet S, and thus does meet hypersurface $t = 0$ outside S at some event e' simultaneous with e_1 and e_3 (i.e., not $e' \gamma e_1$ and not $e' \gamma e_3$). Hence $e' \in J(e_2)$ but $e' \notin J(e_1) \cup J(e_3)$.

When $I^2(e_1, e_3) = 0$, the set S' of all events causally between e_1 and e_3 is just the corresponding segment of the null line determined by e_1 and e_3. Thus the domain of dependence of S' is just S' itself (Geroch, 1970), and the preceding argument may be applied once more. Done.

Causal connectibility and betweenness also allow us to distinguish the boundary of the causal space $J(e)$ from its interior. We begin by defining the appropriate connectibility relations. The corresponding betweenness relations then follow at once.

Events e_1 and e_2 are said to be *light-connectible* just in case $I^2(e_1, e_2) = 0$. The light-connectibility of two events may be defined in terms of causal connectibility as follows:

Definition 3.3. $e_1 \lambda e_2 = $ df. $e_1 = e_2$, or $e_1 \neq e_2$, $e_1 \gamma e_2$, and $LB = \{e \mid CBe_1ee_2\}$ is causally connectible, i.e., for any two events, e, e', in LB, $e \gamma e'$.

Proposition 3.4. $e_1 \lambda e_2$ iff $I^2(e_1, e_2) = 0$.

Proof: The case of $e_1 = e_2$ is immediate. So suppose $e_1 \neq e_2$. In Minkowski space-time, the null line containing two light-connectible events e_1 and e_2 is a causally connectible set. Hence the same is true of the set of all events causally between e_1 and e_2, since they all lie on this line. However, if e_1 and e_2 are causally connectible but have timelike separation, the intersection of the causal future of one with the causal past of the other will contain at least two events which are not causally connectible. Done.

Events e_1 and e_2 are chronologically connectible just in case they have timelike separation, i.e., $I^2(e_1, e_2) < 0$. Clearly the following definition succeeds:

Definition 3.5. $e_1 \tau e_2 =$ df. $e_1 \gamma e_2$ and not $e_1 \lambda e_2$.

Notice that since $e \lambda e$, chronological connectibility is defined so as to be irreflexive (not $e \tau e$), while causal and light connectibility are both reflexive.

With timelike and lightlike connectibility on hand, the causal betweenness relation may be refined further to yield chronological betweenness and null betweenness, the latter being the order of events along null lines. The intent and adequacy of the definitions below are both obvious, so justifying theorems are left as exercises.

Definition 3.6. (a) *TB* $e_1e_2e_3 =$ df. *CB* $e_1e_2e_3$ and $e_1 \tau e_2, e_3$.
    (b) *LB* $e_1e_2e_3 =$ df. *CB* $e_1e_2e_3$ and $e_1 \lambda e_3$.

The above constructions show that, although we start with causal connectibility alone, we may nevertheless distinguish events that are "inside" the causal space of e ($\{e_1 \mid e_1 \tau e\}$) from those that are on its "surface" ($\{e_1 \mid e_1 \lambda e\}$). It should be noted, however, that these constructions rely upon global features of Minkowski space-time and will fail to be adequate in Lorentz space-times in general. For example, in the two-dimensional Minkowski space-time depicted in Figure 3.1, if segment K is removed, definition 3.1 of causal betweenness fails, since the causal space of e_2 is a subset of the causal space of e_1, yet e_2 is not causally between e_1 and e_3.

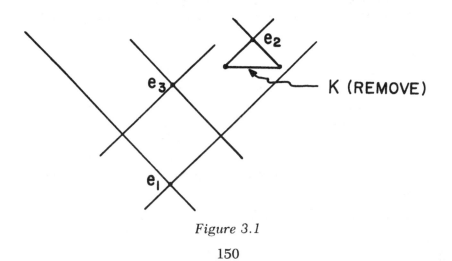

Figure 3.1

150

At this point, it is possible to proceed at once to the development of synthetic causal geometry. However, it is technically convenient to work with asymmetrical causal relations such as causal "precedence," and much contemporary causal research—also Robb—employs such a relation from the outset. While the use of such relations does not presuppose a direction (a privileged future) to causality, it does assume that space-time is *causally orientable*, that is, an arbitrary but consistently assigned causal direction may be selected. Hence, since Minkowski space-time *is* causally orientable, the question might just as well be faced now so that we may freely use the result in later developments.

The causal space of an event *e* breaks up into two components: its causal "future" J^+ and its causal "past" J^-. The event *e* is here considered to be a member of both J^+ and J^-. Now one cannot speak of a "future" or "past" without having first oriented the space-time. Nevertheless, without orientation, one can say that one or the other component of the causal space *J* of *e* is the component in which a given event e_1 resides. This may be accomplished using the causal betweenness relation as follows:

Definition 3.7. Where $e_2 \gamma e_1$, and $e_2 \neq e_1$, the e_2-*component of* e_1 is just the set of all events *e* in the causal space of e_1 such that e_1 is not causally between *e* and e_2, together with e_1, i.e.,

Comp $e_2 (e_1) =$ df. $\{e \mid e \in J(e_1)$ & not $CB \, ee_1e_2\} \cup \{e_1\}$.

Proposition 3.8. Let $\{x^\alpha\}$ be a Lorentz coordinate system for a Minkowski space $\mathfrak{M} = (M, I^2)$. Let e_1 be an event in *M*, with e_2 in the causal future (past) of e_1. (Here $\{x^\alpha\}$, as usual, is used to orient $\mathfrak{M}$.) Then for any event *e*, *e* is in the causal future (past) of e_1 if and only if $e \in$ Comp $e_2 (e_1)$.

Proof: If $e = e_1$, the result is immediate. So assume $e \neq e_1$. Suppose *e* is in the causal future of e_1, yet $CB \, ee_1e_2$. Then there would be a causal curve from e_1 to *e* and from *e* to e_1 which, in Minkowski space-time, is impossible. Next, suppose that $e \in$ Comp $e_2 (e_1)$, i.e., $e \gamma e_1$ and not CB ee_1e_2. Since $e \gamma e_1$, there is either (a) a causal curve from *e* to e_1 or (b) a causal curve from e_1 to *e*. If (a), since there is a causal curve from e_1 to e_2, there is a causal curve from *e* to e_1 to e_2, so we would have $CB \, ee_1e_2$, contrary to our assumption. Hence (b) must hold, so *e* is in the causal future of e_1. Done.

We now *causally orient* Minkowski space-time by selecting an ordered pair of events $\mathcal{O}_{ij} \equiv \, <e_i, e_j>$ which are causally connectible $(e_i \gamma e_j)$, but not identical. Such an event-pair will be called an *orientation* of $\mathfrak{M}$, and

John A. Winnie

$\mathcal{M}' = (M, I^2, \mathcal{O}_{ij})$ is called an *oriented Minkowski space-time*. Let F_{ij} be the e_j component of e_i, with P_{ij} the other component of e_i (i.e., those events in $J(e_i)$, which fail to be in the e_j component of e_i, together with e_i itself). We now define the *causal precedence* relation as follows:

Definition 3.9. Let events e_1 and e_2 be causally connectible. Then e_1 *causally precedes* e_2 *with respect to orientation* $\mathcal{O}_{ij} = <e_i, e_j>$, i.e., $e_1 < e_2$ (reference to $\mathcal{O}_{ij}$ omitted) iff (see Figure 3.2):

   (1) $e_1 = e_i$ and $e_2 \in F_{ij}$, or $e_1 \neq e_i$ and either

   (2) not $e_1 \, \gamma \, e_i$ and Comp $e_2 \, (e_1) \cap F_{ij} \neq \{ \quad \}$, or

   (3) $e_1 \in F_{ij}$ and Comp $e_2 \, (e_1) \subseteq F_{ij}$, or

   (4) $e_1 \in P_{ij}$ and Comp $e_2 \, (e_1) \nsubseteq P_{ij}$.

We show the adequacy of this definition as follows. Let $\mathcal{M}' = (M, I^2, \mathcal{O}_{ij})$ be an oriented Minkowski space-time with orientation $\mathcal{O}_{ij} = <e_i, e_j>$. Now let $K \equiv \{x^\alpha\}$ be a Lorentz coordinate system for $\mathcal{M}'$. If $\{x^\alpha\}$ is such that $t(e_i) < t(e_j)$, then $\{x^\alpha\}$ will be said to be *compatible* with $\mathcal{M}'$; otherwise K and $\mathcal{M}'$ will be called *incompatible*. Clearly, if $\mathcal{M}'$ and K are compatible, the function t makes F_{ij} the causal future of e_i, rather than its causal past. That this works out for all events in M is shown by the following proposition:

Proposition 3.10. Let $\mathcal{M}' = (M, I^2, \mathcal{O}_{ij})$ be an oriented Minkowski

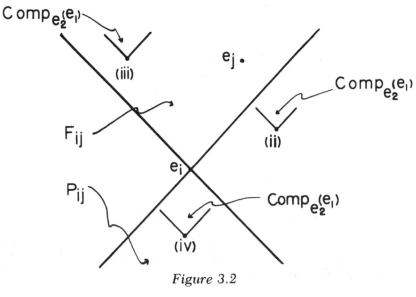

Figure 3.2

152

space-time, with $\{x^\alpha\}$ a Lorentz coordinate system compatible with $\mathfrak{M}$ '. Then $e_1 < e_2$ iff e_1 is causally connectible with e_2 and $t(e_1) \leqslant t(e_2)$.
Proof: We consider the cases as in definition 3.9. Case (1): immediately follows from the adequacy of the definition of components. Case (2): the F_{ij} component of e_i must intersect some component of e_1. If this were its past component, there would be a causal curve from e_i to e_1, contradicting the assumption that not $e_i \gamma e_1$. Hence F_{ij} must intersect the future component of e_1, which is the intended component. Case (3): in Minkowski space-time, when $e \neq e'$, e' is in the causal future of e iff the causal future of e' is a subset of the causal future of e. Case (4): similar to (3). Done.

Given an orientation $\mathcal{O}_{ij}$, we may now define the causal future of e $(J^+(e))$ as the set of all events which e causally precedes, and the causal past of e $(J^-(e))$ as the set of all events that causally precede e. That is,

Definition 3.11. (a) $J^+(e) =$ df. $\{e_1 \mid e < e_1\}$
(b) $J^-(e) =$ df. $\{e_1 \mid e_1 < e\}$

Chronological precedence $(<<)$ and precedence along a null line $(\rightarrow)$—sometimes called the *horismos* relation—are now definable using causal precedence and the appropriate connectibility relation in an obvious way.

Definition 3.12. (a) $e_1 << e_2$ (e_1 chronologically precedes e_2) $= df.$ $e_1 < e_2$ and $e_1 \tau e_2$
(b) $e_1 \rightarrow e_2$ (horismos relation) = df. $e_1 < e_2$ and $e_1 \lambda e_2$.

These definitions provide the basic causal notions to be used throughout this account. Although these constructions are mere preliminaries to the construction of Minkowski geometry proper, it is easy to see that, even at this preliminary level, not all of the preceding constructions succeed in arbitrary Lorentz space-times. This tendency of Minkowski space-time to cooperate with causal constructions will become even more evident as we proceed to construct its linear and metrical structure causally. However, before examining the causal basis of Minkowski metrical geometry, there is an interesting preliminary matter to consider: the causal basis of the *topology* of Minkowski space-time.

IV. Causality and Space-time Topology

According to the causal theory, the entire geometry of Minkowski space-time may be derived from its causal structure. Not only space-time metric or congruence, but the linear structure and topology of Minkowski space-time, must be definable in terms of causality. For some geometries (metric spaces), the problem reduces to defining the metric, since the linear structure and topology of such spaces is contained in their metric. However, this is not the case for the indefinite metrics of relativity theory. For these spaces, the connection between metric and topology is more indirect. Nevertheless, for many of these spaces, Minkowski space-time included, their causal structure permits a simple and direct definition of their topology. So while special relativity is the main concern here, the following causal construction of standard space-time topology succeeds in many of the space-times encountered in general relativity as well.

In the preceding section Minkowski space-time was defined as a pair (M, I^2), where the semi-Riemannian metric I^2 is specified in terms of a bijective coordinate mapping ϕ from M ro R^4. However, the mapping ϕ was also used to define the standard R^4 topology on M by simply calling a subset N of M open just in case $\phi [N]$, its image, is open in the standard topology of R^4. Thus the mapping ϕ serves two independent functions: it provides M with a "metrical" structure by specifying the interval I^2, and, in addition, ϕ is used in a quite different way to induce the R^4 metric topology on M. Naturally, the question arises: what is the relation between these two structures on M? In particular, can the *topology* of Minkowski space-time be derived from the space-time interval I^2?

The problem of the independence of the topological and metrical structures of space-time was clearly recognized by early writers on relativity such as Russell (1954) and, of course, Eddington:

. . . the statement that the world is four-dimensional contains an implicit reference to some ordering relation. This relation appears to be the *interval*, though I am not sure whether that alone suffices without some relation corresponding to *proximity*. It must be remembered that if the interval *s* between two events is small, the events are not necessarily near together in the ordinary sense. (1959, pp. 186–187.)

Eddington's *proximity* is just the local topological structure of space-time. Whereas Russell (1954, p. 56 ff.) held that space-time topology is interval-independent, Eddington did not rule out interval-dependence altogether, although he did recognize that the topology of space-time is not derivable from its interval in the standard way. As will become clear soon, Eddington's caution was justified. For the topology of many space-times (surely all those Eddington and Russell were considering) *is* obtainable from their space-time metric tensor field, although it is the *causal* content of the metric tensor which is exploited.

To begin with, let us be clear about just what aspect of relativistic space-times causes the difficulty. Consider a set N and a function d from $N \times N$ into R_0^+ (the set of nonnegative real numbers). If d is such that:

$$(1)\ d(x, x) = 0;\ (2)\ \text{if } d(x, y) = 0, \text{ then } x = y;$$
$$(3)\ d(x, y) = d(y, x), \text{ and } (4)\ d(x, y) + d(y, z) \geq d(x, y),$$

then (N, d) is said to be a *metric space*. When (1), (3), and (4) hold, the result is called a *pseudo-metric space*. If we let N be the point-set of either a Euclidean space or any of the classical non-Euclidean spaces, and take d to be the corresponding distance function on these point-sets, then each of these geometries may be shown to be a metric space in the above sense.[15] Furthermore, when a space (N, d) is a metric space, there is a natural way of defining a topology on that space: a subset A of N is open just in case, for every point x_0 in A, there is a *ball* $B_\epsilon(x_0) \equiv \{\, y \mid d(x_0, y) < \epsilon \}$, $\epsilon > 0$, containing x_0, and at the same time wholly contained in A $(B_\epsilon(x_0) \subseteq A)$. The fact that the open sets so defined indeed yield a topology on N is easy to prove, but the proof does require that the space (N, d) be a pseudo-metric space. For metric spaces, and thus for the classical geometries, their metric yields their standard topology when the above definition is used. More generally, it can be shown that for any Riemannian space with positive definite metric, the metric topology and manifold topology coincide (see Hicks, 1971, pp. 70–71).

However, Minkowski space-time is neither a metric nor a pseudo-metric space. The Minkowski interval yields distinct events with zero separation, and the triangle inequality fails. It is then easy to show that if "open" sets are defined in terms of balls as above, they do not determine *any* topology on M, much less the standard (R^4) topology. For the indefinite space-times of relativity theory, metrical nearness and topological nearness do not coincide. These facts appear to be the basis for Russell's conclusion that the topology of a space-time cannot be derived from its interval.

In retrospect, it is easy to see that this conclusion was too hasty. For although the *usual* definition of a topology in terms of balls given by the space-time interval will not do, it does not follow that some *other* definition in terms of the interval cannot succeed. Indeed, we now know that for a wide class of relativistic space-times (Minkowski space-time included), another type of construction does yield their standard topology. Moreover, this definition utilizes only that content of the interval which relates to the causal structure of space-time. In brief, the standard topologies of Minkowski and other space-times may be derived solely from their causal structure. Such a derivation is the first step on the way to a full-fledged causal theory of Minkowski geometry.

Like any topology, a topology for Minkowski space-time may be defined by specifying just which sets are to constitute its basis. The causal theory demands that these basis sets be constructed out of causality relations on M. These chosen basis sets are usually called *Alexandroff* intervals and are causally defined as follows:

Definition 4.0. For any events e_1, e_2 in M, *the Alexandroff interval of* e_1
and e_2, Alex (e_1, e_2), is just $\{e \mid TB(e_1ee_2)\}$: the set of all events chronologically between e_1 and e_2 (see Figure 4.0).

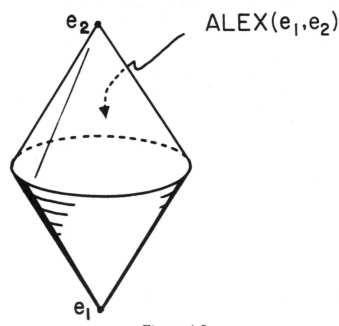

ALEX(e_1, e_2)

Figure 4.0

When e_1 and e_2 are chronologically connectible, then an Alexandroff interval consists of those events in the interior of the cone depicted in Figure 4.0; otherwise Alex (e_1, e_1) is simply the empty set. It is now easy to show that the Alexandroff intervals do provide a basis for a topology on M, i.e., for any two such intervals A_1 and A_2, there is another which is a subset of their intersection. A subset A of Minkowski space-time is now said to be open in the *Alexandroff topology* on M just in case every event e in A is an element of an Alexandroff interval which is itself a subset of A. Clearly in Minkowski space-time, any Euclidean 4-ball contains an Alexandroff interval, and conversely. Thus we have the result: *the Alexandroff topology and the standard E^4-topology on M coincide* (Hawking and Ellis, 1953, p. 196 ff). Since an Alexandroff interval is defined causally, this straightforward result establishes the adequacy of a causal theory of the standard topology of Minkowski space-time.[16]

Thus Russell and Eddington's questions concerning the physical significance of the topology of space-time are answered by Alexandroff's simple construction, at least for the space-time of special relativity. Russell's conclusion that space-time topology is independent of its metric was not justified, although he saw quite clearly that the topology of space-time does not have the *customary* relation to the space-time metric, and also realized that the space-time interval has an intimate connection with causality (1954, chap. 35). However, Russell's work on the foundations of relativity theory was seriously flawed by his being misled by some of the accidental features of the interval, such as its assignment of zero distances to all events on the same light path.[17] In general, Russell read the space-time interval too literally, and as a result never became clear about its relation to causal structure.

While the Alexandroff construction succeeds at once in Minkowski space-time, the situation in general relativity is not so straightforward. In some cases, the counterpart to the Alexandroff topology will be coarser than (have "fewer" open sets than) the manifold topology of general relativistic space-times. However, the two topologies do coincide in all Lorentz space-times that satisfy the condition of *strong causality*. This condition says, roughly, that every event e has arbitrarily small neighborhoods to which no causal curve returns once having left that neighborhood (Hawking and Ellis, 1973, p. 192ff). Since space-times that violate strong causality are in some sense "pathological," the physical significance of these exceptions to a causally definable topology is by no means clear at present.

157

John A. Winnie

V. Synthetic Causal Geometry

A. A. Robb appears to have been the first to suggest that the geometry of the space-time of special relativity can be developed on a purely causal basis. He was certainly the first to work out the details of this idea, and the result was a profound work of considerable magnitude and difficulty (1914).[18] Given Robb's ambitious aims, and the conceptual economy of his foundations, the complexity of his constructions was inevitable. For Robb's account was axiomatic, and the causal theorist who wishes to axiomatize his theory cannot avoid this consequence of having but a single primitive relation at his disposal. Robb's choice was the relation $<$ of causal precedence introduced in the preceding section, and twenty-one postulates are laid down for this relation. Only after a few hundred theorems do coordinates appear, and with them the Minkowski metric in its standard form. On the other hand, the definition of Minkowski space-time in section III above used Lorentz coordinates to achieve its result with a single postulate. Nothing could be a better example of the power of coordinate methods to introduce in a single step a wealth of structure which we do not begin to understand!

Rather than attempt a detailed survey of Robb's constructions, this section will provide a more concise development of the affine geometry of Minkowski space-time. The transition from affine to metric geometry is developed in the following section, although a few of the most important of Robb's metrical constructions are sampled here. The philosophical interest of these constructions lies in the fact that they show how the causal theory solves the central problem of any relational theory: the construction of quantitative (or metrical) comparison out of a qualitative causal relation. As we have seen, most contemporary causal theorists (Robb excepted) have turned to extra-causal sources as the basis of congruence and metric, and furthermore, they have provided epistemological and ontological arguments that purport to demonstrate the necessity for introducing extra-causal considerations. The following developments show how the causality relation alone permits the construction of the affine (linear) structure and interval congruence in space-time. As a result, these same constructions show how Reichenbach's arguments for the necessity of coordinative definitions of congruence (1957, chap. 1) are circumvented by the causal theory. The same is true of arguments that appeal to the homogeneity and continuity of a manifold as demanding an

158

extrinsic basis for congruence in opposition to the intrinsic temporal order afforded by causal relatedness (as in Grünbaum (1969), sec. 3). Both types of argument fail as a result of having overlooked the possibility of a purely causal construction of spatial, temporal, and space-time congruence. The account of the preceding section shows that this underestimation of the resources of causality is understandable; nevertheless, the results of the succeeding sections show this to have been a serious philosophical error.

The constructions that follow outline three areas of space-time geometry: linearity, orthogonality, and congruence. The development of the linear (affine) structure of Minkowski space-time is a considerable simplification of Robb's approach; proofs of the adequacy of its definitions are reasonably direct, and are either cited or provided. The account of orthogonality and congruence follows Robb closely; however, it is incomplete, and the definitions are not accompanied by adequacy proofs. This is remedied in section VI. With few exceptions, the constructions are not *obviously* successful—in general, their adequacy depends upon the world's causal structure in a nontrivial way. The remarkable fact is that, in the case of special relativity, the right sort of structure is forthcoming.

We begin with the linear structure of Minkowski space-time, i.e., the construction of its straight (geodesic) lines, planes, parallelism, and hyperplanes.

Lightlike (null) lines are the easiest to construct. If events e_1 and e_2 are light-connectible, i.e., $e_1 \lambda e_2$, the *null line containing e_1 and e_2 ($L(e_1, e_2)$)* is just the set of all events e which are light-connectible to both e_1 and e_2; in general, a set is a null line just in case it is a set of events such that for some distinct events e_1, e_2, this set is just $L(e_1, e_2)$.

Definition 5.0. When $e_1 \lambda e_2$, and $A \subseteq M$,
   (a) $L(e_1, e_2) = $ df. $\{e \mid e \lambda e_1 \ \& \ e \lambda e_2\}$
   (b) A is a null line $=$ df. for some distinct e_1, e_2 such that $e_1 \lambda e_2$, $A = L(e_1, e_2)$.

Although the adequacy of this definition is fairly obvious, a proof is provided for the sake of completeness.

Proposition 5.1. The null line L determined by two distinct events e_1 and e_2 with zero separation is just the set of events light-connectible to both e_1 and e_2.

Proof: Choose a Lorentz frame in which the coordinates of e_1 and e_2 (0, 0,

0, 0) and (1, 1, 0, 0), respectively. Those events light-connectible with e_1 have coordinates satisfying:

(1) $-(x^0)^2 + \Sigma\ (x^k)^2 = 0,\ k = 1,\ 2,\ 3,$

while those events light-connectible to e_2 satisfy

(2) $-(x^0 - 1)^2 + (x^1 - 1)^2 + (x^2)^2 + (x^3)^2 = 0.$

Subtracting (1) and (2) gives

(3) $x^0 = x^1,$

and substituting (3) in (1), we obtain

(4) $x^2 = x^3 = 0.$

Hence the points satisfying (1) and (2) are those of the form $(a, a, 0, 0)$, i.e., just those points on the null line L determined by e_1 and e_2. Done.

The construction of spacelike lines is not so obvious. Robb's approach is to use null lines to construct planes of a special kind, which he then intersects to recover non-null lines. The following idea is due to Latzer (1972), and is much simpler.

Suppose that the separation of e_1 and e_2 is spacelike (i.e., not $e_1\ \gamma\ e_2$). Then we say that e_3 *is S-collinear with* e_1 *and* e_2 just in case e_3 has a spacelike separation to both e_1 and e_2 (not $e_3\ \gamma\ e_1$ and not $e_3\ \gamma\ e_2$), and the light cones of e_1, e_2, and e_3 fail to have a common point of intersection (there is no event e such that $e\ \lambda\ e_1, e_2, e_3$). The *line* $S(e_1, e_2)$ *determined by* e_1 *and* e_2 is now just the set of all events S-collinear with e_1 and e_2. That is,

Definition 5.2 (a) $Se_1e_2e_3$ = df. (1) not $(e_1\ \gamma\ e_2,\ e_3)$ & not $(e_2\ \gamma\ e_3)$, and (2) there is no event e_4 such that $e_4\ \gamma\ e_1, e_2, e_3$.

　　　　　(b)　A is a spacelike line = df. $A. \subseteq M$ and for some events e_1 and e_2, $A = \{e_3\ |\ Se_1e_2e_3\}$.

The adequacy of this construction is shown by Latzer (1972, proposition 2.11).

Timelike lines may be constructed using the following approach. Let e_1 and e_2 be chronologically connectible ($e_1\ \tau\ e_2$). Next, consider the intersection of their light cones; this is defined as just $LC(e_1, e_2) \equiv \{e\ |\ e\lambda\ e_1$ and $e\ \lambda\ e_2\}$, and in a Lorentz coordinate system in which $x^k(e_1) = x^k(e_2) = 0\ (k = 1, 2, 3)$, $t(e_1) = -2$, $t(e_2) = 0$, this intersection will be a spacelike three-sphere defined by $(x^1)^2 + (x^2)^2 + (x^3)^2 = 1$ in the $t = -1$ hypersurface. (See Figure 5.0, one dimension suppressed.) Connect any two points in this sphere by a spacelike line. The set of all such lines yields a spacelike hypersurface $H(e_1, e_2)$, which is just the $t = -1$ surface of Figure 5.0. This hypersurface now being fixed, any other event e_3 *is said* to be *temporally collinear with* e_1 *and* e_2 ($T(e_1, e_2, e_3)$) just in case the hypersurface $H(e_3, e_2)$ (or $H(e_3, e_i)$) does not intersect the canonical hypersurface $H(e_1, e_2)$. In the figure, e_3 is thus collinear with e_1 and e_2 whereas e_4 is not. The following definition recapitulates this construction, and the theorem which follows establishes its accuracy.

Definition 5.3. Where $e_1 \tau e_2$:

(a) $LC(e_1, e_2) = $ df. $\{e \mid e \lambda e_1, e_2\}$

(b) $H(e_1, e_2) = $ df. $\{A \mid A$ is a spacelike line such that there are at least two events in $LC(e_1, e_2) \cap A\}$

(c) $T(e_1, e_2, e_3) = $ df. (1) $e_3 = e_1$ or $e_3 = e_2$ or (2) $e_3 \neq e_1, e_2$, and $e_2 \tau e_3$ & $H(e_1, e_2) \cap H(e_2, e_3) = \{ \ \}$.

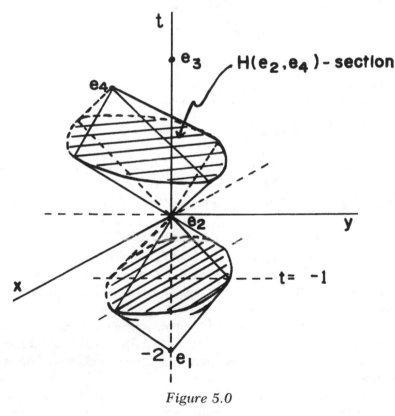

Figure 5.0

161

(d) A is a timelike line = df. $A \subseteq M$ and for some $e_1, e_2 \in M$, such that e_1 τ e_2, $A = \{e_3 \mid T(e_1, e_2, e_3)\}$.

Proposition 5.4. If e_1 and e_2 have timelike separation, then e_3 is collinear with e_1 and e_2 iff e_2 τ e_3 and Hyp (e_1, e_2) $\cap$ Hyp (e_2, e_3) = { }.

Proof: Let e_1 and e_2 have timelike separation, and e_3 be any event such that e_2 τ e_3. Choose a Lorentz frame $K = \{x^\alpha\}$ in which: (1) $x^0(e_1) = -2$, $x^0(e_2) = 0$; and (2) e_3 is in (say) the first quadrant of the $x^0 - x^1$ plane, and thus has coordinates $(a_0, a_1, 0, 0)$ (see Figure 5.0). If e_3 is identical with either e_1 or e_2, we are done. Otherwise, let K' be the Lorentz frame with relative speed $v = a_1/a_0$ to K in the positive direction along the x^1-axis of K, K' being thus the frame in which e_3 lies along the $x^{0'}$-axis. By the Lorentz transformations $x^{0'}(e_3) = ((a_0)^2 - (a_1)^2)^{1/2}$ and thus hypersurface $H(e_2, e_3)$ is given by

$$(1) \quad x^{0'} = \frac{x^{0'}(e_3)}{2} = \frac{((a_0)^2 - (a_1)^2)^{1/2}}{2} \, ,$$

in terms of the K' coordinates. Transforming back to K we obtain:

$$(2) \quad a_0 x^0 - a_1 x^1 = \frac{((a_0)^2 - (a_1)^2)}{2}$$

The $H(e_1, e_2)$ hypersurface is given by $x^0 = -1$, so the intersection of the two hypersurfaces is given by substituting in (2) to obtain

$$(3) \quad x^1 = \frac{((a_1)^2 - (a_0)^2 - 2a_0)}{2a_1}$$

which will have a solution just in case $a_1 \neq 0$, i.e., just in case e_3 fails to be on the x^0-axis of K. Done.

Having constructed the straight lines of Minkowski geometry, we may now go on to define its other linear objects in the standard way. A line ℓ_0 and an event e_0 not in ℓ_0 determine the plane which is the union of all lines containing e_0 and some point in ℓ_0. Parallel lines are those that are coplanar and fail to intersect. Finally, linear hypersurfaces are defined as maximal sets of lines intersecting a plane and a given point outside that plane.

With these definitions, the causal construction of the affine structure of Minkowski space is essentially complete. Since the definitions of planes and hyperplanes are standard, they require no additional justifying theorems.

The causal construction of the metric in all its details is a more complicated matter, and is carried out more efficiently within the mathematical framework of section VI below. Hence the following account is merely a sample of the techniques Robb employed. Nevertheless, the examples chosen (orthogonality and congruence) are of central metrical importance, and the following constructions illuminate their causal-geometrical bases.

Let ℓ_t and ℓ_s be timelike and spacelike lines, respectively, which intersect at an event e_0. When are we to say that ℓ_t and ℓ_s are orthogonal?

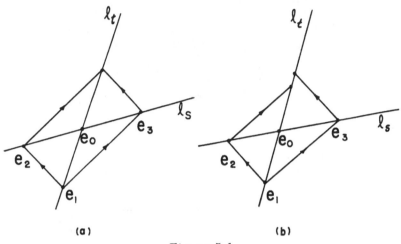

(a)   (b)

Figure 5.1

Robb's definition is simplicity itself (1921, p. 46ff), and incidentally, helps in reading orthogonality from of a space-time diagram. Choose an event e_1 on line ℓ_t ($e_1 \neq e_0$), as in Figure 5.1(a). The light cone of e_1 will now intersect ℓ_s at events e_2 and e_3, and each of their cones will in turn intersect ℓ_t once more. If these last intersections are the *same point-event* on ℓ_t (as in (a)), then ℓ_t and ℓ_s are said to be orthogonal.[19] Otherwise (as in (b) below), ℓ_s and ℓ_t are nonorthogonal. If in Figure (a) ℓ_t and ℓ_s are the x^0 and x^1-axes of an inertial frame, then they do indeed come to "look" orthogonal using this definition. It is not that orthogonal lines in Minkowski space fail to look orthogonal, in contrast to Euclidean lines, but that Minkowskian "orthogonality" bears only a loose family resemblance to its Euclidean counterpart.

Nevertheless, Euclidean spatial orthogonality is now forthcoming. For let ℓ_s^1 and ℓ_s^2 be two spacelike lines which intersect at e_0. We now determine (define) their orthogonality as follows. Let ℓ_t be the timelike line through e_0 and orthogonal (in the sense just defined) to *both* e_1 and e_2 (see Figure 5.2). Then ℓ_s^1 is orthogonal to ℓ_s^2 just in case every timelike line ℓ_t' in the plane determined by ℓ_t and ℓ_s^2 is orthogonal to ℓ_s^1. Remarkably enough, this definition is not only intuitively satisfactory, but can be shown quite rigorously to succeed in reconstructing standard Euclidean orthogonality. (A proof and further discussion are provided in the following section.)

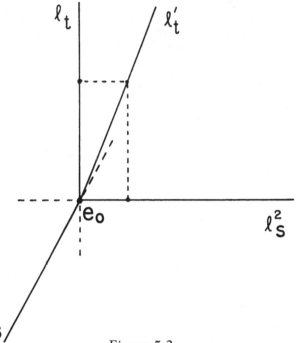

Figure 5.2

The basic idea of this construction (the comparison of a pair of spacelike lines or intervals using a timelike line as an intermediary) is the key to understanding the metrical constructions of causal geometry; the same device is used in the definition of spatial congruence which will soon follow.

Assume now that lines, planes, their parallelism, and orthogonality are given. How does Robb now proceed to construct a theory of congruence from these relations? First of all, consider classical (Euclidean) affine geometry. Within this theory, congruence between segments lying on *parallel* lines is easily established. Thus if ℓ_1 is parallel to ℓ_2, AB is a segment on ℓ_1, and A' is a point on ℓ_2, then the point B' (on a given side of A') which is such that AB is congruent to $A'B'$ is obtained by projecting parallels AA' and BB' from line ℓ_1 to line ℓ_2 in the usual way. Furthermore, by reflecting parallels from another parallel line (say ℓ_2) the interval AB may be "transported" along its own line as well.

Now just as parallelism allows the definition of a *partial* congruence relation in Euclidean affine geometry, the same is true in space-time geometry. Having developed a theory of parallelism, we define intervals which are the opposite sides of chains of parallelograms to be congruent in space-time.

But what of nonparallel line segments? In Euclidean affine geometry, it is time to add more structure: a congruence predicate and the "compass" axiom, the rotation group, or an inner product, depending upon the general approach adopted. In each case, a new *primitive term and axioms* are added, and in this straightforward sense, the theory of the total congruence relation is *external* to affine geometry (see Klein, 1939, p. 162ff).[20]

In the causal theory of congruence, there is no need to add new primitives or coordinative decrees, since constructions involving parallelism do not fully exhaust the metrical resources of a causal theory. In addition, there exists a distinguished set of lines, the *causal boundary lines* (the null lines), and these may be used to extend the theory to a full theory of space-time congruence.[21] Perhaps the best way to see this is to consider some examples.

Let e_1 and e_2 be two events with spacelike separation in two-dimensional Minkowski space-time. Then the light cones of e_1 and e_2 will intersect at two events A and B such that (by Robb's theory of parallelism) $e_1 B e_2 A$ is a parallelogram. (See Figure 5.3.) Such a parallelogram will be called a "light parallelogram" (Robb's "optical" parallelograms). Spacelike segments $e_1 e_2$ and timelike segment AB will intersect at some event O, which will be defined as the center of the light parallelogram. The theory of parallelism now allows a proof that $e_1 O$ is congruent to $O e_2$, and OA is congruent to OB. In other words, the diagonals of a light parallelogram bisect each other. So far, then, only parallelism is involved.

But now consider four-dimensional Minkowski space-time (Figure 5.4, drawn with one dimension suppressed). Again let e_1 and e_2 have a spacelike separation. Now there are many light parallelograms with diagonal e_1e_2. Choose two with diagonals which are both orthogonal to e_1e_2, with the result as shown in Figure 5.4. The light parallelograms $e_1 B$ $e_2 A$ and $e_1 B' e_2 A'$ share the diagonal e_1e_2 and center O. The remaining half-diagonals OB and OB' are now *defined* as congruent. This idea may now be extended to provide a general definition of congruence of timelike intervals along the following lines.

Suppose we are given intersecting timelike lines ℓ_1 and ℓ_2, with segment OB on ℓ_1, and we wish to locate B', i.e., the point on the "B-side" of ℓ_2 such that OB' is congruent to OB. The line with segment e_1e_2 (see Figure 5.4) can be shown to be uniquely determined by the condition that it be orthogonal to both ℓ_1 and ℓ_2. Point-events e_1 and e_2 are now determined by the "backwards" light cone of event B. The "forward" cones of e_1 and e_2 (actually only one is necessary) now determine point B' on ℓ_2; in effect, the "top half" of the intersecting light parallelograms is determined.

With this construction, the problem of the congruence of arbitrary

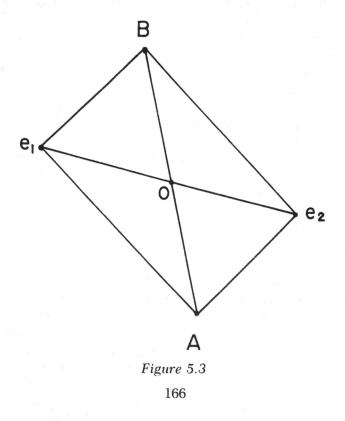

Figure 5.3

timelike intervals is all but solved. Timelike intervals AB and CD on *intersecting* lines are compared by using the above construction to transfer segment AB from its line to the other. Parallelism now allows the transported segment to be compared with segment CD. If intervals AB and CD lie on *nonintersecting* lines, we merely use a third line intersecting the second (CD line) and parallel to the first (AB line) as an intermediary and proceed as before.

An attractive feature of this construction is that it reveals the causal basis of the hyperbolas which form the contour lines of the Lorentz metric. Thus in Figure 5.5(a), let OA be the unit timelike interval. Then the locus of all events later than O and also a unit distance from O is given by the familiar hyperbola shown in the figure. The causal theory of congruence provides some geometrical insight into the situation. Thus let OA be

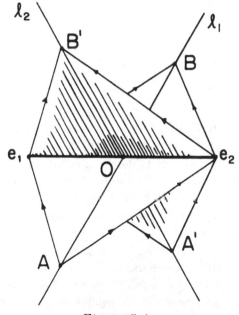

Figure 5.4

a unit interval and consider the problem of finding A' on an arbitrary timelike line ℓ through O and in the $t - x$ plane (see Figure 5.5(b)). By construction, the y-axis is orthogonal to both the t-axis and line ℓ. The light cone of A now determines e_1 and e_2, and so the diagonal of the top half of a light parallelogram. But the e_1 and e_2 cones will both intersect line ℓ in a unique point A', which, by our earlier definition, yields OA' congruent to OA.

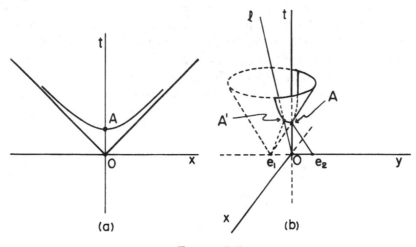

Figure 5.5

Now consider *all* timelike lines like ℓ that intersect O and lie in the $t - x$ plane. Use only the light cone of e_1 to determine the unit distance on this line, since the e_2 cone gives the same result. Then the intersection of the e_1 cone with such lines is the intersection of the e_1 cone with the $x^0 - x^1$ plane, which is just the Minkowski hyperbola.[22]

Finally, let us consider the congruence of spacelike intervals. These are compared by first associating them with a timelike interval, and comparing the latter as above. Once again, the general idea is to use light parallelograms, and then compare diagonals. For example, let AB and $A'B'$ below be spacelike intervals (see Figure 5.6). Let $ADBC$ be a light parallelogram with AB as diagonal (there will be many such; pick any one), and let $A'D'B'C'$ be chosen similarly. Then diagonals DC and $D'C'$ will be timelike (theorem), and AB is congruent to $A'B'$ if and only if (definition) DC is congruent to $D'C'$. The unique congruences thus specified provably yield a Euclidean geometry for both spacelike planes and linear three-dimensional spacelike subspaces of space-time (Robb, 1921, p. 67ff).

Of course, this last group of constructions requires supporting proofs of accuracy; these are supplied by Robb (1914). Nevertheless, with these constructions in mind we may stand back and ask: how do they manage to succeed?

Perhaps the two most remarkable features of the causal geometry of flat space-time are: first, the ability of the causal relations to yield the affine structure of space-time; and second, the lack of necessity for additional primitives (or coordinative definitions) in order to obtain the congruences of the Lorentz metric. From the Euclidean, or even Riemannian, viewpoint, this feat appears almost magical. How does causal geometry do the trick?

To answer this question we must go back to the conical structure of the causal space of each event. The set of events causally connectible with a given event e (the causal space of e) has a boundary or shell, the causal cone of e. The causal cone of e is definable in terms of $<$ alone, and light lines (better called "causal boundary lines" since we need not suppose them to be occupied) are then defined as above. The light lines form the foundation for the linear structure of Minkowski space-time.

Similarly, the *light* parallelograms form the basis for the comparison of nonparallel segments. A comparison with Euclidean space may be useful. Let AB be a segment in Euclidean three-space. Then the Euclidean parallelograms which have AB as their common diagonal form a motley

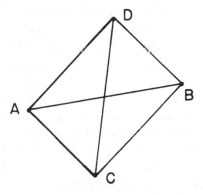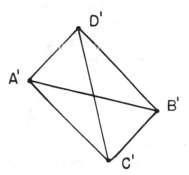

Figure 5.6

169

set whose non-*AB* diagonals have little in common aside from the condition just used to define them. But the *light* parallelograms of Minkowski space with a given diagonal *AB* form a severely restricted set. For one thing, they all have a common center (1914, theorem 59; 1921, p. 31). Hence the initial *plausibility* of defining corresponding half-diagonals of such light parallelograms as congruent. What is remarkable is that this definition is compatible with the affine congruence relations so that both taken together yield a congruence relation which is provably an equivalence relation on its domain.

Next consider the question of spatial geometry from the standpoint of the causal theory. As we have just seen, the congruence of spacelike intervals is defined uniquely in the causal theory by comparison with timelike intervals. The spatial geometry of a spacelike plane (or spacelike, three-dimensional, linear subspace of space-time) is then determined by the space-time congruences and linear structure of the events in the plane (subspace). The resulting geometry is then provably Euclidean.

At this point, an interesting issue arises. As Grünbaum has pointed out, although a metric yields a unique geometry on a space, the converse is, in general, false. In particular, the same set of points may be provided with two Euclidean metrics which nevertheless differ in their congruence verdicts (1973, p. 98 ff.).[23] However, the causal constructions just outlined yield a Euclidean spatial geometry by *first* delivering the congruence verdicts of that geometry! Thus, although the existence of alternative Euclidean metrics is a mathematical possibility, it is not, within special relativity, a physical possibility!

There is, of course, an important gap in this argument. For it has only been shown (or more precisely, claimed) that *given* the space-time geometry, spatial congruence is uniquely determined. Does Minkowski geometry itself allow alternative congruences? If so, we might expect this fact to reinstate derivatively alternative spatial congruences as well.

Clearly the preceding constructions show that this situation does not arise. For space-time congruence is *explicitly definable* in terms of causal structure. Given that this (causal) structure is fixed, space-time congruence—and hence spatial congruence—are uniquely determined by the preceding constructions.[24] Thus *the causal structure of Minkowski space-time allows the unique determination of spatial congruence within its spatial linear hyperspaces.* As a result, no extra-causal resources such

170

as extended rods are needed in order to specify spatial congruence. On the contrary, the causal definition of spatial congruence may be used to provide an ontological basis for the assessment of the rigidity or accuracy of material congruence standards.

The preceding constructions were not, in general, accompanied by justifying theorems. However, the development of the linear theory was reasonably rigorous. The metrical constructions (orthogonality and congruence) were not provided with proofs of adequacy, and were, in any case, incomplete. The following section repairs this defect by defining the Minkowski metric in terms of its linear and causal structure, with proofs of adequacy provided. Fortunately, this development may be carried out in a reasonably direct way within the framework of modern metric affine geometry.

VI. From Affine Structure to the Minkowski Metric

In the preceding section it was shown that the construction of a space-time metric falls naturally into two parts: (1) the theory of linear or affine structure; and (2) the theory of congruence and metric. Constructions that yield the affine structure of Minkowski space-time have been provided and justified. However, the causal construction of the Minkowski metric was merely outlined. In this section, the construction of the Minkowski metric from parallelism will be developed in detail along lines proposed by Robb, but using the techniques of the modern theory of metric affine spaces. The aim is to show that, given the usual linear structure of affine spaces, the mere addition of the Minkowski null cone makes possible the construction of the metric structure of Minkowski space-time: orthogonality, congruence, and interval. For Minkowski space-time, the transition from parallelism to metric—which, in Euclidean spaces, is carried out by adding congruence postulates or an inner product—may be achieved merely by adding the null cone structure. We begin by outlining the general structure of affine metric spaces.[25]

Let V be a vector space over the field of real numbers. A (nonsingular) *metric vector space* is obtained by providing V with an *inner* product ($<$, $>$) which is just a function that assigns a real number to every pair of vectors in V, and is, in addition, linear, symmetric, and nonsingular.[26] Thus a metric vector space[27] may be regarded as a pair $V_I = (V, <,>_I)$ with V a real vector space, and $<,>_I$ an inner product on V.

171

The choice of this inner product now determines the geometry of V_I. For example, if we now require that the inner product be *positive definite*, i.e.,

for all $v \in V$, $<v, v> \geq 0$, with
$<v, v> = 0$ iff $v = 0$

the resulting metric vector space is said to be Euclidean.[28] *Minkowski vector space* (V_m^4) may be defined as a four-dimensional (dimension here is that of the vector space V) real metric vector space whose inner product satisfies the following condition:

(M) There are four linearly independent vectors

$v_0, \ldots, v_3$ such that
$<v_0, v_0> < 0$, and $<v_k, v_k> > 0$, $k = 1, 2, 3$.[29]

The *light cone* or *null cone* of V_m^4 consists of all vectors v, called *null vectors*, whose length[2] $(<v, v>)$ is zero. Clearly the zero vector, $\vec{0}$, is in the null cone of V_m^4, and it is easy to show that the null cone consists of a three-dimensional surface (not subspace, however) in V_m. A vector v is said to be *timelike* if $<v, v> < 0$ and *spacelike* if $<v, v> > 0$. Two vectors u and v are *congruent* if and only if they have the same length, i.e., $<u, u> = <v, v>$. Notice that only timelike vectors are congruent to timelike vectors, and so too for spacelike vectors.[30] All null vectors are hereby made congruent, but this is merely a convenience. Null vectors may be compared *along a given null line* by means of the scalars of V^4. (Thus av and $-av$ might be held congruent, av and $2av$ not congruent, even when v is null.)

The above account outlines the standard approach to Minkowski metric vector spaces. A specific inner product is added to V^4, and then used to define vector length[2], the light cone, spacelike and timelike vectors, and, if we like, congruence of vectors. The analogue of Robb's approach is quite different. Here we begin with the vector space V^4 and its light cone, which will be called "Null." Congruence, orthogonality, and finally, an inner product are now defined in terms of this structure. In other words, where the standard account begins with a structure of type $(V^4, <, >)$, and goes on to define Null, a subset of V^4, as:

(1) Null $\equiv \{v \mid v \in V^4 \text{ and } <v, v> = 0\}$,

the following account defines congruence and the inner product function in terms of Null (and the vector space structure of V^4).[31] However, before proceeding, a few remarks are in order concerning the relationship of the above Minkowski metric *vector* space to affine Minkowski space-time, the universe of the latter consisting of points (or point-events), that of the former being a set of vectors.

Let X be a nonempty set of points. Recall that $V_I = (V, <,>_I)$ is a metric *vector* space. We now make the set of points X into a geometrical space (*a real affine metric space*) by using the metric *vector* space V_I to induce a metric on X. The obvious way to do this is to associate each pair of points in X with a vector in V. Thus if $<x, y>$ is an ordered pair of points in X, the vector in V associated with $<x, y>$ is designated by "$<\overrightarrow{x, y}>$." The distance[2] between x and y in X is now defined in a natural way as the length[2] of $<\overrightarrow{x, y}>$ in V_I (i.e., as $= <<\overrightarrow{x, y}>, <\overrightarrow{x, y}>>$). The straight line through x and y is given by the set of points associated with all vectors $a <\overrightarrow{x, y}> \epsilon V$, $a \epsilon$ Reals, etc.

Now the association of point-pairs in X with vectors requires an *association function A*, usually called the *action* (function) of V_I on X. Of course, A must induce the structure of V_I on X in a "natural" way, and is accordingly subjected to the following requirements:

(2) A is a function from $V \times X$ to X, such that:
   (a) $A(\overrightarrow{0}, x) = x$, for all $x \epsilon X$,
   (b) If $A(v, x) = A(w, x)$, then $v = w$, and
   (c) $A(u + v, x) = A(u, A(v, x))$. (Cf. Snapper and Troyer, 1971, p. 6.)

A *real affine metric* space is now defined as a set X, and a real metric vector space V_I together with an action function A.

Now consider two such affine spaces, (X, V_I, A) and (Y, V'_I, A'). It is not difficult to show that these spaces are isometric if and only if V_I and V'_I are isometric (Snapper and Troyer, 1971, p. 387, ex. 12). This amounts to saying that the study of an affine metric geometry essentially reduces to the study of its metric *vector* space V_I. In short, while an affine space (say Minkowski space-time) and its associated metric vector space (say V_m^4) are not identical, the geometric structure of the affine space derives entirely from its associated metric vector space, so the geometry of the former essentially reduces to that of the latter. Hence Minkowski space-time, i.e., affine Minkowski metric space, is defined as a triple (M, MV^4, A), where M is the set of its "points" (or point-events), $MV^4 = (V^4, <, >_m)$, and A specifies the action of MV^4 on M in accord with the requirements of (2) above.

173

The above isomorphism theorem shows that, having distinguished Minkowski space-time from its metric vector space, we now may proceed to concentrate on the latter. The first task will be to use the Null vectors of V^4 to define congruence of its vectors, and then to go on to show how to define orthogonality and a Minkowski inner product on V^4.

The central notion in the following is that of the *conjugate class*, Conj (v), of a vector v. A vector u is in the conjugate class of v if and only if both u and $-u$ yield a null vector when added to v. It turns out (proposition 6.1 below) that the conjugate class of v contains just those vectors that are orthogonal to v and have length^2 equal to negative length^2 of v.

Definition 6.0. Where $v \in V^4$,

Conj (v) = df. $\{u \mid u \in V^4$ and $v + u, v - u \in$ Null$\}$.

Proposition 6.1. For all $u, v \in V^4$,

$u \in$ Conj (v) iff $<u, v> = 0$ and $<u, u> = -<v, v>$.

Proof: (a) Suppose $u \in$ Conj (v). Then (1) $<u + v, u + v> = 0$ and (2) $<u - v, u - v> = 0$. Expanded, (1) and (2) become:

(3) $<u, u> + 2<u, v> + <v, v> = 0$ and (4) $<u, u> - 2<u, v> + <v, v> = 0$.

Adding and subtracting (3) and (4) yields:

$<u, u> = -<v, v>$ and $<u, v> = 0$.

(b) As in case (a). Done.

The following two propositions will be used later in proving the accuracy of the forthcoming congruence definition:

Proposition 6.2. (a) Conj $\vec{(0)}$ = Null.

(b) If $u \in$ Null and $u \neq \vec{0}$, then Conj (u) = $<<u>>$, the subspace of V^4 spanned by u.

Proof: (a) Immediately from definition 6.0.

(b) Suppose $v \in$ Conj (u). Then by proposition 6.1, $v \in$ Conj (u) iff (1) $<v, v> = -<u, u> = 0$ and (2) $<u, v> = 0$. Hence if $v = \vec{0}$, $v \in$ Conj (u), and thus in $<<u>>$. Suppose then that $v \in$ Conj (u), $v \neq \vec{0}$ and $v \notin <<u>>$. Then u and v are linearly independent, and so $<<u; v>>$, the subspace spanned by u and v, is a plane in V_m^4. Let w be in $<<u; v>>$. Then $w = au + bv$, for some $a, b \in$ Reals. However, by virtue of (2), $<w, w> = 0$. Hence $<<u; v>>$ is a null plane in V_m^4, which contains no such planes (see Snapper and Troyer, 1971, corollary 195.2). Thus we must have $v \in <<u>>$.

On the other hand, if $v \in <<u>>$, $v = au$, so $<v, v> = a^2<u, u> =$

0, and $<u, v> = a<u, u> = 0$. Hence by proposition 6.1, $v \in$ Conj (u). Done.

Next it is shown that a vector and its negative have the same conjugates, and that every vector has some conjugate.

Proposition 6.3. (a) For all u, Conj (u), $\neq \{ \quad \}$
                 (b) For all u, Conj (u) = Conj $(-u)$.
Proof: (a) If $u \in$ Null, then the result follows from proposition 6.2. Suppose then that u is timelike. Let v be spacelike and hence linearly independent of u. Then $<<u, v>>$ is a Lorentz plane and contains a spacelike vector w orthogonal to u. Multiply w by the appropriate scalar a to obtain aw: $<aw, aw> = -<u, u>$, and $<u, aw> = 0$. Proposition 6.1 now yields $aw \in$ Conj (u). If we suppose u to be spacelike, we choose a timelike vector v' and proceed similarly.
    (b) Immediately from definition 6.0. Done.

We are now in a position to define congruence of vectors in V^4. Note that by proposition 6.1, conjugate vectors are not in general congruent (their lengths being of opposite sign), but for any vector u, the vectors in Conj (u) are all congruent to each other. Still, if u and v are, say, congruent timelike vectors, can we count on there being a spacelike vector w such that $u \in$ Conj (w) and $v \in$ Conj (w), thereby yielding their congruence? The following definition relies on this being the case, and the propositions which follow justify this confidence.

Definition 6.4. $u \simeq v$ (u is congruent to v) = df.
    Conj $(u) \cap$ Conj $(v) \neq \{ \quad \}$.
Proposition 6.5. For all u, v,
    (1) $u \simeq u$ (reflexivity),
    (2) if $u \simeq v$, then $v \simeq u$ (symmetry).
Proof: (1) Immediately from definition 6.4 and proposition 6.3.(a).
    (2) Immediately from definition 6.4.

The following proposition now establishes the accuracy of the congruence definition.

Proposition 6.6. $u \simeq v$ iff $<u, u> = <v, v>$.
Proof: (a) Assume $u \simeq v$. Then for some vector w_0, $w_0 \in$ Conj (u) and $w_0 \in$ Conj (v). Hence by proposition 6.1, $<u, u> = -<w_0, w_0> = <v, v>$.
    (b) Suppose $<u, u> = <v, v>$.

(1) $u, v \in$ Null. By proposition 6.2, $\vec{0} \in$ Conj (u) and $0 \in$ Conj (v), whether or not u and v are the zero vector.

(2) $u, v \notin$ Null.

(α) u, v linearly dependent. Then $u = kv$, and thus $<u, u> = k^2<v, v>$. But $<u, u> = <v, v>$, so $k = \pm 1$.

Hence $u = v$ or $u = -v$. In either case, by proposition 6.3, Conj $(u) \cap$ Conj $(v) \neq \{ \quad \}$.

(β) u, v linearly independent. Since $<u, u> = <v, v>$, and $u, v \notin$ Null, u and v are either both timelike or both spacelike. Hence $<<u; v>>$ is either a Lorentz plane or a Euclidean plane. In either case $V^4 = <<u; v>> \oplus <<u; v>>^*$, where $<<u; v>>^*$ is the orthogonal complement of $<<u; v>>$. (See Snapper and Troyer, 1971, p. 155, ex. 4b.) Thus $<<u; v>>^*$ is either a Euclidean plane or a Lorentz plane, respectively. Hence when u, v are both timelike, a spacelike vector w_0 is in $<<u; v>>^*$ and is thus orthogonal to both u and v. Multiplication by an appropriate scalar will now yield an aw_0, $<aw_0, aw_0> = -<u, u> = -<v, v>$. Hence by proposition 6.1, $aw_0 \in$ Conj $(u) \cap$ Conj (v). When u, v are both spacelike, a similar argument succeeds. Done.

Now that it has been shown that congruence of vectors is definable using the null cone of V^4_m, the orthogonality of vectors in V^4_m is defined next. We want our definition to yield u orthogonal to v just when $<u, v> = 0$. With congruence and orthogonality defined, the construction of the Minkowski inner product on V^4 is simple and direct.

As an aid to the constructions to follow we now define what it means for two vectors u and v to be of the same type ($u \longleftrightarrow v$)—intuitively, to be either both null, both timelike, or both spacelike. The definition is obvious. The proposition that follows provides justification.

Definition 6.7. $u \longleftrightarrow v =$ df. for some $a \neq 0$, $au \simeq v$.

Proposition 6.8. $u \longleftrightarrow v$ iff $u, v \in$ Null, or u, v timelike, or u, v spacelike.

Proof: Immediately from proposition 6.6 and from the fact that if u and v are of the same type, a suitable choice of $a \neq 0$ will yield $<au, au> = <v, v>$.

By the *type of a vector* v we shall mean the set of all vectors of the same type as v. However, the above definition has the zero vector different in type from any spacelike or timelike vector; so for the sake of convenience we add it to each class.

Definition 6.9. Where v is any vector, the type of v (Type (v)) is just the set of all vectors of the same type as v, together with the zero vector, i.e.,

Type (v) = df. $\{u \mid u \longleftrightarrow v \text{ or } u = \vec{0}\}$.

The construction of orthogonality is helped by being able to distinguish timelike and spacelike vectors. It is well known (cf. Noll, 1964) that the set of spacelike vectors contains a subspace of dimension three, whereas the timelike vectors contain subspaces of dimension one at most. We use this fact below to distinguish the two sorts of vectors.

Definition 6.10. Where u is any vector,
   (a) u *is timelike* = df. Type (u) contains subspaces of at most dimension one;
   (b) u *is spacelike* = df. Type (u) contains subspaces of dimension three.

We are now in a position to define orthogonality of vectors in v.

The definition of orthogonality is made easier by recalling that in Minkowski spaces: (1) timelike vectors are only orthogonal to spacelike vectors, (2) two (nonzero) null vectors are orthogonal if and only if they are linearly dependent, and (3) the zero vector is orthogonal to every vector (cf. Trautman, 1965, section 3.2). Hence only three interesting cases of orthogonality remain: spacelike vectors orthogonal to (nonzero) null vectors, timelike vectors, and other spacelike vectors. The case of the orthogonality of a spacelike and a timelike vector amounts to a simple extension of the idea of conjugate vectors introduced above. The geometrical significance of the other two cases of orthogonality is quite different, however, and each case merits individual treatment. First, the case of two orthogonal *spacelike* vectors.

The intuitive idea is this. Let u and v be orthogonal spacelike vectors. Then there will be a plane $<<w_0, v>>$, containing v, all of whose vectors (and hence all of whose timelike vectors) are orthogonal to u (see Figure 5.2).

The orthogonality of a (nonzero) null vector and a spacelike vector has still another significance. Let u and v be the null and spacelike vector, respectively. Then u and v may be such that their "tips" are not causally connectible, i.e., $u - v$ is spacelike. If this is so *no matter how* v *is extended* (for all av, $a \neq 0$), then and only then are u and v orthogonal. In other words, an intersecting spacelike and a null line are orthogonal just in case no point on the spacelike line, save the intersection point itself, is

177

causally connectible with any point on the null line (cf. Robb, 1921, p. 46, definition B).

These intuitive explanations are stated more precisely in the following definition. The succeeding proposition shows that the construction is successful.

Definition 6.11. $u \perp v = $ df.

   (1) $u = 0$ or $v = 0$, or

   (2) $u, v \in$ Null, $u \neq \overrightarrow{0} \neq v$, and $u = av$, for some $a \in$ Reals, or

   (3) u is spacelike and:

      (a) v is timelike and for some $a \in$ Reals, $a \neq 0$, $au \in$ Conj (v); or

      (b) v is spacelike and for some $w_0 \in$ Conj (v), if t is timelike and in $<<w_0; v>>$, then $u \perp t$, i.e., t satisfies (a); or

      (c) v is null, $v \neq 0$, and $v - au$ is spacelike for all $a \in R$, $a \neq 0$; or

   (4) as in (3), with u and v interchanged.

Proposition 6.12. $u \perp v$ iff $<u, v> = 0$.

Proof: (A) Suppose $<u, v> = 0$. By the remarks above, definition 6.11 covers all possible cases. Hence we need only show its accuracy in each case. Cases (1) and (2) are trivial.

Case (3) (a): Choose $a \neq 0$ such that $<au, au> = -<v, v>$. Then by proposition 6.1, $au \in$ Conj (v).

Case (3) (b): Since $<u, v> = 0$, u and v are linearly independent, so $<<u; v>>$ is a Euclidean plane. Let w_0 be a timelike vector, $<u, w_0> = 0$, $<v, w_0> = 0$, suitably chosen so that $<w_0, w_0> = -<v, v>$. By proposition 6.1, $w_0 \in$ Conj (v). Now suppose t timelike and $t \in <<w_0; v>>$. Then $t = aw_0 + bv$.

Hence $<u, t> = a<w_0, u> + b<v, u> = 0$. Choose $d \neq 0$, so that $<dt, dt> = -<u, u>$. Then $<dt, u> = 0$ and by proposition 6.1, $dt \in$ Conj (u). Case (3) (c): By a standard theorem, (i) $<v - au, v - au> = <v, v> - 2<v, au> + a^2<u, u>$.

Since $<v, v> = <v, u> = 0$, (i) becomes (ii) $<v - au, v - au> = a^2<u, u>$.

Since $<u, u> > 0$, (ii) shows that $v - au$ is spacelike for all $a \neq 0$. Furthermore, (i) shows that unless $<v, u> = 0$, some choice of $a \neq 0$ will make $<v - au, v - au> < 0$, proving sufficiency as well.

Case (4): as in case (3).

   (B) Suppose $u \perp v$. Cases (1) and (3) are trivial. Case (3) (a) follows from proposition 6.1. Case (3) (c) was shown under (A) above. Case (4) is treated like case (3). This leaves only Case (3) (b). Choose $t_1, t_2 \in <<w_0;$

$v >>$ (a Lorentz plane) to be timelike and linearly independent. Then by (3) (b), $u \perp t_1$, $u \perp t_2$, so by case (3) (a), $<u, t_1> = <u, t_2> = 0$. Now $v = at_1 + bt_2$ (since t_1, t_2 a basis of $<<w_0; v>>$), so $<u, v> = a<u, t_1> + b<u, t_2> = 0$. Done.

A Minkowski inner product is now defined as follows. First, choose a timelike vector t_0 as a unit. Let s_1 be any vector in Conj (t_0). Hence $t \perp s_1$ and $<s_1, s_1> = -<t_0, t_0>$. We now complete the set by adding s_2, s_3, both spacelike, congruent to s_1, and such that the vectors $B = \{t_0, s_1, s_2, s_3\}$ are pairwise orthogonal. (Since this is V_m^4, we know that such exist.) The set B is thus a Lorentz basis of V_m^4, so we may define the inner product $<,>_m^*$ in the usual way:

Definition 6.13. $<u, v>_m \equiv -u^0 v^0 + u^k v^k$, k = 1, 2, 3, where u^α and v^α are the components of u and v in any basis $B = \{t_0, s_1, s_2, s_3\}$, t_0 is time-like, $s_1 \in$ Conj (t_0), $s_3 \simeq s_2 \simeq s_1$, and the members of B are pairwise orthogonal.

The following theorem follows automatically:

Proposition 6.14. Let $<,>^*$ be a Minkowski inner product on V^4, with null cone = Null and $<t_0, t_0>^* = -1$. Then $<u, v>^* = <u, v>_m^*$.
Proof: Since the null cone of $<,>^*$ is Null, all previous theorems apply. Let $B = \{t_0, s_1, s_2, s_3\}$ be as specified in definition 6.13. Its members are mutually orthogonal $\perp$, and so, by theorem 6.12, have pairwise inner products $(<,>^*)$ equal to 0. But $<t_0, t_0>^* = -1$, and so by propositions 6.1 and 6.6, $<s_k, s_k>^* = +1$, k = 1, 2, 3. Thus B is a Lorentz frame and $<u, v>^* = -u^0 v^0 + u^k v^k \equiv <u, v>_m^*$. Done.

So, given the vector space V^4 and the Minkowski null cone in V^4, the Minkowski metric may be uniquely defined (up to a unit of length). In Euclidean geometry, on the other hand, the choice of a unit vector does not suffice for the construction of the metric. In addition, a set of $n - 1$ linearly independent vectors must be exhibited and defined to be mutually congruent and pairwise orthogonal. The arbitrariness of the resulting metric, from the physical standpoint, will reside in the extent to which the choice of these other vectors is based upon factual physical relations. When Minkowski geometry is seen in a similar way, the choice of s_1, s_2, and s_3 comes under scrutiny, and issues involving the conventionality of the spatial length of transported spatial measuring standards will naturally arise. The above construction shows how this difficulty is eliminated

179

within Minkowski space-time. In addition to the vector t_0, we obtain the metric by exhibition, but now by the exhibition of the null cone in V^4. However, the null cone, now interpreted as the boundary of the causal connectedness relation, is free from the conventions that attach to the choice of various spatial and temporal measuring devices.

The last theorem (proposition 6.14) is also a step on the way toward fulfilling the claim made earlier (section V) that two Minkowski metrics on the same set M which have identical null cones are identical up to a constant scale factor. However, proposition 6.14 assumes that the two metrics yield the same linear structure for space-time, and this is an unnecessary restriction, since two Minkowski metrics with the same null geodesics agree on all geodesics. Hence the following theorem is just what is needed:

Proposition 6.15. Let $\mathfrak{M}_1 = <M, \eta>$ and $\mathfrak{M}_2 = <M, \eta'>$ be Minkowski space-times [32] with identical null cone structures. Then $\eta' = k\eta$, where k is a positive constant.

Proof: Since $\mathfrak{M}_1$ and $\mathfrak{M}_2$ have the same null cone structure, they are conformal, i.e., $\eta' = \sigma\eta$, where σ is a positive real-valued, smooth function defined on all of M. Since $\mathfrak{M}_1$ and $\mathfrak{M}_2$ are both flat, it can be shown—as in Haantjes (1937), p. 702—that σ must satisfy
(1) $2S_{\lambda,\mu} = S_\mu S_\lambda - (\eta_{\mu\lambda}/2)(\eta^{\alpha\beta}S_\alpha S_\beta)$, where $S_i \equiv \partial \ln\sigma/\partial x^i$. Equations (1) may be solved, giving $\sigma = $ constant as a solution, along with solutions of two other types. However, for each of these types, σ is not defined on *all* of M (Haantjes, 1937, section 2). Hence $\sigma = $ constant. Done.

It now follows that the causal structure of Minkowski space-time fixes its metric up to a constant factor. As a result the spatial metric of the inertial frames of special relativity is determined (up to a constant factor) as well.

VII. The Causal Definition of Congruence

While the constructions of synthetic causal geometry move deeply into the structure of Minkowski space, the homogeneous geometry of Minkowski space-time permits the effective use of group-theoretic methods in its study. These methods provide a considerably easier way to show the causal definability of congruence in Minkowski space. For E. C. Zeeman has shown (1964) that the causal group of flat space-time and its "Lorentz" group are one and the same; in this section it will be shown that the causal definability of Minkowski space-time congruence is a relatively easy consequence of Zeeman's result. To begin with, let us be quite clear about what Zeeman's theorem asserts.

Recall that "$TB(e_1 e_2 e_3)$" asserts that event e_3 is *chronologically between* events e_1 and e_3, and "$CB(e_1 e_2 e_3)$" asserts that e_2 is *causally* between events e_1 and e_3. Now a *symmetry* (or automorphism) of a given structure, say $<M, TB>$ or $<M, CB>$, is a bijective (one-one onto) mapping of its universe M to itself which, in addition, preserves certain (perhaps all) of that structure's relations. The set of all such symmetries is easily shown to be a group when functional composition is taken as the group operation, and so is called a "symmetry group" of that structure. The symmetry groups of $<M, TB>$ and $<M, CB>$ will henceforth be called the *chronology group* (*TG*) and the *causality group* (*CG*) of Minkowski space-time, respectively. Thus *TG* consists of all one-one mappings of Minkowski space-time onto itself that preserve chronological betweenness, while *CG* contains just those mappings that preserve causal betweenness.[33]

However, in Minkowski space-time, the chronology relation $<<$ and the causality relation $<$ are interdefinable (see Kronheimer, 1967, section 2). It follows immediately that a mapping which preserves the one will preserve the other, and so the chronology group and causality group are identical ($CG = TG$).

Of course, this easy result is not Zeeman's theorem. What Zeeman has shown is that the causality-chronology group is in turn identical with the set of all mappings that preserve the Minkowski metric up to a scale factor. This group thus contains the metrical symmetries together with those maps that uniformly expand or contract Minkowski space-time ("dilatations," "magnifications"). For the sake of clarity, let us define the groups involved more carefully.

Let (M, I^2) be the points of Minkowski space-time together with the interval function $I^2: M \times M \rightarrow R$, where $I^2(e_1, e_2)$ is the interval squared between e_1 and e_2. Then an *isometry* of M onto M is a one-one mapping of M onto M which preserves I^2. The set of all such isometries forms a symmetry group, called the Poincaré Group (*PG*) of M.[34]

It would be too much to expect the causal group *CG* to be identical with the Poincaré group of flat space-time, since the Poincaré group fixes the scale of space-time geometry, and this would seem to require extra-causal considerations or conventions. In any case, the geometrically fundamental group is not the isometry group, but rather the *similarity group*, of a geometry: the set of all one-one mappings of the space onto itself that preserve the metric up to a scale factor.[35]

Thus a mapping h *is a similarity* of Minkowski space if and only if h is a bijection of M to M such that:

$$I^2(e_1, e_2) = j \cdot I^2(h(e_1), h(e_2)),$$

for some positive real number j. The group of all similarities of Minkowski space includes the Poincaré Group as a subgroup (take $j = 1$), and will henceforth be called the Extended Poincaré Group (*EPG*). A theorem on metric affine spaces (Snapper and Troyer, 1971, theorem 416.1) now tells us that every similarity is the product of a magnification and an isometry, so we may characterize *EPG* as obtained from the Poincaré group by adding magnifications and closing under functional composition.[36]

We are now in a position to state Zeeman's theorem:

(Z) The causality (= chronology) group of Minkowski space is just the Extended Poincaré Group; i.e., $CG = TG = EPG$.

In other words, any one-one mapping of Minkowski space-time onto itself that preserves its causal structure also preserves its metric up to a scale factor. (The converse is obvious since $<$ and $<<$ are definable in terms of I^2.)

Clearly Zeeman's theorem shows that the metrical and causal structure of space-time are very closely related. According to the theorem, the similarity group of the structure $<M, I^2>$ and the symmetry group of $<M, CB>$ are identical. However, exactly what does this tell us about the connection between the relation CB of causal betweenness and the space-time interval I^2? In particular, may we now conclude that these two relations are interdefinable? As a first step towards considering this question, let us consider the relation between symmetry groups and definability in a more general setting.

Let $Th(R, S)$ be a theory containing just two relation-terms R and S. Then a possible model of $Th(R, S)$ will be a structure (A, R, S), where A is a nonempty set and R and S are relations on A of the appropriate types. Now let $M = (A_1, R_1, S_1)$ be a model of $Th(R, S)$, and consider the reduced structures $K = (A_1, R_1)$ and $J = (A_1, S_1)$. Both K and J have their associated symmetry groups Sym(K) and Sym(J). Suppose now that these two groups are *not* identical. Then R and S are *not* interdefinable in $Th(R, S)$. This result is an immediate consequence of the following theorem.

Theorem 7.0. Let Th(R, S_i), $i = 1, 2, \ldots, n$ be a theory containing the relation terms R, S_1. Let $A = (M, R, S_1, \ldots S_n)$ be any model of Th(R, S_i). Then if R is definable in Th(R, S_i) we must have Sym(($M, S_1, \ldots, s_n$) $\subseteq$ Sym((M, R)). (Here Sym((M, R)) and Sym((M, S_i)) are the symmetry groups of (M, R) and (M, S_i), respectively.)

Proof: Suppose $h_0 \in$ Sym((M, S_i)) but $h_0 \notin$ Sym((M, R)). Now consider the structure ($M, h_0[R], h_0[S_1], \ldots, h_0[S_n]$) $\equiv A'$. By construction, h_0 is an isomorphism of A and A'; thus A' is also a model of Th(R, S_i). Since $h_0 \in$ Sym((M, S_i)), $h_0[S_i] = S_i$; but $h_0 \notin$ Sym((M, R)), so $h_0[R] \neq R$. Thus $A' = (M, R', S_1, \ldots, S_n)$, where $R' = R$. Hence, by Padoa's method, R is not definable in Th(R, S_1).[37]

In order to apply this result to the causal definability of congruence, it is only necessary to note that the mappings in the Extended Poincaré Group are just the pairwise *congruence-preserving* mappings of Minkowski space-time, since *EPG* is just the similarity group of Minkowski space-time. This may be seen as follows. As in affine metric spaces in general, congruence of point-pairs in Minkowski space-time may be defined in this way:

Definition 7.1. Let $A = <u, v>$ and $B = <x, y>$ be pairs of points in M. Then A *is congruent to* B iff there is an isometry h of M such that $h(u) = x$ and $h(v) = y$.[38]

A four-placed congruence relation $C(u, v, x, y)$ (read "the pair u, v is congruent to pair x, y" or "u is just as far from v as x is from y") may now be defined as obtaining just in case $<u, v>$ and $<x, y>$ are congruent according to the above definition.

Now suppose h is a *similarity* of M (in particular, $h \in EPG$), and $C(u, v, x, y)$. Then we must also have $C(h(u), h(v), h(x), h(y))$, since the similarity h magnifies both $<u, v>$ and $<x, y>$ by the same factor. Thus it is easy to see that, quite generally, the similarities of a geometry are just those mappings that preserve the congruence relation.[39] So, for Minkowski space, the symmetry group of (M, Con) is just the Extended Poincaré Group (*EPG*).

We are now in a position to apply the earlier theorem. Let Th(CB, C) be a theory containing the three-place relation term CB (to be interpreted as causal betweenness) and the four-place relation term C (to be interpreted as the point-pair congruence relation), and containing all sentences true of Minkowski space under this interpretation. Is C definable

John A. Winnie

in Th(CB, Con)? The above theorem now tells us that this will be the case *only if*

(1) Sym(<M, CB>) $\subseteq$ Sym(<M, C>), when (M, CB, C) is any model of Th(CB, C). But, as we have just seen, Sym(<M, C>) is just *EPG*, and so by Zeeman's theorem

(2) Sym(<M, CB>) = Sym(<M, Con>).

Thus Zeeman's theorem establishes a *necessary* condition for the causal definability of congruence in Minkowski spaces.

Naturally, we now ask: is Zeeman's theorem also a sufficient condition for causal definability of congruence? Or, more generally, does the equality of two symmetry groups (as in (2)), both models of a theory Th, insure the interdefinability in Th of the involved relations?

In general, the answer to this question is negative. The smaller the symmetry groups involved, the less likely it is that their identity will be of any logical importance. It is just this fact about symmetry groups that lies behind the failure of Klein's *Erlangen Program* to provide a comprehensive foundation for geometry. Spaces of highly variable curvature have relatively small symmetry groups (in some cases, the trivial group), and so the structure of these groups provides little or no information about the underlying geometry.

However, the Extended Poincaré Group is not the "small" symmetry group of an inhomogeneous space, but a "large" symmetry group of a strongly homogeneous structure: Minkowski space-time. So although the causal definability of congruence does not follow *in general* from a result such as Zeeman's theorem, it does indeed follow *in this instance* owing to the homogeneity of Minkowski space. This may be shown as follows.

First of all, although Zeeman's theorem merely provides us with a causal characterization of the similarity group of Minkowski space-time, let us suppose that we have managed to provide, somehow or other, a purely causal definition of the *isometry* group *PG* of Minkowski space-time. We may then define congruence as in definition 7.1 above, and we are done. However, it should be noted that even this last step is not entirely trivial, since the homogeneity of Minkowski space enters here in an important way. Consider, for example, the following alternative approach to congruence in metric spaces:

(**) *Definition.* Let A = <u, v> and B = <x, y> be ordered pairs of points in M. Then A *is congruent to* B iff there exists a bijection h *from* A *to* B which preserves the metric on M.

Notice that this last definition requires only that h map A onto B, rather than *all* of space M onto itself as well. Thus it may serve to define congruence for all metric spaces, even those with trivial symmetry groups. In addition, this last definition of congruence captures the intuitively desirable feature that two figures should not be deemed incongruent solely on the basis of remote inhomogeneities of the space.[40] Fortunately it can be shown that, for a wide class of geometries—Minkowski space-time included—the two definitions of congruence are equivalent.[41] The proof of this theorem depends heavily on the metrical homogeneity of the spaces involved, and thus the success of the above definition of congruence in terms of the Poincaré Group is not a trivial matter.

It now remains to provide a causal definition of the *isometry group PG* of Minkowski space-time. First of all, Zeeman's theorem tells us that the causality group CG is the similarity group of Minkowski space-time, so the following succeeds as a causal definition of that group (*EPG*).

Definition 7.1. EPG = df. $\{h \mid h: M \to M, h$ bijective, and for all x, y, z, ϵ $M, CB(x, y, z)$ iff $CB(h(x), h(y), h(z)).\}$

Now we need to recover the isometry group *PG* from the similarity group *EPG*. The following method succeeds in solving this problem for all (nonsingular) real affine metric spaces.[42]

Call the subset of *EPG* whose members each fix at least two points "*TF(EPG)*." More precisely,

Definition 7.2. TF(EPG) = df. $\{h \mid h \epsilon EPG,$ and for some $s, y \epsilon M, x \neq y,$ $h(x) = x$ and $h(y) = y\}$.

Clearly *TF(EPG)* contains only isometries, since it contains only similarities, all of which have ratio 1. Thus *TF(EPG)* $\subseteq PG$. Now close *TF(EPG)* under finite functional composition, labeling the result $\overline{(TF(EPG))}$. That is,

Definition 7.3. $\overline{TF(EPG)}$ = df. $\{h \mid h = h_n \circ h_{n-1} \circ \ldots \circ h_1, h_i \epsilon TF(EPG),$ $i = 1, \ldots, n\}$.

Since the composition of two or more isometries is again an isometry, we still have that $\overline{TF(EPG)} \subseteq PG$. However, we can now say more: $\overline{TF(EPG)}$

185

contains *all* the isometries, i.e., $TF(EPG) = PG$. This is shown in the proof of the following theorem:

Proposition 7.4. Let M^d be any (nonsingular) affine metric space of dimension $d \geq 2$, with Sim the similarity group of M^d. Then $\overline{TF(\text{Sim})}$ is the isometry group Is of M^d.

Proof: The above argument shows that $\overline{TF(\text{Sim})} \subseteq Is$. Suppose then that $h \in Is$. According to the Cartan-Dieudonné theorem (see Snapper and Troyer, 1971, section 69), there exist at most $d + 2$ symmetries[43] of M^d such that $h = s_1 \circ s_2 \circ \ldots \circ s_{d+2}$. Since any symmetry of an affine metric space pointwise fixes a $d - 1$ dimensional subspace of M^d, and $d \geq 2$, every symmetry is in $\overline{TF(\text{Sim})}$. Since h is a finite functional composition of symmetries, $h \in \overline{TF(\text{Sim})}$. Done.

Hence the Poincaré Group is just $\overline{TF(EPG)}$, and the accuracy of the following definition is established:

Definition 7.5. $PG = $ df. $\overline{TF(EPG)}$.[44]

Definition 7.1 above may now be used to define the congruence of point-pairs in Minkowski space-time. With this construction, congruence in Minkowski space-time is shown to be ontologically independent of conventions concerning the isochrony of periodic processes and the congruence of transported spatial units. Congruence in Minkowski space-time is reducible to its causal structure.

The availability of the isometry group PG (or even EPG) of Minkowski space-time makes the definability of its linear structure obvious, since these transformations are all linear. Thus, beginning with two point-events x and y, first consider all transformations in EPG that leave x and y fixed. Let $L(x, y)$ be the set of all points that are left fixed by *all* of the above transformations. When x and y are not light-connectible (not $x \lambda y$), $L(x, y)$ will be the straight line in Minkowski space connecting x and y. If x and y are light-connectible ($x \lambda y$), then $L(x, y)$ may be defined (as in section V) as the set of all events z, such that z is light-connectible to both x and y ($x \lambda z$, $y \lambda z$). Lines are classified as spacelike, timelike, or lightlike, of course, just in case x and y have the corresponding relation. (Of course, any pair of events on a given line may be chosen.)

The above constructions use the similarity group of Minkowski space together with the causality relation $<$, and, for present purposes, this is perfectly appropriate. Let me, however, put forth the following conjecture, which is of some independent geometrical interest: given *only* the similarity group (*EPG*) of Minkowski space-time, it is possible to construct the entire space-time geometry *including the order of points within its lines*. For Euclidean and classical non-Euclidean geometries, their similarity group does not suffice; additional axioms of order must be provided.[45] However, regardless of the truth of this last claim, the preceding constructions clearly suffice for present purposes, and thus the causal definability of the geometry of Minkowski space-time is established.

Epistemological issues are not hereby resolved, for the above constructions are profoundly nonoperational. Space-time pairs are congruent if (and only if) a global map of *M* onto itself of the right sort exists; but how are *we*, in concrete measuring situations, to determine the existence or nonexistence of such a map? At first sight, the synthetic constructions of congruence outlined earlier might seem to provide a local method, since fortuitous light parallelograms of finite dimension might then be used to determine interval-congruence. But this turns out to be illusory. These constructions depend upon a prior determination of the collinearity (spacelike and timelike) of at least three events. However, the definitions of spacelike collinearity given by Robb and Latzer, and the definition of timelike collinearity given here (section V) are essentially global. Of course, this does not show that a local causal definition is impossible, and it would be of considerable interest to see this matter resolved.

It should not be supposed, however, that even if Minkowski space-time congruence is bound to global causal structure in an essential way, we are thereby left with the conventionality of congruence for all practical purposes, or worse still, the view that even if *STR* congruence is not conventional, we cannot know it to be so. For auxiliary hypotheses may now serve to provide us with a multiplicity of ways to ascertain global causal structure, and with it, space-time congruence.

VIII. Foundations of Space-time Theories

The reducibility of the metric of Minkowski space-time to its causal structure has not gone unchallenged. Thus Grünbaum, endorsing the views of Reichenbach (1957, section 27) and L. L. Whyte (1953), has held that

Using only light signals and temporal succession without either a solid rod or an isochronous material clock, it is not possible to construct ordinary measures of length and time (Grünbaum, 1973, chap. 13, p. 414).

187

The basis for this claim is a result that emerges in the course of Reichenbach's axiomatization of special relativity. Reichenbach (1969) attempts to define the Minkowski metric by first constructing the inertial frames of special relativity, then showing these frames to be related by the Lorentz transformations, and, finally, defining the space-time interval as the invariant $-(\Delta x^0)^2 + \Sigma (\Delta x^k)^2$, where the coordinates $\{x^\alpha\}$ are those of any inertial (Lorentz) frame. The construction begins with the causal precedence relation, together with what he calls "real particles," the latter being the trajectories of continuous timelike paths. Reichenbach's postulates now enable him to "define" (conventions are adopted along the way) the class of all frames of reference S'' in which a light signal propagates according to

(1) $-(\Delta x^0)^2 + \Sigma (\Delta x^k)^2 = 0 \ (k = 1, 2, 3)$.

The question now arises: is the class of frames S'' just the class of inertial frames of special relativity?

The answer to this question is negative; for it can be shown that if K is an inertial frame, then any other frame K' with coordinates $\{x^{\alpha'}\}$ related to those in K by a transformation such as the inversion

(2) $x^{i'} = \dfrac{x^i}{-(x^0)^2 + \Sigma (x^k)^3}$

will also have an equation of form (1) as its description of the propagation of a light signal (cf. Reichenbach, 1957, p. 172). Fortunately, all of these non-Lorentz frames are essentially of the same sort: some of their spatial points are moving with respect to each other so as to preclude certain sorts of light signal communication. Thus, in Figure 8.0, world lines P' and Q are at rest in K' at $x' = 1$, $y' = z' = 0$, and $x' = 1/2$, $y' = z' = 0$, respectively. Any signal sent from P' after event e^* fails to reach world line Q. Why then not use such facts, or the fact that the transformations such as (2) above are not one-one, in order to characterize the inertial frames of S''?

This problem was touched upon earlier in the proof of proposition 6.15. There we were considering those transformations that preserve light paths "actively," as conformal mappings of space-time onto itself, rather than "passively," as coordinate transformations. However, the main point remains: there is no bijective mapping (or coordinate transformation) of M (Lorentz frame $\{x^\alpha\}$) onto M (another frame $\{x^{\alpha'}\}$) that preserves light-propagation. In this and a number of other ways, inertial paths may be causally distinguished from accelerated paths, as the constructions of section V above demonstrate. However, as was mentioned there, these constructions are global: they involve reference to all of the space-time M. As a result, the definitions do not provide operational procedures for determining whether or not the defined relations obtain.

This brings us to the source of Reichenbach's and Whyte's reluctance to rule out the accelerated reference frames in terms of their global properties. Whyte admits that global restrictions are admissible *logical* devices, but rejects the resulting distinction by calling it "a fact of no operational significance" (Whyte, 1953, p. 161); Reichenbach, in a similar vein, remarks:

. . . since no unlimited spaces can be utilized for a decision, this method is not fruitful. We can always describe systems T [of the accelerated sort] that deviate from systems of class I [inertial systems] only outside the space we have at our disposal. (Reichenbach, 1957, p. 173.)

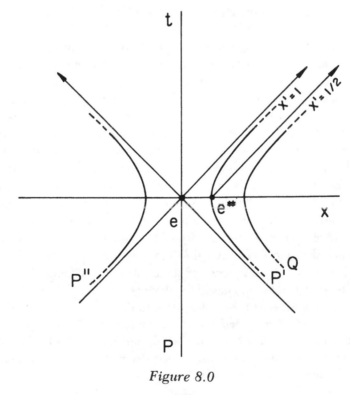

Figure 8.0

189

Reichenbach's solution is to introduce *rigid material bodies* and use them to determine the accelerations of the points in the unwanted frames. Since the "rigidity" of such bodies is not subject to prior physical determination, it can be only a matter of convention and utility just which class of material bodies is so-taken. Reichenbach realizes that material clocks may be used instead of "rigid" rods, but maintains that essentially the same problem arises all over again for material clocks, now in the guise of attributing congruence to their successive periods. However, as the preceding remarks show, this appeal to 'rigid rods' or 'isochronous clocks' as supplementary to the causal structure of special relativity is needed for purposes of *verification* only, and this is explicitly admitted by Reichenbach (1969, p. 88) as well as by Whyte. Thus writers such as Grünbaum who are concerned to establish the conventionality of Minkowski spacetime as a matter of *ontological* fact gain no support for their views from the Reichenbach-Whyte contention. On the contrary, a closer look at the context of Reichenbach's contention reveals that his constructions tend to support, rather than detract from, the causal theory of Minkowski spacetime.

Naturally, rejecting operationism, or the verifiability theory of meaning, does not amount to maintaining that global hypotheses about causal structure cannot be confirmed by experiment. To be sure, the results of any set of confirming experiments will also be compatible with a different causal structure in regions not yet investigated, but this is an inductive problem of the sort that inevitably arises when any cosmological hypothesis is under consideration. Furthermore, as the causal theory is extended by adding hypotheses to the effect that light rays travel on causal boundary lines (null lines), free massive particles travel on timelike lines, etc., the ways of obtaining indirect confirmation for global hypotheses of causal structure will multiply indefinitely. In this connection, it is important to note that the equality of the successive periods of atomic clocks has the status of a hypothesized physical fact within the causal theory, as does the congruence of standard rigid rods and clocks under gentle transport. Rods and clocks are devices that survey a region's causal structure, and are accurate or inaccurate to the extent that they yield the causally-grounded congruence classes. Just as space-time has now replaced space and time as the fundamental geometrical object, "etiometry," the measurement of causal structure, replaces and explains both chronometry and geometry.

Still, Reichenbach's failure to distinguish causally the inertial frames of the special theory is only partly explained by his adherence to the verifiability theory of meaning. More fundamental was his failure to recognize the geometrical power and complexity of the causal structure of relativistic space-time. Reichenbach saw clearly that the new causal structure lies behind the relativity of simultaneity, and he saw the demise of absolute *time*; but (unlike Robb) he continued to minimize the differences between classical and relativistic causal structure.[46]

. . . the light-axioms in the theory of relativity do not differ from those in classical theory except for the assertion of the limiting nature of the velocity of light. (1957, p. 175.)

And in referring to the experimental confirmation of his axioms:

Axioms I to II,4 and Axiom II,6 express merely facts known before the theory of relativity. Only axiom II,5 and axiom III are new. These [last] axioms formulate the limiting character of the velocity of light (1969, p. 92).

Unfortunately, one reason so few of Reichenbach's light axioms are new is that not all of them are true. Thus his Axiom II,1 asserts that if we choose an arbitrary event e on any timelike world-line P, and P' is any other timelike world-line, a signal may be sent from e on P to P'. Reichenbach himself recognizes that this is false in relativistic space-times (as P and P' in Figure 8.0 illustrate), but proceeds to use it in any case (1969, p. 31, fn. 9)!

Axiom II,2, however, fares no better. According to this axiom, if e'' is any event on particle world-line P'' and P' is any other world-line, then a signal may be sent from P', arriving at e'' on P''. Once again, this axiom is false, as P' and P'' in Figure 8.0 illustrate.[47]

Another aspect of Reichenbach's axiomatization is worth a final note. Being convinced that spatial and temporal congruence were infected with conventions, Reichenbach proceeded to axiomatize relativity theory so as to *exhibit* these conventions clearly. In this way, when statements involving temporal or spatial congruence are derived, the conventions that are presupposed by this assertion may be traced, thereby clarifying the physical content of the original congruence claim. Indeed, by changing *conventions*, and by investigating the effect of these changes on the form of our assertions, we are better able to recognize the "accidental" conventional aspects of a given physical claim.[48]

191

However, the method has its disadvantages. The most serious difficulty is that the process of conventional variation leads to a set of apparently contradictory statements all having the same physical content. Hence none of them mean what they apparently say. We thereby learn what parts of a locution ought *not* to be taken as an assertion of physical fact, but we may remain as ignorant as ever of that statement's "core" of nonconventional physical content. An appeal to Reichenbach's axioms is of no help here, for, having adopted some particular set of conventions, we find succeeding axioms couched in terms of those conventions. For example, Reichenbach's Axiom V reads:

It is possible to choose the static system relative to A so that the defined spatial geometry will become three-dimensionally Euclidean. (1969, p. 53.)

There is no need to know what a "static system" is in this axiom to appreciate its power *and* its opacity. For the physically interesting question is: *by virtue of what non-convention-laden features of the world is such a choice possible?* The way to exhibit the physical content of our theories is not to *exhibit* their implicit conventions, but to *eliminate* them. Coordinate-free methods are a step in this direction; Robb's axiomatization exhibits the method in its purest form, along with its attendant technical difficulties.[49]

While Reichenbach regards the introduction of metrical conventions as importantly involving epistemological issues, Adolf Grünbaum is most concerned with the ontological status of spatial, temporal, and space-time congruence. For Grünbaum, the philosophical importance of metrical conventions results from their being a symptom of the factual *underdetermination* of physical metrics. The theorist, not nature, forces metric closure by the implicit adoption of physically arbitrary but useful conventions. The congruence of two disjoint spatial, temporal, or space-time intervals is a matter of convention, not because we are unable to ascertain their congruence, but because there *is* no physical relation of metrical equality among the world's spatial, temporal, or space-time intervals.

One moral of this view is clear, and well worth heeding. The standard formulations and practices surrounding a physical theory do not provide clear and explicit guides to its ontological claims. At best, they furnish preliminary clues to be used by the critical scientific realist in his efforts to create a reconstructed version of the theory that will exhibit its physical commitments with greater perspicuity. In particular, the physical claims of a theory may become especially problematic when the theory is both recent and conceptually revolutionary. For example, it is only by virtue of the relatively recent development of invariant methods that it has become clear (to most relativists) that the relativistic notion of mass is just as much an invariant as was its Newtonian counterpart.

Yet to applaud the merits of Grünbaum's critical approach to the problem of congruence and conventions is not necessarily to agree with his assessments of conventionality or with his diagnosis of its sources. It is well known that Grünbaum locates the source of the conventional ingredients in assertions of spatial, temporal, or space-time congruence in the postulated continuity (locally Euclidean topology) and homogeneity of the underlying manifolds.[50] However, before such a claim may be established, it is necessary to state explicitly just what relations on a manifold M are to be taken as nonconventional, and why. In a recent work (1973, chap. 16) Grünbaum has attempted to answer the latter question in his general account of the "intrinsicality" of a relation on a set. Only a narrower question will be considered here: namely, given that a set of relations $\{R_i\}$ *are* held to be "intrinsic" or "objective" on a set M, how are we to assess the physical objectivity of a relation S_0 not in $\{R_i\}$ *with respect* to the set $\{R_i\}$?[51]

Following Klein and Weyl (1949, par. 13), a natural group-theoretic condition, which I propose be regarded as necessary, is the following:

Condition 8.0. Let $(M, R_1, \ldots, R_n)$ be such that M is a nonempty set and the R_i are relations on M. Then the relation S on M is objective with respect to (M, R_i) only if the symmetry group of (M, R_i) is a subset of the symmetry group of (M, S).[52]

When this condition obtains, S is sometimes said to be an *invariant* of (M, R_i). Intuitively, the condition amounts to saying that a permutation of M that leaves all of the relations $\{R_i\}$ fixed does not alter the "R_i-objective structure" of M. Should such a permutation then alter another relation S, this would indicate that S brings additional structure to M over and above that already guaranteed by the relations $\{R_i\}$; S is "extrinsic" *with respect to* (M, R_i).

We may now apply this condition to the Leibniz space-times of section II. Let (M, CB) be a temporally continuous Leibniz space-time with CB as its causal betweenness relation. Let d_t be any nontrivial temporal metric on the set of instants of (M, d) and d_s be any nontrivial spatial metric on any of its simultaneity slices. Then, by proposition 2.10, d_t and d_s are not invariant with respect to (M, CB).

In many cases, Grünbaum contrasts the extrinsicality of the congruence relation on a set M with the relation of 'causal connectibility.' Thus in his discussion of simultaneity in Newtonian mechanics (1969, section 3, subsection 1), he asserts that the arbitrarily fast causal chains of the Newtonian world establish the simultaneity of distant events "*as a matter of ordinal temporal fact*," whereas "there is no ordinal or topological basis that would yield" the congruences of *successive* temporal intervals. These remarks imply that Grünbaum holds causal connectibility between events to obtain or not as a matter of fact, with the causal relations of a Newtonian world sufficing only for the determination of the temporal order of its events.

As we have seen, the results of section II confirm this *conclusion*; however, they do *not* confirm Grünbaum's diagnosis of the *source* of this causal impotence. For Leibniz space-times, the source is *twofold*: the Leibniz postulate precludes nontrivial *spatial* metrics which are causally invariant, and continuity has nothing to do with this matter. On the other hand, it is the Leibniz Postulate *together with* the postulate of temporal continuity that precludes nontrivial, causally invariant, *temporal* metrics.

From this standpoint it becomes obvious that if we now begin with the relation of causal connectibility γ on *Minkowski space-time*, the question of the objectivity or causal invariance of the congruence relations of this space-time needs to be completely reexamined because of the failure of the Leibniz Postulate in these space-times. The fact that the spatial, temporal, and space-time intervals or paths of special (and general) relativity remain continuous fails to provide even a plausibility argument for their causal noninvariance. Indeed, given Grünbaum's causal nonconventionalism, the causal definability of flat space-time congruence would appear to provide a clear example of what he regards as a nontrivial intrinsic metric.

Furthermore, as we have seen, causal connectibility need not be regarded as merely supplementing a previously supplied topology on M; but we may use the causal connectibility relation to construct that topology as well. In this sense, the causal theory undercuts questions of metrical intrinsicality or extrinsically by providing a single relation from which *all* the various "layers" (topology, affine geometry, metric) of the geometry of Minkowski space-time may be constructed. Yet there is a sense in which these affine and metrical structures are not unique, for there are space-times that have neither the same affine nor metrical structure as Minkowski space-time, yet they have the same causal structure. How is this possible?

Suppose (M, g) is a semi-Riemannian manifold with g as its metric tensor field. Then (M, g') is said to be *conformal* to (M, g) just in case $g' = \phi g$, where ϕ is some positive, real-valued, smooth function on M. For our purposes, the important fact about conformal Lorentz space-times is that they have the same null trajectories (see, e.g., Hawking and Ellis (1973), p. 42) and thus the same causal structure. Now if (M, g') is conformal to Minkowski space-time (M, η), we shall call it *globally conformally flat* (*gcf*). Unless $g' = k\eta$, where k is a constant, (M, g') will not be isometric to (M, η) nor have the same similarity group. But, being conformal to (M, η), such a nonflat space-time will have the same causal group as (M, η). The problem posed above is this: how is it possible that the metric η is (up to a constant factor) definable in terms of the causal structure of (M, η), while there exist semi-Riemannian manifolds (M, g') *not* isometric to (M, η), but conformal to (M, η) and hence sharing its causal structure?

If we apply the invariance condition C.8.0 above, we obtain the answer to this question at once. First of all, notice that when (M, g') is (globally) conformally flat but not flat, its similarity group will be a proper subgroup of the similarity group (EPG) of (M, η) (cf. Levine, 1936), i.e.,

(1) $\text{Sim } (M, g') \subset EPG = \text{Sym}(M, \eta) = CG.$

Hence there must be a causal automorphism h_0 which is not a similarity of (M, g'),[53] so

(2) $CG \nsubseteq \text{Sim}(M, g'),$

i.e., g' is not invariant with respect to (M, γ).

It now follows from theorem 7.0 above that g' (up to a constant factor) is not causally definable. As Howard Stein has noted, the problem comes down to this. If $g' = \phi_0 \eta$, then a mapping which preserves η up to a constant factor will not, in general, likewise preserve the function ϕ_0: $M \to R^+$. In fact it is easy to see that if ϕ_0 is any nonconstant function on M, *some* function in the extended Poincaré group (*EPG*) will fail to preserve ϕ_0 up to a constant factor. Hence the only way to obtain the invariance of

g' is to add a distinguished function ϕ_0 to (M, γ), obtaining (M, γ, ϕ_0). Now the metric $g' = \phi_0\eta$ is an invariant of (M, γ, ϕ_0), and may be defined in a theory $T(\gamma, \phi_0)$ which incorporates an *additional* primitive term ϕ_0. In terms of the intuitive explanation provided earlier for the invariance condition, the permutation h_0 of M which leaves causality unaltered does change a global conformally flat metric of nonzero curvature, thereby showing the latter to be sensitive to extra-causal features of the space-time structure.

Such cases of the failure of causal invariance in conformally flat spaces lead naturally to consideration of the status of the causal theory with respect to the space-times of general relativity. And if general relativity is considered "liberally," the causal theory is clearly false. For there are models of general relativity, such as Gödel's (1949), such that for every point-event e in M, there is a closed causal curve through e. In such a universe, every event is causally connectible to every other event; hence γ is the universal relation, so the automorphisms of (M, γ) or (M, CB) are just *all* the diffeomorphisms of M onto M. Clearly, no nontrivial semi-Riemannian metric on M is invariant with respect to (M, γ) or (M, CB), and a causal theory of such space-times must fail.

Suppose, however, that we restrict our attention to relativistic space-times in which the condition of strong causality obtains.[54] Indeed, the interpretation of general relativity so as to rule out such causality violations was suggested by Einstein in his comments on the Gödel solution (1959). In this case, the failure of the causal theory is not so clear. For example, if we consider all vacuum solutions to Einstein's field equations ($\tau = 0$, where τ is the matter-energy tensor) these reduce to

(3) Ricci $= 0$,

where 'Ricci' is the Ricci tensor (cf. Schouten, 1954, p. 148). Such spaces are sometimes called "Einstein spaces," or "special Einstein spaces." We now have the following result which generalizes Zeeman's theorem.

Theorem 8.1. If (M, g) is a geodesically complete four-dimensional Lorentz space-time in which Ricci $= 0$, then either: (1) every local conformal transformation of (M, g) is a similarity or (2) every global conformal transformation of (M, g) is a similarity.

Proof: Case (1): (M, g) is not conformally flat. By a theorem of Brinkman (Schouten, 1954, p. 314), every local conformal transformation is homothetic.

Case (2): (M, g) is conformally flat. Then since Ricci $= 0$, (M, g) is locally flat (Schouten, 1954, p. 314). But all global conformal transformations of a complete locally flat Lorentz space-time are similarities (cf. Appendix B for an outline of the proof).

Since strong causality ensures that the causality-preserving transformations of a Lorentz space-time be conformal diffeomorphisms,[55] it follows that the metric of (up to a scale factor) of a large class of relativistic space-times is a causal invariant. The extent to which this result may be generalized poses an interesting question for further investigation. It should be noted that the constructions of Ehlers, Pirani, and Schild (1972), Woodhouse (1973), and Marzke (1964) do not resolve the issue, since they rely on adopting "free particles" in addition to causality.[56]

While the existence of space-times conformal to but not isometric to Minkowski space-time does not impugn the causal theory of Minkowski space-time, the existence of such space-times as models of *general* relativity appears to be a fatal blow to a causal account of these space-times. For if $\mathcal{M} = (M, \eta, \gamma)$ is Minkowski space-time, and $\mathcal{M}^* = (M, \phi^2\eta, \gamma)$ is nontrivially conformal to $\mathcal{M}$, then $\mathcal{M}$, $\mathcal{M}^*$ are a pair of models of general relativity having the same causal structure yet metrics which differ nontrivially. They are analogous to a Padoa pair of models in general model theory whose existence suffices to show the nondefinability of the (in this case) metric tensor. Equivalently, the similarity group of $\mathcal{M}^*$ is a *proper* subset of the similarity group of $\mathcal{M}$ and so (by Zeeman's theorem) a proper subset of the automorphisms of (M, γ). Hence there is a causal automorphism of $\mathcal{M}^*$ which is not a similarity of $\mathcal{M}^*$. Thus the metric $\phi^2\eta$ (up to a constant factor) is not a causal invariant of $\mathcal{M}^*$. Since invariance is a necessary condition of definability in any reasonable sense of that term, the failure of causal structure to determine the metric of $\mathcal{M}^*$ (up to a constant) would appear to be demonstrated.

It should also be noted that examples like $\mathcal{M}^*$ cannot be ruled out by imposing further restrictive causality conditions on the models of general relativity since $\mathcal{M}^*$ has the same causal structure as Minkowski space-time—itself a paradigm of good causal behavior. More tellingly, if $\mathcal{M}$ is any admissible model of general relativity, then a conformally equivalent $\mathcal{M}^*$ as above will also be a model, and the argument proceeds as before.

A causal theory of space-time is not necessarily a "relational" theory, as philosophers have used this term. While results such as those discussed here bear upon the reducibility of space-time structure to causal structure, they do not thereby resolve the issue of the ontological status of these causal relations. Briefly: are events causally related because of the obtaining of relations between material particles or radiation, or do the latter, by their presence, merely modify an independently existing "absolute" causal structure? One reason for regarding standard relativity as endorsing the absolutist interpretation is that there are empty space-times ($T_{\mu\nu} \equiv 0$; "vacuum" solutions to the field equations) which nevertheless have a determinate causal and metrical structure; furthermore, these solutions are not all causally isomorphic or isometric. How can causal or geometrical structure be a function of matter-radiation relations, when there is no matter or radiation present? The difficulty becomes even more striking when it is realized that these empty space-times are not necessarily devoid of "activity." They may, for example, contain pure gravitational waves (plane waves) propagating along some of their null geodesics.[57]

Appendix A. The Two-Point Homogeneity Property of Real Metric Affine Spaces

A discussion of the general structure of real affine metric spaces is given in section VI above. First, the two-point homogeneity property is defined.

Definition A.1. Let $A = (X, (V, <,>))$ be a nonsingular, real, affine metric space. Then A *is two-point homogeneous* if and only if for any pairwise distinct points a, b, c, $d \in X$ such that $d^2(a, b) = d^2(c, d)$, there is an isometry $j: X \rightarrow X$ such that $j(a) = c$ and $j(b) = d$.

In other words, for every congruent pair of point-pairs, there is an isometry of the entire space that maps the first pair to the second. The following theorem now asserts that all such spaces (which includes Minkowski spaces) are two-point homogeneous.

Theorem A.1. All nonsingular, real, affine metric spaces are two-point homogeneous.

Proof: Let $A = (X, (V, <, >))$, with a, b, c, d distinct, in X, and such that $d^2(a, b) = d^2(c, d)$. First we construct an isometry $i: V \rightarrow V$ such that $i(<\overrightarrow{a, b}>) = <\overrightarrow{c, d}>$, where $<\overrightarrow{a, b}>$ and $<\overrightarrow{c, d}>$ are the vectors in V associated (by the action function) with $<a, b>$ and $<c, d>$, respectively.

198

Let $<<\ <a,\vec{b}>\ >>$ and $<<\ <c,\vec{d}>\ >>$ be the one-dimensional subspaces of V spanned by vectors $<a,\vec{b}>$ and $<c,\vec{d}>$. Define i_0: $<< <a,\vec{b}>\ >> \to\ << <c,\vec{d}>\ >>$ by: (1) $i_0(v) \equiv i_0(k<a,\vec{b}>) \equiv k<c,d>$, where k is a real constant $\neq$ zero. Clearly i_0 is an isometry, and so by the Witt theorem (Snapper and Troyer, 1971, p. 202) may be extended to an isometry $i: V \to V$.

Finally, we compose i with a translation $T<a,\vec{c}>$ of A by the vector $<a,\vec{c}>$, obtaining (in the notation of Snapper and Troyer, 1971) $J \equiv T$ $<a,\vec{c}> L(a, i)$. By a standard theorem (Snapper and Troyer, 1971, prop. 381.1), J is an isometry of A. Clearly $J(a) = c$ and $J(b) = d$. Done.

Appendix B. The Causal Group of Locally Flat Space-times

Although a Lorentz space-time (M, g) may be everywhere locally flat, it need not be globally isometric to Minkowski space-time (M, η). The reason for this is that local flatness does not imply that M has a globally Euclidean topology. Even compact flat space-times are possible, although these are ruled out below by the requirement of strong causality (see section VIII above). It will now be shown that for all complete, flat space-times, every global conformal mapping is a similarity. (A Lorentz space-time is said to be *complete* just in case every affine geodesic may be extended to arbitrarily high (or low) values of its parameter.)

The theory of covering spaces is used freely below.[58] The general idea is this: the universal covering manifold of a complete, flat space-time $\mathcal{M} \equiv$ (M, g) is just Minkowski space-time $\mathcal{\tilde{M}} = (\tilde{M}, \eta)$. Global conformal mappings of M onto itself lift up to global conformal mappings of $\tilde{M}$ onto $\tilde{M}$ in a natural way.[59] Hence a conformal mapping of M onto M which was not a similarity would lift to an analogous mapping of $\tilde{M}$ (Minkowski space-time) onto itself. But, by Zeeman's theorem, or proposition 6.15 above, there can be no such mapping of $\tilde{M}$ onto $\tilde{M}$.

To begin with let $(\tilde{M}, p)$ be the universal covering manifold of a C^∞ manifold M. Let $\phi: M \to M$ be a diffeomorphism. We lift ϕ to a diffeomorphism $\tilde{\phi}: \tilde{M} \to \tilde{M}$ as follows. The function $\phi \circ p$ is a smooth mapping from $\tilde{M}$ onto M. Since $\tilde{M}$ is simply connected, by the unique lifting theorem (Massey, theorem 5.1), if $\tilde{x}_0, \tilde{x}_1 \epsilon \tilde{M}$, $x_1 \epsilon M$, and $\phi \circ p\ (\tilde{x}_0) = x_1 = p(\tilde{x}_1)$, there is a unique smooth mapping $\tilde{\phi}: \tilde{M} \to \tilde{M}$ such that $\tilde{\phi}$ is a lift of $\phi \circ p$, i.e.,

(1) $p \circ \tilde{\phi} = \phi \circ p$

John A. Winnie

and $\bar{\phi}(\bar{x}_0) = \bar{x}_1$. That $\bar{\phi}$ is a diffeomorphism follows from: (a) $(\bar{M}, \phi \circ p)$ covers M; and (b), from (1), $\bar{\phi}$ is a homomorphism of $(\bar{M}, \phi \circ p)$ into $(\bar{M}, p)$. Hence (Massey, lemma 6.7) $(\bar{M}, \bar{\phi})$ covers $\bar{M}$. Since $\bar{M}$ is a universal covering manifold, ϕ is a diffeomorphism.

Suppose that $\mathfrak{M} \equiv (M, g)$ is a complete, flat Lorentz space-time, and let $(\bar{M}, p)$ be the universal covering manifold of $\mathfrak{M}$. The manifold $\bar{M}$ is supplied with a unique metric by using p to lift g to $\bar{M}$; explicitly,

$$(2) \quad \bar{g}(u_x, v_x) \equiv g(p_*(u_x), p_*(v_x)),$$

where p_* is the differential mapping of p. Since M is flat and complete, so is $\mathfrak{M} \equiv (\bar{M}, \bar{g})$; but $\bar{M}$ is also simply connected, and so by the Cartan-Ambrose-Hicks theorem (see Wolf, *Spaces of Constant Curvature* (Boston: Publish or Perish, Inc., 1974), section 1.9) is isometric to Minkowski space-time. Hence we may let $(\bar{M}, \bar{g}) = (\bar{M}, \eta)$, i.e., *the universal covering manifold of any complete, flat, Lorentz space-time is just Minkowski space-time.*[60]

A smooth diffeomorphism $\phi: M \to M$ is said to be *conformal* just in case

$$(3) \quad g(\phi_*(u_x, v_x)) = e^{\psi(x)} g(u_x, v_x),$$

for some smooth function $\psi: M \to R$. When ψ is a constant function, ϕ is a *similarity* or *homothetic*. A conformal mapping that is not a similarity is a *properly conformal* mapping; if $\psi \equiv 0$, ϕ is an *isometry*.

We have already shown how to lift any diffeomorphism $\phi: M \to M$ to a diffeomorphism $\bar{\phi}$ of $\bar{M}$ onto $\bar{M}$. From (1) and (2) it follows that if $\phi: M \to M$ is a proper (homothetic, isometric) conformal diffeomorphism of M, then $\bar{\phi}$ is a proper (homothetic, isometric) conformal diffeomorphism of $\bar{M}$. We now have our result at once; for if ϕ were a proper conformal diffeomorphism of M, then $\bar{\phi}$ would be a proper conformal diffeomorphism of $\bar{M}$, Minkowski space-time, and this, by Zeeman's theorem, we know to be impossible. *Hence every conformal mapping of a complete flat space-time is a similarity.* Since for strongly causal space-times, the conformal group and causal group are identical (see section VIII), we have the generalization of Zeeman's theorem: *For every strongly causal, complete, flat space-time, its causal group and its similarity group are identical.*

Notes

1. As in Tarski (1959), where the Euclidean line is asymmetrically ordered by using betweenness and an arbitrary pair of points. In section III below, this procedure is spelled out in detail for Minkowski space-time.

THE CAUSAL THEORY OF SPACE-TIME

2. Causal connectedness vs. causal connectibility as suitable interpretations of γ will be discussed below. For now, either construal will do.

3. The earlier comments on the *nondirectionality* of causal precedence thus apply derivatively to temporal precedence.

4. Compare this postulate and proposition 2.6 with Leibniz's argument: "And since my prior state, by reason of the connection between all things, involves the prior state of other things as well, it also involves a reason for the later state of these other things and is thus prior to them. *Therefore whatever exists is either simultaneous with other existences or prior or posterior.*" (1956), p. 1083; italics in text.

5. See van Fraassen (1970), chap. 6, section 6.

6. As A. Grünbaum has pointed out (in conversation), we have the following plausible options for the pair (U, γ):

    (1) U is the set of actual events, γ is 'causal connectedness';

    (2) U is the set of actual events, γ is 'causally connectible'; or

    (3) U is the set of possible events, γ is 'causally connectible.'

The third option is, to my mind, the most plausible, with the physical claim being that the set described in (1) is embeddable in this structure.

7. A formulation of the theory of real order is given in Tarski (1961), p. 214; connectedness (Tarski's Axiom 1') requires the Leibniz postulate.

8. It is important to note, however, that such considerations have clear-cut significance only within the context of a general program (such as the causal theory) of *reduction* for space and time. Often this reduction has been undertaken for operational motives, but this *need* not be its rationale. Reduction may also be desired for the sake of theoretical *explanation*, in which case it is to no avail to reply to the conventionalist that we may *postulate* an intrinsic spatial or temporal metric without scientific embarrassment. Of course; but it is just the postulated metric which is the object of the causal theorist's attempted reduction. It will become clear in the following that such a reduction succeeds provably for special relativity and fails for classical physics.

9. For more on automorphisms, see Weyl (1949), section 13, and also (1952) for developing the right intuitions.

10. Zeeman's theorem and its implications are discussed more fully in section VII below.

11. Throughout this section, I am greatly indebted to the aid of Geoffrey Matthews, Indiana University.

12. In Latzer's interesting paper (1972), chronological betweenness is constructed from the symmetrical relation (λ) of light (null) connectibility.

13. See section IV following for a definition of Alexandroff intervals.

14. A curve in M is here taken to be a smooth mapping from I into M, where I is an open interval of real numbers. A causal curve through e_1 and e_2 is said to go *from* e_1 *to* e_2 when e_2 has the higher parameter value.

15. See Blumenthal (1970), section 9, for these and other examples.

16. Interesting nonstandard space-time topologies are also possible, as was shown by Zeeman (1967).

17. He was by no means alone in this. See, for example, Lewis (1926), and the admittedly speculative ideas of Bohm (1965). The point is that intervals along a given light path *may* be compared metrically by using an affine parameter along that path, although intervals belonging to distinct nonparallel paths may not be compared meaningfully.

18. Robb (1914) is outlined with proofs omitted in Robb (1921); this is the best introduction to Robb's work. The later Robb volume (1936) is essentially the same as (1914), with some of its constructions and proofs simplified. The introduction to (1936) is a clear statement of Robb's views about the foundations of geometry.

19. A more rigorous formulation is this: let ℓ_t and ℓ_s intersect at e_0, ℓ_t timelike, ℓ_s spacelike. Then ℓ_t and ℓ_s are orthogonal iff for any event $e_1 \in \ell_t$, $e_1 \neq e_0$, there are events e_2, $e_3 \in \ell_s$ with $e_1 \lambda e_2$, e_3, such that if $e_4 \lambda e_2$ and $e_4 \in \ell_t$ and $e_5 \lambda e_2$ and $e_5 \lambda \ell_t$, then $e_4 = e_5$.

20. Let T_1 be a theory and T_2 an extension of T_1. Then T_2 may be obtained from T_1 by: (1)

John A. Winnie

adding postulates which involve no new primitives, or (2) by adding postulates which involve new primitive terms not then definable in T_2. The above example is of type (2).

21. Not quite a "full" congruence relation in that comparisons are made between (1) timelike intervals, (2) spacelike intervals, and (3) lightlike intervals on the same or parallel light lines. This is more than enough to give us the Minkowski metric.

22. Contrast this with the misleading account given by Reichenbach (1957), section 28.

23. A simple way to see this is to consider an n-dimensional vector space V^n, choose n linearly independent vectors $<v_1, \ldots, v_n>$, and decree them orthogonal and of unit length. Define their Euclidean inner product as $<u^\alpha v_\alpha, w^\beta v_\beta> \equiv \Sigma\, u^\gamma v^\gamma \gamma = 1, \ldots, n$. To obtain another Euclidean inner product differing in its congruence verdict, simply choose, say, $<2v_1, v_2, \ldots, v_n>$ as a basis and define the inner product similarly. Clearly v_2 is congruent to v_1 in the first case and not in the second. (An account of metric vector spaces is provided in the following section.)

24. This issue is discussed in more detail in section VI, proposition 6.15, where it is shown that no two Minkowski geometries (dim > 2) with the same causal structure may have metrics that deliver differing congruence verdicts.

25. For the general theory of such spaces, see Snapper and Troyer (1971); this work is used as a reference throughout this section.

26. These conditions are defined as follows: (1) (linearity) $<aw + bv, w> = a<u, w> + b<u, w>, <w, au + bv> = a<w, u> + b<w, v>$, (2) (symmetry) $<u, v> = <v, u>$, and (3) (nonsingularity) if $<u_0, v> = 0$, for all v, then $u_0 = \bar{0}$ (the zero vector).

27. In what follows, the fact that the vector space V is real and its inner product nonsingular will be assumed throughout, and thus not always stated explicitly.

28. It should be noted that some writers include positive definiteness in the definition of inner product. The approach taken here follows Snapper and Troyer (1971), chap. 2.

29. The customary formulation of (M) is: there exist basis vectors $e_0, e_1, \ldots, e_3$ of V such that for all $u, v \in V$, $<u, v> = -u^0 v^0 + u^k v^k$ (sum on $k = 1, 2, 3$), where u^i and v^i are the components of u and v in basis $e_0, \ldots, e_3$. However, it is easy to show (using the Gram-Schmidt process) that the two formulations are equivalent. The former (M) has the advantage of obvious basis independence. Note that, again, a signature $(-, +, +, +)$ is used.

30. Nothing of importance in what follows depends upon this. Thus we might equally well have defined vectors with equal $\left|\text{length}^2\right|$ as congruent.

31. More precisely, let $V(V_m^4)$ be the theory of Minkowski metric vector spaces formulated so as to contain both $\underline{<.>}$ and N (interpreted as 'Minkowski inner product' and 'null vector,' respectively) as primitives. Statement (1) above will now be a theorem of $T(V_m^4)$ showing the definability of N in terms of $\underline{<,>}$. The construction to come shows the definability of $\underline{<,>}$ (under to an arbitrary constant) in terms of N.

32. For generality's sake, a Minkowski space-time is here defined as a four-dimensional differential manifold M, together with a metric tensor field η on M, such that there is a coordinate system $\{x^\alpha\}$ on M onto R^4 in which the components of η are everywhere diag $(-1, 1, 1, 1)$.

33. More precisely: $TG = $ df. $\{h \mid h: M \to M, h$ a bijection, and for all $e_1, e_2, e_3 \in M$, $TB(e_1, e_2, e_3)$ iff $TB(h(e_1), h(e_2), h(e_3))\}$; similarly for CG.

34. PG is often called the "inhomogeneous Lorentz group." The Lorentz groups are subgroups of PG, each consisting of rotations about a given point. By virtue of the homogeneity of Minkowski space-time, these groups are isomorphic, hence *the* Lorentz group.

35. For more discussion of the relationship between similarities and isometries in geometry, see Weyl (1949), section 14; and Freudenthal and Bauer (1974), section 22.

36. A magnification of M with center e and ratio r $(r \neq 0)$ may be defined as a bijective mapping of M onto itself which (1) has exactly one fixed point e, and (2) maps every other point e' onto the "tip" of vector $k(\overrightarrow{e, e'})$. For a precise account, see Snapper and Troyer (1971), section 11.

37. For an account of Padoa's method, see Beth (1962).

38. It should be noted that such a definition succeeds only in spaces that are sufficiently

homogeneous to have enough isometries to "move" all congruent pairs onto each other (pairwise free mobility). This property is sometimes called (Birkhoff, 1944) "two-point homogeneity." Euclidean, hyperbolic, and spherical spaces all possess this property. The cylinder is a simple example of a flat space which does not. Nor do Riemannian and semi-Riemannian spaces of variable curvature. However, all real affine metric spaces, and hence Minkowski space, are two-point hmogeneous. See Appendix A for a proof of this; also the discussion later in this section.

39. In group-theoretic terms, the similarity group is the normalizer of the isometry group (cf. Weyl, 1949, section 14).

40. Thus consider a Euclidean plane $<E, g>$, with g a Euclidean metric tensor field on E. Let A and B be two nearby point-pairs which are congruent (according to both definitions) in $<E, g>$. Now choose a new metric tensor g' so as to agree with g on a possibly quite "large" connected region U containing A and B, yet have g' place a single "bump" in $E - U$ somewhere. Isometries of $<E, g'>$ must now map the bump onto itself, and thus there may be none which also map A onto B.

41. The theorem, mentioned above, is that all (nonsingular) real affine metric spaces are two-point homogeneous. A proof is given in Appendix A below.

42. For definitions, see the account in Snapper and Troyer (1971).

43. A symmetry of M^d is defined as a reflection of M^d about a fixed hypersurface ($d - 1$-dimensional subspace) of M^d. See Snapper and Troyer (1971), pp. 219, 386.

44. Professor L. Janos, University of Montana, has pointed out that the involutions of EPG could be used in place of $TF(\text{Sim})$, since every such involution is an isometry and a symmetry.

45. See the work of Bachman for a group-theoretic approach to the classical geometries, especially Bachman, Pejas, Wolff, and Bauer (1974).

46. Lest this judgment seem overly harsh, it should be noted that Robb's initial confidence in the causal theory seems to have been based on the unwarranted conviction that scientific realism *alone* demanded the truth of the causal theory! (Robb, 1921, p. v. ff.)

47. In his footnote commenting on Axiom *II, 1* he writes: "There are singular cases in which this axiom does not hold; *cf.* the example on p. 80. Axiom *II, 2*, however, holds without exception. This fact indicates a fundamental difference between the two axioms." (1969, p. 31, note 9).

48. Perhaps the best example of the method is its application to the conventionality of simultaneity within the inertial frames of special relativity. For recent analyses, see Grünbaum (1969), Salmon (1969), Winnie (1970), and the critical article by Friedman in this volume.

49. An analysis of length-contraction and time-dilation in special relativity which *eliminates simultaneity* conventions is given in Winnie (forthcoming).

50. Cf. Grünbaum (1968), chap. 3, par. 2.9, 2.10; Grünbaum (1973), chap. 16, is a more recent and detailed account.

51. "Relation" is here used in the broad sense to include functors, sets of relations proper on M, etc.

52. Symmetry groups are explained more fully in section VII above.

53. In the general context of semi-Riemannian spaces, h is *homothetic* or a similarity of (M, g) iff $h^*(g) = kg$ ($k = $ constant), where h^* is the differential of map h (cf. Hicks, 1971, pp. 9, 73). Similarities preserve arc length up to a constant factor.

54. See section IV, p. 45 above for a rough characterization of this condition and a reference.

55. Cf. Hawking, King, and McCarthy, *Orange Aid Preprint*, OAP-405 (1975), or David Malament's dissertation referred to in the next note.

56. For an excellent critical survey of these constructions, see Grünbaum (1973, chap. 22). For some additional results along these lines, see chap. 3 of David Malament's (Rockefeller University) doctoral dissertation (1975).

57. Such cases and others are discussed penetratingly and in detail by Adolf Grünbaum in his contribution to this volume.

John A. Winnie

58. For the basic theory and relevant results, see, e.g., R. Geroch, "Topology in General Relativity," *Journal of Mathematical Physics* 8, no. 4 (1967):782–786.

59. I am indebted to Professor John Ewing, Department of Mathematics, Indiana University, for helpful comments.

60. It is assumed that, by definition, all Lorentz space-times are connected.

References

Bachman, F., W. Pejas, H. Wolff, and A. Bauer. (1974). "Absolute Geometry," in Behnke, H., F. Bachman, K. Fladt, and H. Kunle, eds., *Foundations of Mathematics: II: Geometry*. Cambridge, Mass.: MIT Press.

Behnke, H., F. Bachman, K. Fladt, and H. Kunle, eds. (1974). *Foundations of Mathematics: II: Geometry*. Cambridge, Mass.: MIT Press.

Beth, E. W. (1962). *Formal Methods*. Dordrecht: Reidel.

Birkhoff, G. (1944). "Metric Foundations of Geometry, I," *Transactions of the American Mathematical Society*, vol. 55, pp. 465–495.

Bishop, R. C. and S. I. Goldberg. (1968). *Tensor Analysis On Manifolds*. New York: Macmillan.

Blumenthal, L. (1970). *Theory and Applications of Distance Geometry*. 2nd ed. Bronx, N.Y.: Chelsea.

Bohm, D. (1965). "A Proposed Topological Formulation of the Quantum Theory," in I. J. Good, ed., *The Scientist Speculates*. New York: Capricorn Books.

Earman, J. (1972). "Notes on the Causal Theory of Time," *Synthèse*, vol. 24, pp. 74–86.

Eddington, A. S. (1959). *Space, Time, and Gravitation*. New York: Harper.

Eddington, A. S. (1965). *The Mathematical Theory of Relativity*. 2nd ed. Cambridge: Cambridge University Press.

Ehlers, J., F. A. E. Pirani, and A. Schild. (1972). "The Geometry of Free Fall and Light Propagation," in L. O'Raifeartaigh, ed., *General Relativity*. Oxford: Oxford University Press.

Einstein, A. (1959). "Reply to Criticisms," in P. Schilpp, ed., *Albert Einstein: Philosopher-Scientist*. New York: Harper.

Fine, A. (1971). "Reflections on a Relational Theory of Space," *Synthèse*, vol. 22, pp. 448–481.

Freudenthal, H. and A. Bauer. (1974). "Geometry—A Phenomenological Discussion," in Behnke, H., F. Bachman, K. Fladt, and H. Kunle, eds., *Foundations of Mathematics: II: Geometry*. Cambridge, Mass.: MIT Press, pp. 3–28.

Geroch, R. (1970). "Domain of Dependence," *Journal of Mathematical Physics*, vol. 11, pp. 437–449.

Gödel, K. (1949). "A Example of a New Type of Cosmological Solution of Einstein's Field Equations of Gravitation," *Review of Modern Physics*, vol. 21, pp. 447–450.

Grünbaum, A. (1968). *Geometry and Chronometry in Philosophical Perspective*. Minneapolis: University of Minnesota Press.

Grünbaum, A. (1969). "Simultaneity by Slow Clock Transport," *Philosophy of Science*, vol. 36, no. 1, pp. 5–43.

Grünbaum, A. (1973). *Philosophical Problems of Space and Time*. 2nd ed. Dordrecht: Reidel.

Haantjes, J. (1937). "Conformal Representations of an n-dimensional Euclidean space with a Non-definite Fundamental Form on Itself," *Koninklijke Nederlandse Akademie van Wetenschappen. Proceedings. Series A, Mathematical Sciences*, vol. 40, pp. 700–703.

Hawking, S. W. and G. F. R. Ellis. (1973). *The Large Scale Structure of Space-Time*. Cambridge: Cambridge University Press.

Hicks, N. (1971). *Notes on Differential Geometry*. London: Van Nostrand.

Hocking, J. G. and G. S. Young. (1961). *Topology*. Reading, Mass.: Addison-Wesley.

Klein, F. (1939). *Elementary Mathematics From An Advanced Standpoint: Geometry*. New York: Dover.

Kronheimer, E. H. and R. Penrose. (1967). "On the Structure of Causal Spaces," *Proceedings of the Cambridge Philosophical Society*, vol. 63, pp. 481–501.

Latzer, R. (1972). "Nondirected Light Signals and the Structure of Time," *Synthèse*, vol. 24, pp. 236–280.

Leibniz, E. W. (1956). "The Metaphysical Foundations of Mathematics," in L. Loemker, ed., *Philosophical Papers and Letters*, vol. II, Chicago: University of Chicago Press, pp. 1082–1094.

Levine, J. (1936). "Groups of Motions in Conformally Flat Spaces," *Bulletin of the American Mathematical Society*, vol. 42, pp. 418–422.

Lewis, G. N. (1926). "Light Waves and Light Corpuscles," *Nature*, vol. 117, pp. 236–238.

Marzke, R. and J. Wheeler. (1964). "Gravitation as Geometry-I," in Chiu and Hoffman, eds., *Gravitation and Relativity*. New York: Benjamin.

Noll, W. (1964). "Euclidean Geometry and Minkowskian Chronometry," *American Mathematical Monthly*, vol. 71, pp. 129–143.

Penrose, R. (1972). *Techniques of Differential Topology in Relativity*. Philadelphia: Society for Industrial and Applied Mathematics.

Reichenbach, H. (1969). *The Axiomatization of the Theory of Relativity*. Los Angeles: University of California Press.

Reichenbach, H. (1957). *Philosophy of Space and Time*. New York: Dover.

Riemann, B. (1973). "On the Hypotheses which Lie at the Basis of Geometry," in, e.g., C. W. Kilmister, ed., *General Theory of Relativity*. Oxford: Pergamon, pp. 101–122.

Robb, A. A. (1914). *A Theory of Time and Space*. Cambridge: Cambridge University Press.

Robb, A. A. (1921). *The Absolute Relations of Time and Space*. Cambridge: Cambridge University Press.

Robb, A. A. (1936). *The Geometry of Space and Time*. Cambridge: Cambridge University Press.

Russell, B. (1954). *The Analysis of Matter*. New York: Dover.

Salmon, W. (1969). "The Conventionality of Simultaneity," *Philosophy of Science*, vol. 36, pp. 44–63.

Schouten, J. A. (1954). *Ricci-Calculus*. 2nd ed. Berlin: Springer-Verlag.

Snapper, E. and R. J. Troyer. (1971). *Metric Affine Geometry*. New York: Academic Press.

Tarski, A. (1959). "What is Elementary Geometry?" in Henkin, Suppes, and Tarski, eds., *The Axiomatic Method*. Amsterdam: North-Holland.

Tarski, A. (1961). *Introduction to Logic*. 2nd ed. Oxford: Oxford University Press.

Trautman, A. (1965). "Foundations and Current Problems of General Relativity," in Trautman, Pirani, and Bondi, eds., *Brandeis 1964 Summer Institute On Theoretical Physics*, vol. 1. Englewood Cliffs, N.J.: Prentice-Hall.

Trautman, A. (1973). "Theory of Gravitation," in J. Mehra, ed., *The Physicist's Conception of Nature*. Dordrecht: Reidel, pp. 179–198.

Van Fraassen, B. (1970). *An Introduction to the Philosophy of Time and Space*. New York: Random House.

Weyl, H. (1949). *Philosophy of Mathematics and Natural Science*. Princeton: Princeton University Press.

Weyl, H. (1952). *Symmetry*. Princeton: Princeton University Press.

Whyte, L. L. (1953). "Light Signal Kinematics," *British Journal for the Philosophy of Science*, vol. 4, pp. 160–161.

Winnie, J. (1970). "Special Relativity Without One-Way Velocity Assumptions, Parts I and II," *Philosophy of Science*, vol. 37, nos. 1 and 2.

Winnie, J. "Length-Contraction and Time-Dilation in Special Relativity" (forthcoming).

Woodhouse, N. M. J. (1973). "The Differentiable and Causal Structures of Space-time, *Journal of Mathematical Physics*, vol. 14, pp. 495–501.

Zeeman, E. C. (1964). "Causality Implies the Lorentz Group," *Journal of Mathematical Physics*, vol. 5, pp. 490–493.

Zeeman, E. C. (1967). "The Topology of Minkowski Space," *Topology*, vol. 6, pp. 161–170.

——————LAWRENCE SKLAR——————

Facts, Conventions, and Assumptions in the Theory of Space Time

I. Introduction

Given a physical theory, there are always those who will try to formalize it. At a minimum this entails an attempt to present the theory informally in terms of a number of basic propositions about the world from which all the consequences of the theory in question can be derived informally. At a maximum such an undertaking involves an exact and rigorous presentation of these fundamental propositions in some formal language (first order quantification theory or set theory, say), with the implicit claim that all the consequences of the original theory could, if they themselves were suitably "formalized," be derived formally from these rigorous fundamental propositions by the mere application of the formally specified rules of the logic in question.

Clearly such "axiomatization" can be a useful (or even crucial) component of ongoing science or the philosophy of science. Axiomatization need not be merely a more or less careful display of some degree of technical ingenuity applied to no apparent purpose, even if that is what it frequently turns out to be. But just *how* axiomatization can be applied to scientific or philosophic purpose is not always completely clear from the axiomatizations themselves. The scope and limits of formalization as an aid to science, and especially to philosophy, are the subject of this paper. But I shall treat these general questions within the limited context of some formalizations of the theory of space-time, since these particular cases provide highly illuminating examples of the general issues.

NOTE: I am grateful to the John Simon Guggenheim Memorial Foundation for their generous support during the period in which this paper was written. I am also grateful to David Malament, John Stachel, and Robert Geroch for their advice on some points of physics and mathematics.

II. Formalizations and Their Differences

All standard axiomatizations have some familiar components. First of all there is the accepted logical framework, which may be restricted to "purely logical" terms or may encompass a vocabulary of mathematics that is in general taken for granted. Then there are the *primitive descriptive terms* of the theory in question. They are just "givens," and although they may be "explicated" informally outside of the formalization, nothing within the formal presentation pretends to "define" them or otherwise specify their "meaning." There are the *axioms*, statements accepted as true and containing in their vocabulary only the "logical" and primitive descriptive terms—at least initially.

Next there are the *definitions*, and over their status much philosophical debate can rage. In the definitions we have, standardly, new descriptive terms introduced on one side of an equation or equivalence, and phrases involving only the primitive vocabulary or previously defined terms on the other side. Not all equations can serve as definitions. They must meet the conditions of eliminability and noncreativity. If a definition is legitimate, one must be able to replace the defined term in whatever context it appears in further on in the formalization by some expression using primitive terms alone. The theory that results when the definition is included must not contain any consequences phrasable in terms of the primitive vocabulary alone that would not be consequences were the definition eliminated and the defined term replaced everywhere by its defining expression.[1]

Let us note some important ways in which formalizations may differ from one another:

(1) Two formalizations may have all the same consequences, but differ in that one and the same term appears as a primitive in one formalization and as a defined term in the other. Naturally this will force the formalizations to differ also in which consequences they count as definitions and which as postulates.

What we count as a primitive term and what as defined are in some sense not "internal" to the formalization but imposed upon it by us. That is, this differential attribution of status to terms is not forced upon us in any way by the consequences of the formalization. Let us call formalizations that differ only in regard to which terms are considered primitive and which defined, formalizations differing merely in the *status attributed to terms*.

207

(2) Two formalizations may agree as to which terms are primitive and which defined. And they may agree in that the consequences of each framable in primitive terms alone are the same. Yet they may still differ in at least one consequence which is phrased using a defined term. Let us say that such formalizations differ merely in *definitional consequence*.

(3) Suppose two formalizations differ in what they both agree to be a consequence framable in primitive terms. Then, as we shall see, all parties are likely to agree that the formalizations are based on different theories. But suppose two formalizations differ in some consequence. And suppose one formalization so categorizes terms as primitive and defined that it declares the differences to be only in definitional consequence. Suppose further that the other formalization declares some of the differential consequences to be framable in primitive terms only, given its classification of terms as primitive or defined. Let us call formalizations that differ in this way, formalizations differing only in *one-way definitional consequence*.

(4) Two formalizations may agree as to which terms are primitive and which defined; and they may have exactly the same consequences. But one formalization may declare, "in the margin," some of these consequences to be definitions that the other calls postulates or axioms, and vice versa. Let us say that two such formalizations differ only in the *attribution of definitional status to consequences*.

The relevance and importance of these distinctions will become clear as we proceed. I shall be concerned primarily with two fundamental questions:

(1) Given two formalizations that differ in one of the ways noted above, what good scientific and/or philosophical grounds could there be for preferring one such formalization to another?

2) What are the scientific and/or philosophical consequences of opting for one such formalization over an alternative that differs from it in one of the ways noted above?

III. Formalizations and Minkowski Space-time

A. Robb's Formalization of Minkowski Space-time

Instead of continuing the discussion at this abstract level, let us descend to a concrete case. I shall start with Robb's ingenious, elegant, and under-appreciated axiomatization of Minkowski space-time.[2] Most of us would view this as a formalization of special relativity; but since Robb preferred to *contrast* his theory with that of Einstein, let us keep the more neutral form above. In Robb's formalism one takes as primitive a class of events and one primitive relation on them: *a* is after *b*. Intuitively *a* is after *b* if and only if it is possible to propagate a causal signal from *b* to *a*, and *a* and *b* are distinct events. Robb imposes twenty-one axiomatic conditions on the 'after' relation, sufficient to guarantee that the structure of causal connectibility among events is that among events in four-dimensional Minkowski space-time.

Robb then shows that it is possible to introduce definitions, in terms of the 'after' relationship, that are sufficient to capture such well-known features of Minkowski space-time as timelike connectivity, null connectivity, spacelike connectivity, etc. More surprisingly, he shows that one can offer definitions of 'is a timelike inertial path,' and 'is an interval congruent to a given interval,' even for spacelike intervals, such that the space-time structures so defined will again correspond in all their geometric features to the usual such structures of Minkowski space-time.

The temporal asymmetry of the 'after' relationship seems irrelevant to his main task, and indeed it can be shown that if one starts with the symmetric relationship corresponding to 'either *a* is after *b* or *b* is after *a*,' then, following out a Robb-like construction, one can construct all those features of Minkowski space-time that are themselves independent of a choice for the "direction of time."[3]

Now Robb, being concerned only to construct a formalized theory of space-time, has little to say about the relationship between the observable behavior of material objects like clocks, rods, and free particles and the space-time structures so constructed. He does tell us, however, that it is an empirical fact that light rays travel null straight lines, as he constructs them, and I assume he would take it that in our full physics we would simply add more axioms to tell us that atomic clocks measure time-like intervals relative to a reference frame, and rigid rods spacelike intervals relative to a reference frame; and that free particles travel timelike straight lines. These are totally nondefinitional on his view. We shall discuss this in more detail shortly.

B. A Philosophical Thesis Drawn from Robb's Formalization

Let us look at one aspect of an answer to our second question above: What important philosophical consequences can be drawn from opting for a given formalization of a theory? John Winnie has argued as follows:

Grünbaum and others have asserted that the metric structure of space-time is conventional since, even given the topology of space-time, no particular set of metric relations between the events is singled out as preferred. Therefore the choice of metric is, in some sense, arbitrary. Hence the metric of space-time is conventional.[4]

While Robb did not discuss the problem of formalizing the topology of space-time, it is clear that for Minkowski space-time, a formalization is available that takes only the 'after' relationship between events as primitive, and in which we can fully define all the topological features of the space-time. This is certainly true in the following sense: given the "after" relationship, and the Robbian axioms governing it, we can define what it means for an event to be in the interior of the forward (respectively backward) light cone of the event. Then we can take as the topology of the space-time the Alexandroff topology, i.e., the coarsest topology in which all such interiors of light cones are open. This will agree in all its structure with the usual manifold topology on Minkowski space-time.[5]

But, Winnie continues, Robb has shown that the metric congruence structure of the space-time is definable by the 'after' relationship alone. The same basic relation defines simultaneously the topological and metric features of the space-time. Thus we cannot really first fix the topology and still have free choice as to the metric; hence the metric is not really conventional. If we opt for Robb's formalization we see that, at least in the context of Minkowski space-time, the metric of the space-time is not really a matter of "convention."

A full critique of this argument will reveal much that is of general philosophical importance. But rather than attack this thesis directly, I shall take a more circuitous route. I hope the reader will bear with me through a lengthy digression.

C. General Relativistic Considerations

In the first edition of *Raum, Zeit, Materie* Weyl suggested that we could, assuming space-time to be a four-dimensional pseudo-Riemannian manifold, completely map out the metric of the space-time in which we live by using light rays alone. The fact that the light cone structure of the space-time is the structure of causal connectibility suggests that 'after' is a sufficient primitive on which to found the definitions of all the metric concepts we use to characterize the space-time. But, as Lorentz pointed out to Weyl, and as Weyl noted in later editions of his book, this thesis is wrong. For any two nonisometric space-times that are *conformally* equivalent (i.e., such that there is a one-to-one angle-preserving transformation from one to the other) will have the same light cone structure.

Weyl then pointed out that if we added to our body of observational data, so far consisting of the paths of light rays, the paths of free particles, and if we assumed these free particles traveled timelike geodesics in the space-time, then we could indeed fully determine the metric. For if there is a one-to-one mapping from one space-time to another that preserves the paths of material particles (i.e., the timelike geodesics), then the space-times are isometric, at least up to a constant factor.[6, 7]

Robb's results on the definability of the metric by causal connectibility in Minkowski space-time, and those that tell us that the metric is most certainly not uniquely fixed by the causal structure in Riemannian space-times, are, of course, completely compatible with each other. Suppose we know that space-time is Minkowskian. Robb has shown us that we can determine the interval separation of any pair of points (relative to a given separation taken as unit) by means of light rays alone, light rays being taken to demarcate the boundary of causally connectible sets of events. But if we know only that the space-time is Riemannian, then the results noted above tell us that we could not even determine that the space was flat (Minkowskian) using light rays alone, much less measure relative interval separations fully between all its pairs of points.

To clarify this point further, we might reflect for a moment on an important result of Zeeman. He shows that if space-time is Minkowskian, any automorphism of the space-time onto itself that preserves, in both directions, all the causal connectibility relations and the time-ordering of events that are causally connectible (i.e., if b is after a in Robb's sense, and if f is the mapping, $f(b)$ is after $f(a)$; and if d is after c, $f^{-1}(d)$ is after $f^{-1}(c)$) is a member of the group G, where G is the group generated by the orthochronous Lorentz group, translations, and dilations of the space-time. In other words, if the space-time is Minkowskian, "causality implies the Lorentz group."[8]

But Zeeman most certainly does *not* show us that any mapping between two pseudo-Riemannian space-times which is a one-to-one mapping and which preserves 'after' in both directions is such an extended Lorentz transformation. I think this is parallel to the fact that, while Robb shows

211

us that in Minkowski space-time, the 'after' relation defines inertial lines and both spatial and temporal congruence, he has certainly not shown us that we can tell which pseudo-Riemannian world we are in by using the data on 'after' relations alone. For, once again, even if Robb's axioms for 'after' are satisfied, the space-time need not even be Minkowskian.

Now if space-time is Riemannian, one of two possibilities holds: either the causal connectibility (or 'after') structure obeys Robb's axioms or it does not. In general it does not. In fact, unless the space-time is conformally flat, the causal structure of the space-time will not even be *locally* like that of Minkowski space-time at each point.[9] And even if the space-time is conformally flat, i.e., locally conformal to Minkowski space-time, Robb's axioms still need not hold, for they are of a global nature; they will hold only if the space-time is conformal to Minkowski space-time in a global way. It is, in fact, the global nature of Robb's axioms that allows him to define the metric structure in terms of the causal when his axioms are satisfied. For, as Weyl showed, the *local* causal structure is insufficient to fix the metric even in the Minkowskian case.[10]

Now then, Robb's axioms may not hold. We allow this as a possibility, believing as we do in general relativity and therefore in the possibility that space-time is not globally conformal to Minkowski space-time. Indeed, when we look at the almost inevitably singular nature of cosmological solutions to the general relativistic field equations, it seems most probable that Robb's axioms do not hold. And if Robb's axioms fall, so does his program of defining the congruence of the metric in terms of causal connectibility.

But suppose we did believe Robb's axioms hold. Should we accept his definitions of the metric congruence in terms of the causal connectibility relation? Two possibilities arise: (1) While we believe Robb's axioms hold, we believe special relativity in general does not. From the general relativistic point of view this corresponds to a theory of a space-time which, while globally conformal to Minkowski space-time, is not flat. Empirically this is revealed to us by the fact that free particles do not travel timelike straight lines as causally defined in Robb; rigid rods do not measure Robbian spatial intervals; and atomic clocks do not measure Robbian timelike intervals. (2) Not only do we believe Robb's axioms hold, but we believe that the special relativistic predictions about free particles, rods, and clocks hold as well.

IV. Some Alternative Formalizations of Robbian but non-Minkowskian Space-time

A. The Maximal Robbian Formalization

Robb's axioms hold; but from the general relativistic point of view, the space-time is believed to be non-Minkowskian. What alternative formalizations could we choose?

Option A: The maximal Robbian choice.

Here we keep as close to the Robbian analysis as possible. His postulates for 'after' remain the same. And we hold to the same definitions. Of course we must now drop such assumptions as that free particles travel timelike straight lines, that rigid rods measure intervals, etc.

Now we have some warrant for taking this approach in the history of the theory after all. In Newtonian space-time, for example, we never assumed that particles acting only under the influence of gravitation followed space-time geodesics. Instead we took these geodesics to be straight lines and said that the trajectories of the particles were distorted from the geodesics by the gravitational *force*. Now, of course, given the assumption that rigid rods measure general relativistic spacelike intervals and that clocks measured general relativistic timelike intervals, we must allow these "forces" (or, better, *potentials*) to influence the behavior of material objects in the space-time in other ways as well. The gravitational force must become also a gravitational "stretching-shrinking" field.

In this light it is interesting to see how Robb himself has responded to the introduction of general relativity. He worries about how his space-time theory is going to fare in the light of general relativity in an appendix to his *The Absolute Relations of Time and Space*. His remarks there are brief and enigmatic, but I think we can see that something like what we are proposing here as Option A, but with differences, is going on.

Robb says that although there might be some reason for speaking of particles as traveling geodesics in one of Einstein's "complicated geometries," as a *façon de parle*, "this does not imply any 'curvature of space.' " His meaning is clear: space-time is still flat, and timelike geodesics are still straight lines. While free particles can be viewed as travelling "geodesics" in a *fictitious* Riemannian space-time, they do not really travel as geodesics at all.

Robb also realized that in general relativity even light rays may fail to travel straight line geodesics or even to satisfy his axioms. Here his reply

213

is hard to comprehend. He seems to argue that there may be other causal influences that will obey his axioms even if light paths do not. In particular he thinks the propagation of gravitational influence might take the place of light in marking out the boundaries of sets of causally connectible events. Then, he says, we could still adopt his space-time theory, taking it as an empirical fact that gravitation selects the extremely causally connectible events, and allowing light, like free particles, to be influenced by the matter-field so as to fail to follow the "real" geodesics of the space-time.[11]

But of course this will not do. If general relativity is correct, there can even be space-times in which extremely causally connectible events simply do not meet Robb's postulates, inasmuch as gravitational force in a vacuum propagates with the velocity of light whether under the influence of "real matter" or in an empty non-Minkowskian space-time. Robb should have faced the fact that in the light of general relativity, even his axioms for causal connectibility might break down, and with it his whole program for a theory of space-time.

But our modest Robbianism is still worth examining. Once again, it states: if at least the Robbian axioms for causal connectibility hold up— i.e., if the space-time is at least globally conformal to Minkowski space-time—we can retain a formalization of our space-time theory that preserves Robb's axioms and definitions, even for such things as timelike geodesics (inertial lines) and spatial and temporal congruences, and make our theory fit the facts by changing the postulates that connect the behavior of free particles, rigid rods, and clocks with the space-time structure. That is, we can introduce gravity as a "potential" superimposed on the space-time.

B. A Formalization Differing in Definitional Consequence Only

Option B: Choose a formalization that agrees with Robb's on the status of terms and on those consequences framable in only the (mutually agreed) primitive terms; but that differs from Robb on the definitions offered, and on definitional consequence generally. What would this look like? First of all the formalization would take as primitive only those terms designated as primitive by Robb. And any proposition involving only the primitive terms would be a consequence of the new formalization if and only if it were a consequence of Robb's formalization. But the formalization would differ from Robb's, for example, by defining 'timelike geodesic' as 'path of a free material particle.' And whereas in Robb's formalization free particles would not travel timelike geodesics, rigid rods would not measure spacelike intervals, and clocks would not measure timelike intervals, in this formalization the usual relativistic assumptions about the relationship between these material objects and these space-time structures might hold.

But the Robbians and the proponents of this new space-time theory *might* very well agree that their two formulations were formalizations of the same theory, although I do not know if Robb himself would have accepted this line. How would they argue? Perhaps as follows: "We agree about the primitive consequences of our formalizations. And the primitive consequences of a formulation contain the totality of its *factual import*. We *choose* to define some nonprimitive terms in different ways, but that is a free choice in any case, subject only to the formal conditions of proper definability. Having chosen these differing definitions, it is hardly surprising that we extract from our axioms *apparently* incompatible consequences in cases in which they are propositions containing defined terms. But this apparent incompatibility is only apparent. To think that there is real incompatibility is to fall prey to fallacies of equivocation induced by failure to pay proper attention to the differing *definitions* we offer of the nonprimitive terms. Our formalizations, despite their differing linguistic appearance, really are just different ways of expressing one and the same theory."

We noted above that what we called the maximal Robbian formalization had the "advantage" that in the case in which Robb's axioms were satisfied but in which not all of special relativity was correct, this formalization made gravity play a role similar in some ways to its role as a force in Newtonian mechanics and in special relativity. The advocate of this new formalization might argue like this about the earlier physics: "It is true that in pre-general relativistic physics we treated gravitation as a force superimposed on the space-time. In my formalization, factually and observationally equivalent to Robb's though it is, we *talk* as if gravity really were curvature of space-time. This is a more natural way of talking than Robb's, for my way maintains the connection between rigid rods and spatial separations, and between clocks and temporal separations, which existed in the pre-relativistic theory and which is preserved both in the usual formalizations of special relativity and in the usual formalizations of general relativity.

"Of course in my formalization we speak of particles acted on only by

215

gravity as following curved timelike geodesics, and we speak of curved space-time, even though we did not talk this way in the pre-general relativistic theory. But we must realize that that older way of talking was *only* a way of talking. For just as we have a choice between the Robbian-Minkowski space-time plus gravitational potential fields, and the curved space-time formalization, to formalize our present theory of space-time, which—in general relativistic terms—is globally conformal to Minkowski space-time but not flat, we had the same options in pre-general relativistic physics. For example, as Trautman has shown, we can formalize Newtonian physics from the standpoint of curved space-time just as the maximal Robbian shows us we can formalize our present theory from the standpoint of flat space-time. In the curved space-time formalization of Newtonian physics, the same natural association of rigid rods and clocks with the appropriate intervals is maintained. But we now talk of free particles—those acted upon only by gravity—as 'following the timelike geodesics of a curved spacetime.'[12]

"But we must note that in both these cases it is only a matter of *how we talk*, not *what we belive*. For in both cases the observational (factual) consequences of the alternative formalizations are the same. This follows from the fact that the formalizations agree on the primitive and defined status of terms and propositions and on primitive consequences, differing only on definitional consequence."

Such a thinker might continue to defend a preference for his way of formalization as follows: "While both the maximal Robbian and my formalization of this space-time theory are based on one and the same theory, mine is superior to his in a number of ways, even though it differs from his merely in the definitions adopted for the (agreed) defined terms. The only advantage of the Robbian's formulation is that in his theory, as in the usual Newtonian theory, particles acted upon by gravity do not travel timelike geodesics, but instead are acted upon by forces. As I have just pointed out, this "continuity" with pre-general relativistic physics is not a critical point, for we could, if we wished, even reformulate Newtonian physics to meet the general relativistic way of talking, in which particles acted upon only by gravity follow timelike geodesics in a curved space-time.

"My way of speaking has other continuities, both with pre-general relativistic physics and with general relativity, which the maximal Robbian's formulation lacks. While Newtonian physics, special relativity, general relativity and my formulation all speak of rods and clocks as correctly measuring space-time intervals, the maximal Robbian drops this common way of speaking for the sole reason that this is the only way he can keep his Robbian definitions of congruence, etc.

"My view is further shown to be preferable by the following reason: I believe Robb's axioms for causal connectibility hold. For they hold both in the maximal Robbian formalization and in mine. But I know about general relativity, so I realize my present theory may be wrong, so wrong that the Robbian axioms really do not hold. Should I discover this, I shall have to change my theory. Now the maximal Robbian will not only have to give up belief in the truth of his causal connectibility axioms, he will have to drop all his definitions of congruence, timelike geodesic, etc., as well, for the very formulation of these definitions depends upon the Robbian axioms holding.

"I, on the other hand, shall, of course, have to drop my Robbian axioms of causal connectibility. But I shall be able to preserve my definitions of timelike geodesics and congruence intact, for they do not depend, in my formalization, on the Robbian axioms, and they correspond exactly to the usual general relativistic definitions of these space-time terms, definitions which are adequate even when the space-time is not globally conformal to Minkowski space-time. My formalization, then, while admittedly differing only definitionally from the maximal Robbian's, is more continuous in its way of talking—both with pre-general relativistic physics and with general relativity in their usual formulations—than is the maximal Robbian's way of speaking. Thus one should opt for my formulation." Here the reader will undoubtedly recall the familiar thesis of Eddington, Schlick, and Reichenbach: as long as two formalizations have all the same "observational consequences," they are the formalizations of one and the same theory, differing only in linguistic formulation.[13]

At this point several crucial issues about definition in formalizations have finally surfaced, and we must now make an initial assault upon them. The reader should be assured that our tentative probes here will hardly be the last moves we shall make on these issues, for we are now about to discuss fundamental problems which will appear again and again as we examine particular formalizations of particular theories, their similarities and differences, the reasons for and against adopting them, and the consequences of doing so.

In the rationalization given above for claiming that the two formalizations were based on the same theory, a fundamental proposition was assumed: identity of primitive consequences constitutes identity of observational and of factual consequences. But why should anyone assume this? I think the answer is clear: in both formalizations the choice of primitive or of defined status for a term, and the resulting classification of consequences of the formalizations into primitive and definitional, were *designed* to relegate the totality of observational and factual consequences of the formalizations—these two classes being *assumed* to be coextensive— to the class of primitive consequences.

First of all, the argument assumes that a consequence of a theory cannot be factual unless it is, at least in principle, observably either the case or not the case. We accept that a theory cannot entail any putative "facts" unless they are putative observable facts; for, in the best verificationist vein, we accept that a "factual" difference with no observable consequence is only an *apparent* factual difference. Second, the argument assumes that the classification of terms and propositions into primitive and defined is not an arbitrary classification, but one designed to mark out in the margins of the formalization those parts of its contents that are accesssible to empirical *test* and those that are open to "arbitrary" linguistic *choice*.

What should count as a primitive term in a formalization? The frequent answer is that primitive terms should be only those whose applicability or nonapplicability to a situation can be determined without in any way presupposing the truth of the theory in question. After all, the fundamental axioms of the formalization are to be presented as framed in the primitive terms only. And it is the empirical test of the correctness of these axioms that will tell us whether or not the theory so formalized is correct. But if we could not determine the applicability of a primitive term to a situation without presupposing the correctness of the theory in question, how could empirical testing ever get under way?

One version of this thesis would have us restrict the primitive terms to those whose applicability or nonapplicability is in some sense a matter of "immediate sensory awareness." Primitive terms are then the basic observation terms of some radically empiricistic philosophies of science. This is certainly Robb's position with respect to his 'after' primitive. For, he says, the grounds for choosing 'after' as a primitive are that we can be *directly conscious* of some pair of events—that is, a single "consciousness" can—and this same single consciousness can immediately, without any theoretical presupposition or inference, tell that one of these events in the pair is after the other.

The reader acquainted with some history of philosophy who reads Robb's introductory sections on the status of 'after' and on the "locality of simultaneity," and who encounters Robb's belief that simultaneity for events spatially separated is nonsensical, will recognize a close similarity between Robb's position and that expressed in the third chapter of Bergson's reply to special relativity, *Duration and Simultaneity*. Bergson's work has been, I think, unjustly neglected. This is largely his own fault, for many readers have been dissatisfied with his account of relativity because of the irrationalism rampant throughout Bergson's work, and because of the latter parts of *Duration and Simultaneity*, which constitute a confused attempt to refute of the so-called clock paradox in special relativity. [14]

This rationale for choice of a primitive leads immediately to philosophical trouble: it looks as though our physics is going to be founded on a solipsistic, private, subjective observation basis. Reichenbach, in his well-known axiomatization of special relativity, was well aware of this problem, even though elsewhere in his work he talks more like Robb and seems to take the primitive as the "directly apprehendable by a consciousness." In his book on special relativity Reichenbach uses, for example, *coincidence* of events as a primitive. But, he says, by this I mean *physical* coincidence, not coincidence of "appearances" in some subjective consciousness. Reichenbach calls those propositions that are framable entirely in terms of his primitives, *elementary facts*. What makes them elementary is not immediate apprehendability by consciousness, but rather the fact that they "remain invariant with respect to a great variety of interpretations." For example, we perform the Michelson-Morley experiment. We might have many different physical interpretations of the results. But all such theoretical accounts will agree as to whether the physical interference lines did or did not remain coincident with designated marks when the interferometer was rotated. [15]

As it stands, this definition of elementarity leaves something to be desired. For surely I could have considered a theory in which I explained away the *apparent* null results of the Michelson-Morely experiment, not as was done, by assuming the null results were a physical reality and by choosing one of the alternative accounts for them (Lorentzian ether theories or special relativity, for example), but by explaining the results away as a subjective illusion somehow to be accounted for on the basis of a theory which no one has ever proposed and which I cannot even now begin to construct.

If primitiveness is to be identified with elementarity in Reichenbach's sense, then, rather than with direct apprehendability in Robb's, I think Reichenbach's definition in terms of "relative invariance with respect to a great variety of interpretations" must be supplemented to read "relative invariance with respect to the great variety of interpretations *which one has in mind as plausible theoretical accounts of the experimental observations.*" The reader may now begin to see the point of the title of this paper. I am suggesting that even to begin to characterize the basis on which we shall choose either to accept a theory or formalization or to reject it in favor of some alternative, we need some preliminary *assumptions* about the range of theories we are going to consider.

Many philosophical readers will at this point recall the perennial doctrine of the "theory-ladenness of all terms including the observational" which recurrently makes its mark on the philosophy of science. Indeed, issues of this kind are what I have in mind. But I shall reserve my more general philosophical arguments about this issue and try to outline those aspects of the general philosophy behind the primitive/defined distinction in formalizations, which we need in order to continue our examination of the concrete dispute we are enmeshed in.

If we are presented, then, with a formalization in which some terms are taken as primitive, we should probably assume these are the terms of the theory whose applicability or nonapplicability to a given physical situation is independent of the acceptance of his theory. Naturally the defined terms are supposed to have just the opposite status. If the axioms of the theory are correct, then certain physical situations exist. We can characterize these in terms of the primitive alone. But it may be more convenient to introduce new single terms to refer to some of these physical entities or relations, so we introduce definitions and defined terms.

Just as restricting the axioms to propositions framable only in the primitive vocabulary was supposed to have the advantage that we could *test* the axioms for correctness or incorrectness without presupposing the theory correct, the definitions are supposed to have the opposite virtue of not being amenable at all to physical test. Since the defined terms are *introduced* by the definition, and *stipulated* to hold only when the defining situation characterized in the primitive vocabulary holds, there can be no question of *testing* the correctness of a definition, although it is, of course, a testable question whether the definition is "admissible," since the existence and uniqueness propositions necessary to legitimatize it as a definition are themselves expressible in the primitive terms alone. And since the primitive vocabulary exhausts those terms whose applicability or nonapplicability is decidable independent of accepting the theory, the totality of empirical tests of the theory resides in drawing from it its primitive consequences and testing them.

We are now in deep philosophical waters indeed, and this will hardly be the last we have to say about these matters. But we have enough, I think, to understand at last why our maximal Robbian and his opponent can agree about a number of issues. They agree as to which terms are primitive and which defined in their formalizations, i.e., their formalizations are alike in the status they attribute to terms. They agree that their formalizations are based on the same theory. They disagree only with regard to definitional consequences of their theories; but they agree that these differences are simply the result of having *chosen* differing definitions for some terms, and that these differences are in no way a mark of *empirical* disagreement.

C. A Formalization Differing Only in the Attribution of Definitional Status to Consequences

Let us now return to our concrete case. We are considering the options available to us in the situation in which, from the general relativistic point of view, we believe that space-time is globally conformal to Minkowski space-time but not flat. Physically this means we believe that whereas Robb's axioms for causal connectibility hold, special relativity does not, because it fails on the "matter" side. Free particles, rigid rods, and clocks do not, we believe, correspond in their behavior to the geometry constructed using light (or causal connectibility or 'after') in Robb's manner as they would if special relativity were correct.

Here is another option we could take informalizing our beliefs:

Option C: We adopt a formalization that agrees with the maximal Robbian about the primitive or defined status of the terms. It also has exactly the same consequences, primitive and definitional, as the maximal Robbian formalization. But it allocates definitional and postulational status to some of these consequences in a different way.

In this formalization one would agree, for example, with the maximal

Robbian that free particles do not travel inertial lines, that rigid rods do not measure spatial intervals, and that clocks do not measure timelike intervals. But the formalization might declare that what *defines* an inertial line is still fixed by that consequence that relates the behavior of free particles to inertial lines. And the formalization would then declare that what the maximal Robbian takes to be a causal *definition* of inertial line is a *postulate* about the relation of inertial line structure to causal structure.

As one might imagine, the proponent of this formalization will usually say he is most certainly formalizing the same theory that the maximal Robbian is formalizing. He is even using the same words, and he believes the same terms to have their criteria of applicability independent of accepting the theory. But, he says, since his definitions are the maximal Robbian's postulates, and vice versa, he is disagreeing with the maximal Robbian about the *meanings* of at least some of the defined terms.

Why would anyone prefer such a formalization to the maximal Robbian's? In this case, I doubt that anyone would, but later we shall see a case in which just such a preference is plausible. The proponent of this formalization might argue that his is preferable to the maximal Robbian's simply because the meanings he gives to the defined terms are closer to "what they ordinarily mean" than are the meanings the maximal Robbian gives to them.

He might argue like this: "Our ordinary meaning of 'inertial line' is that it is something fixed by the motion of free particles. And our ordinary meanings of spatial and temporal 'intervals' is that they are given by rods and clocks. The maximal Robbian instead defines these quantities in a causal way. Now I do not say that he *cannot* so define these terms. Nor do I claim that there is anything physically wrong with the theory he formalizes in the peculiar way that results when one gives such novel meanings to the terms. I am only saying that my definitions, my meanings, are closer to what the terms 'meant all along.'"

This argument is not very persuasive here. For while it may be "giving a novel meaning" to 'inertial line' to define it causally as the maximal Robbian does, the meaning given this expression by the proponent of formalization C is peculiar as well. For according to him, in agreement with the maximal Robbian and in opposition to Option B discussed above and Option D discussed below, free particles do not travel inertial lines. Now if 'inertial line' had an ordinary meaning it was, I suppose, 'path traveled by a free particle.' So the proponent of Option C is offering a definition of 'inertial line' just as peculiar in its own way as the maximal Robbians. As I noted, we shall later see a case in which just such a move from the Robbian causal formalization to an alternative seems more plausible.

Another possibility is this: someone might agree with the maximal Robbian that only causal connectibility should be taken as primitive. He might, however, be reluctant to assign "definitional" or "postulational" status among the sentences of his formalization which involved terms other than the primitive terms. In a sense he would be unwilling to assign propositions in formalizing the role uniquely, either of fixing meanings or of stating facts. Those who eschew the analytic/synthetic distinction would argue in this way, I believe.

Nonetheless such a thinker might still agree with the maximal Robbian that both formalizations have all the same "factual" consequences—assuming, that is, that they have the same primitive consequences. The idea here would be of a formalization that, like that of Option B, was "merely another way of formalizing the same theory the maximal Robbian formalized," superior in that by refusing to allocate definitional vs. postulational status to the propositions of the theory which involved vocabulary other than the primitive one, this formalization permitted us a more flexible way of talking about theory change and a more realistic way of talking about meaning.

Basically this position would look like this: "After all, the real import of a theory is, indeed, its observational consequences. My formalization, like the maximal Robbian's, has all the right observational consequences, and these are captured in the primitive consequences that my formalization shares with the maximal Robbian's. He, however, has taken the further—and gratuitous and misleading—step of allocating "analytic" and "synthetic" status to the propositions of his theory which involve nonprimitive terms. Since this has no empirical consequence, he shouldn't do it."

On this account, while it is important to demarcate the observational import of a theory, further discrimination among its consequences by marginal notes about their "analytic" or "synthetic" status are pointless and, from the point of view of an adequate account of meaning, misleading. But I shall not burden the reader with one more attack on the analytic/synthetic distinction, for I think we have more relevant fish to fry.

D. A Formalization Differing Only in One-Way Definitional Consequence

Suppose someone offered a formalization of a theory to account for the space-time of this world in which he, the maximal Robbian, and the proponent of Option B above all agreed that this new formalization differed from the earlier two in a primitive consequence. Then all would agree there was no sense in saying that this was a new formalization of the theory formalized alternatively by the maximal Robbian and the proponent of Option B. Instead they would agree that this was the formalization of a new theory incompatible with that earlier formalized.

But suppose the following peculiar situation arose: a new formalization is proposed. Let us call it Option D.

Option D: Here we are presented with a formalization, some of whose consequences differ from those of the maximal Robbian. But this formalization also differs from the maximal Robbian's in the status it attributes to some terms. In particular, at least one consequence of this new formalization is incompatible with those that follow from the maximal Robbian's formalization, and it has the following interesting feature: whereas the maximal Robbian declares this consequence to contain a defined term—defined, at least, in *his* formalization—the new formalization, with its new status attribution to terms, declares the consequence to be *primitive*. Further, let us suppose the maximal Robbian declares no consequences of the new formalization primitive but incompatible with the consequences of his own.

Let us look at the situation first from the point of view of the maximal Robbian, and then from the point of view of the exponent of the new formalization. The maximal Robbian says: "This is just like the situation discussed in Option B above. This new formalization differs from mine only in definitional consequences, for the only consequences of it "incompatible" with consequences drawable from my formalization are those containing what *I* take to be defined terms. So this is really a new formalization of the same theory as mine, except that this formulation chooses to express the theory in a different way. If there are grounds at all for choosing among these formalizations, they are just like the grounds for choosing between my maximal Robbian formalization and Option B above. They are choices of expression." (Incidentally, again, I am not sure that Robb himself would have tolerated this move.)

But now let us see what the proponent of this new formalization has to say: "My formalization has consequences incompatible with those drawn from the maximal Robbian formalization. Furthermore, *I* say that these consequences, framed in what *I* take to be primitive terms alone, serve as grounds for testing the theory. They contain no terms whose applicability or nonapplicability requires presupposing the truth of the theory. Clearly what I am proposing is not a new formalization that merely 'rephrases' the theory formalized by the maximal Robbian; I am proposing a new theory to account for the data *incompatible* with the theory formalized by the maximal Robbian."

Why would anyone declare that a term taken by the maximal Robbian to be defined, is really a primitive term? This is a subtler issue than it appears to be at first sight. One reason he might have for disagreeing with the maximal Robbian on this issue might be this: he might be taking primitive terms to be those whose applicability or nonapplicability is determinable by "immediate observation," and then saying there are more features of the world accessible to immediate, totally theoretically unmediated, test than the maximal Robbian allows. He might argue, for example, that the maximal Robbian is wrong in thinking that only facts about causal connectibility are directly accessible epistemically. He might assert, for example, that he can immediately tell whether or not two noncoincident but nearby rigid rods are congruent, just as Robb asserts that we can immediately tell when one of two events in our consciousness is "after" the other. If that is his line of reasoning, we are in for one of those endless debates about just what is accessible to immediate observation, and about whether, if we insist upon immediate apprehensibility as a criterion of primitiveness, we ever can construct a science that breaks the "veil of perception" and becomes more than a lawlike systematization of a solipsistic consciousness.

On the other hand, he might argue like this: "I don't take 'primitive' to mark out 'determinable by immediate inspection and so totally independent of theorizing.' But I do have a notion of primitiveness which is important in that it captures 'factualness' in the following way: propositions framed entirely in primitive terms state *facts*. They are in no way 'conventional.' Of course it is in some sense convention that the words mean what they mean, but when I say that a term in a formalization is primitive, I am indicating that a consequence of this formalization framable in these terms alone is true or false irrespective of the way in which terms gain meaning only from the role they play in *this* theory."

Whichever position he takes about primitive terms, he will continue like this: "While the maximal Robbian, for example, says that free particles do not travel timelike geodesics, I say they do. While the proponent of Option B also says they do, he says he is disagreeing with the maximal Robbian only as to the meaning of 'timelike geodesic,' a term whose meaning, for both of them, is fixed by the differing definitions in their respective formalizations. But I say we should take 'timelike geodesic' as a primitive term in our formalization; I disagree with the maximal Robbian about a *fact* of the world (whether free particles do or do not *really* travel timelike geodesics). Furthermore, while the proponent of Option B *thought* he was disagreeing with the maximal Robbian only about 'meaning of terms,' he too is proposing a theory *incompatible* with the maximal Robbian's. The incompatibility, however, was disguised by his *mistaken* attribution of defined status to 'timelike geodesic.'"

As we know, some would even deny the usefulness or intelligibility of any observational/nonobservational or "factual"/"conventional" distinction altogether. I think we shall be able to consider them, without loss of content, as advocating that all the terms of a formalization of a theory should properly be taken as primitive in our sense.

E. Variations on Option D

The exponent of Option D is most understandable when he takes the first line of reasoning we have proposed for him: he believes in "pure observation languages"; he believes in terms whose applicability or nonapplicability to a situation is totally theory-independent. He disagrees with the maximal Robbian and with the proponent of Option B only about which terms are really in the "pure observation language."

The exponent who takes the other option clearly has his work cut out for him. Just what is his theory of meaning, and just how do words get their meaning according to it? How does he rationalize a useful distinction between primitive and defined terms in a formalization if he can't pick out the former by their "pure observational status?" Perhaps he has some kind of theory in mind which, like that implicit in the Reichenbachian notion of "elementary fact" noted above, rests upon a view of the role of antecedent theories (and theories one is entertaining but not believing) in fixing the meaning of the terms in the theory we are as a matter of fact proposing. I shall return to these questions, if not answer them. But let me note here a few of the issues that arise when one reasons in this way.

226

While I shall not try to develop the theory of *meaning* required by this position, I shall try to say something about another problem faced by the individual who takes such a line.

This individual proposes that his formalization is based on a theory *incompatible* with that of the maximal Robbian. But he agrees that both his formalization and the maximal Robbian's give rise to all the same consequences about causal connectibility. Now he might consider that Robb is even wrong in thinking that the facts about causal connectibility were somehow "immediately accessible to observation in a totally theory-independent way." We shall, indeed, explore this issue later in some detail. But for the moment suppose he grants that at least those propositions are open to immediate empirical determination. Still, he says, I take our theories to be incompatible because they conflict on other propositions which, while I admit them to be *not* directly testable empirically without presupposing the correctness of the theory, I still take them, unlike the maximal Robbian, to be *factual* and not in any sense a matter of conventional choice about the meaning of words.

But, one immediately asks, how could one possibly decide which theory was correct? For example, how could one decide whether or not free particles follow timelike geodesics, given that this individual proposes that this is a *factual* question—not in any way a definitional one—and yet tells us we cannot decide its correctness by any "immediate empirical" means? Of course our maximal Robbian has no problems here, for he says these are really just different formalizations of one and the same theory. But our individual continues to deny this.[16]

One thing he could do is opt for skepticism: "I don't really believe free particles travel timelike geodesics, unlike the maximal Robbian, who says they do not. All I wanted to do was demonstrate to you that there were other theories besides maximal Robbianism that were true to all possible observational facts (if maximal Robbianism is) and yet incompatible with maximal Robbianism. Actually I believe you haven't the faintest reason for choosing one over the other of these theses about free particles. Of course only one of these theses can be true, but you simply can never know which." Skepticism.

Another thing he could do is opt for Poincaréan conventionalism, whatever that is. He could say: "Here we have two incompatible theories with all the same possible observational consequences. No empirical test could ever induce preference for one over the other. So we must *conventionally choose* which is true. We might even go so far as to say 'Truth just *is* a matter of convention.' Notice, however, that I am not claiming that our choice is merely a matter of choosing 'how we talk.' That is what Eddington, Schlick, and Reichenbach believe. I am claiming it is a choice among *incompatible*, but empirically indiscriminable, beliefs."

Another position he could opt for is a version of a priorism: "In choosing which theory to adopt we must always use, above and beyond the empirical evidence in favor or against a theory, an initial a priori plausibility of the theory. Without such 'believabilities,' intrinsic to the theory in question and independent of the empirical evidence for or against it, theoretical decision-making could never get under way. So even if two incompatible space-time theories have all the same observational consequences, there may still be good reason, on an a priori basis, to adopt one rather than the other."

Or he might assert that the choice of theory is always to some degree motivated by "continuity with previous theory," and that considerations of this kind could motivate his choice. He might say: "Suppose I have two incompatible space-time theories that are observationally indistinguishable. One of these theories may still be more "continuous" with the older theory it replaces. That is, it makes less of a change in our beliefs to drop the old theory for this new alternative than for the other alternative. In this circumstance it is more reasonable to believe the new hypothesis, which is minimally different from our older rejected theory, than it is to believe the alternative more 'radical' theory."

It has been alleged, for example, that even if general relativity is really incompatible with the Reichenbachian alternative of flat space-time and universal forces, and even if these two alternatives are observationally indistinguishable, we should still opt for general relativity since it is "more continuous with" or "a more conservative change from" the older theories of gravitation. Since this philosophical claim has been made with some frequency, it might be worthwhile to comment on it here.

There are, I think, two problems with this approach:

(1) First of all, each of the new alternatives can be, and usually will be, "more conservative" with respect to the antecedent theory in some respect or other. How can we decide which *way* of being conservative should take precedence or otherwise "weight" the value of being conservative in some particular respect? For example, general relativity is more conservative than the universal force hypothesis in that in the former

case, clocks and rods at different space-time locations are congruent if they are congruent when brought into coincidence. And this is the way things were in Newtonian and special relativistic physics.

But in Newtonian physics, and perhaps in what would have been a "natural" theory of gravitation from the special relativistic point of view, particles acted upon by gravity do *not* travel timelike geodesics. They are not "free" but "forced." This remains true in the universal force hypothesis, but fails to be true in general relativity. Crudely, the universal force hypothesis is more conservative than general relativity in that it, like Newtonian mechanics and special relativity, is a *flat* space-time theory, whereas general relativity is not.

Which way of being conservative with respect to the older theory is more important? Why?

(2) A second difficulty with the approach is this: we are told to be as conservative as possible with respect to antecedent theory. But suppose, when we have considered the options available to us for the new theory, we go back over the older theory and realize that it too could have taken another form observationally indistinguishable from the form it had. We might end up with this view: one option for our new theory is most conservative with respect to antecedent theory as it was. But the other option is most conservative with respect to another theory we could have had earlier, which is observationally indistinguishable from the older theory we did have.

Suppose, for example, we decide that the universal force theory is most conservative with respect to Newtonian mechanics because it, like Newtonian mechanics, is a flat space-time theory. But then we discover, reflecting on general relativity, something like Trautman's curved space-time version of Newtonian mechanics. And suppose general relativity is most conservative with respect to this now revised antecedent theory!

The individual whose "conservative" doctrine we are now examining could avoid this second problem only by becoming more conservative. He would have to say: "I now realize that all along there was an alternative theory that was observationally indistinguishable from my older theory. But I hold to the following conservative maxim: if you believe a hypothesis, there is no good reason to drop your belief just because you discover that there is another hypothesis which is just as good as, but no better than, the one you hold. I now use this retrospectively to justify my taking the older theory, which I actually had as preferable in believability to these newly discovered alternatives to it, and as being the correct starting point for conservative modifications of theory."[17]

But another problem arises for this position in the present context. We may say that if a Trautman had come along in the eighteenth century, a good conservative would have been justified in not dropping his Newtonianism and replacing it with this alternative theory. But now, given that we accept general relativity, might it not be rational for us, even if we are generally conservative, to assert that the theory that should have been held in the nineteenth century was Trautman's curved space-time Newtonian mechanics, and not Newton's flat space-time theory? It is one thing to be conservative with respect to physics as it was, but quite another to be conservative with respect to physics which, we *now* see, would have been as it should have been.

Let us consider a last option for our individual who tells us that theories may be incompatible even if observationally indistinguishable, and who wants to tell us which of such alternative theories we should adopt. Suppose he argues like this: "I believe that Robb's axioms for causal connectibility hold. But I am well aware that I might be wrong about this. Now I believe that free particles do not travel Robbian timelike geodesics, for I believe that the matter axioms of special relativity do not hold. Suppose I found out that, contrary to my present view, Robb's axioms failed. What would I do? The only theory I have available to apply in that case is general relativity. In general relativity Robb's axioms frequently—in fact, usually—fail to hold of the world. But free particles do still travel timelike geodesics in this curved space-time theory. Therefore, looking toward the possible worlds I now envision, and considering the possible theories I have in my repertoire to describe these worlds, I argue as follows: here are two incompatible theories that are observationally indistinguishable. One is the maximal Robbian, in which free particles do not travel timelike geodesics; the other is the version of general relativity suitable to a world which is globally conformal to Minkowski space-time and in which Robb's axioms hold but free particles do travel timelike geodesics. Which should I believe?

"I say I should believe the latter. Why? Because if I discover that, contrary to my present beliefs, Robb's axioms do not hold, I could, with this latter choice move easily and conservatively to a new theory—a general relativistic account of a world not globally conformal to Minkowski space-time. In this theoretical transition I could hold much invariant. I could still believe, for example, that free particles travel timelike geodesics. If I adopted the maximal Robbian theory, however, and then discovered that Robb's axioms do not hold, I would need to change my theoretical beliefs more radically.

"I am assuming the world is one of the possible general relativistic worlds. Which one is it? I believe it is one in which Robb's axioms hold. But I might be wrong about this. So, given that I believe that they do hold, I should believe that theory which is one of a spectrum of possible theories I have in mind, one of which will hold even if some of my present assumptions fail.

"I wish to believe a theory in which Robb's axioms hold. But which of the many incompatible theories in which they hold should I believe, given that I think there are many that are observationally indistinguishable? I should believe that theory which (1) is a specialization of a general class of theories, one of whose members I would adopt if I no longer believed that Robb's axioms do hold; and (2) is selected because it is the one member of this class in which Robb's axioms do hold and which is compatible with the observational data.

"Now the class of theories I have in mind is the class of general relativistic models of the world. The theory in this class in which Robb's axioms do hold is one in which free particles travel timelike geodesics, rigid rods measure spatial intervals, and clocks measure timelike intervals; i.e., given our data, the theory is the general relativistic model of a world globally conformal to Minkowski space-time. Now I believe the world is not flat. So I believe the free particles will not travel the timelike geodesics of the maximal Robbian theory. Therefore I should choose the theory that says free particles do travel timelike geodesics, and in which the maximal Robbian's timelike geodesics are not really timelike geodesics at all; I should not choose the maximal Robbian's theory, in which his timelike geodesics are really timelike geodesics but free particles just don't travel them.

"I make this choice *not* on the basis of my observation, for the maximal Robbian's theory and mine predict the same observational consequences. Rather, I make this choice because my theory is a special instance of a class of similar theories, one of which I would adopt even if I found out, contrary to my present belief, that one of Robb's hypotheses about causal connectibility is wrong; and the maximal Robbian's theory does not naturally fit into a class of theories at least one of which is still available to me

231

even if I were to find out that some of my present assumptions about the world are wrong."

This individual is saying: theories can be genuinely incompatible even if observationally indistinguishable. But I may still have grounds for preferring one over the other. The grounds are not a priori plausibility, nor even conformity with antecedent theory. They are rather the natural or unnatural role which the theory plays in the light of my assumptions about the other possible theories of the world which *might* hold—although I do not believe they do—and which I *would* adopt if I found out that some of my assumptions about the world are false.

Once again, it is asserted that theoretical decision-making depends upon some *assumptions* we make about the spectrum of possible theories we are likely to consider, and not just upon the "observational facts." We shall return to this point in some detail when we consider formalizations of general relativity later in this paper.

F. A Formalization Differing Only in the Status Attributed to Terms

I noted earlier four ways in which formalizations can differ from one another, but I have discussed only three alternatives to the maximal Robbian formalization. I have saved the fourth, Option E, till last, primarily because it is the one option which, to my knowledge, no one has yet espoused. Option B would be espoused by Eddington, Schlick, and Reichenbach, I believe, on the grounds that it is "descriptively simpler" than the maximal Robbian formalization. Option C would be espoused by someone who thinks the consequences of maximal Robbianism perfectly all right as they stand, but who has a different predilection for attributing definitional or postulational status to them. A Quinean, for example, who wishes to eschew such marginal status attribution altogether would adopt a version of Option C. Some version of Option D would probably be espoused by someone who, looking at the data in the light of general relativistic possibilities, and disagreeing with Robb's insistence that only causal connectibility should count as a primitive notion, would interpret the data to mean that the space-time of the world, although globally conformal to Minkowski space-time, was *really* not flat at all.

Here is Option E: Choose a formalization exactly like the maximal Robbian, except that different terms are designated 'primitive' and different terms are designated 'definitional.' This could be done in two ways: either (1) one might take causal connectibility as definitional instead of primitive; or (2) one might take some additional terms over and above 'causally connectible' as primitive. Since the latter is the more likely move, let me discuss what might motivate it.

Someone making this move might argue: "Robb is wrong to take causal connectibility as the only observational (or, perhaps, factual) feature of the world captured by a theory of space-time. For example, congruence for spatial intervals at a distance is also observational (factual), as is being an inertial line. But the maximal Robbian is correct in thinking that the space-time of the world is flat, that free particles do not travel inertial lines, etc. What the maximal Robbian says about the space-time of the world and its connection with material objects is correct; he is only wrong in thinking that some of these consequences of his formalization are matters of definition when they are really matters of observation (or fact)."

It is no surprise that no one has ever offered an approach like this. Who, faced with empirical facts that would normally lead a general relativist to say that space-time was nonflat but globally conformal to Minkowski space-time, would ever assert that the space-time was flat, and not as a matter of "definition" or "convention"—i.e., not because we could, if we wished, reformulate the general relativistic theory as a flat space-time theory, but because we could observe that this was really the case? Or, even if we couldn't decide this observationally, because there were other reasons for thinking this Robbian account both truly incompatible with the general relativistic account and factually correct?

I shall forego consideration of the other approach, with its denial of primitive status to causal connectibility, until later, when we shall examine that approach in another context.

G. Summary of Part IV

Let me summarize what has been going on in the last few sections. I have been exploring the desirability or undesirability of the Robbian definitions of such notions as inertial line (timelike geodesics) and spatial and temporal congruence, by subjecting Robb's theory to stress. Maximal stress consists in saying: suppose Robb's axioms for causal connectibility do not even hold. Under that stress his method of definition clearly breaks down. But then I have said: let us subject Robb's theory to less, but still to some, stress. Let us assume we believe that his axioms of causal connectibility hold, but that the postulates that would supplement his theory if special relativity were correct break down. That is, let us assume we believe that free particles do not travel Robbian inertial lines, rigid rods do not measure Robbian spatial congruence, and clocks do not measure Robbian temporal congruence.

Applying this stress to Robb's theory, we have seen that the following options can be delineated: we could hold to a maximal Robbian formalization in which Robb's axioms for causal connectibility are retained intact, as are his definitions. We would then have to abandon the postulates which would supplement Robb's original theory and tell us how free particles, rods, and clocks behave relative to the space-time structures as defined by Robb.

We could adopt an alternative formalization in which the usual connection between free particles, rods, and clocks and space-time structures is retained, at the same time retaining Robb's axioms for causal connectibility but dropping his definitions of inertial line and of spatial and temporal congruence. We could maintain that this was merely a different formalization of the same theory as the formalization first described, but that it expressed this theory in a different, and preferable, way of speaking.

We could adopt a formalization in which all the consequences of the maximal Robbian formalization were retained intact, and in which the same status—primitive or defined—was attributed to the terms as in the maximal Robbian formalization, but in which the consequences designated "definitions" and "postulates" differed from those so designated by the maximal Robbian formalization. We could maintain that this was a formalization of the same theory as that formalized by the maximal Robbian formalization, but argue that it formalized the theory in a way that gave "more usual meanings" to some of the defined terms. Or we could adopt a formalization exactly like the maximal Robbian's except that in it we refuse altogether to assign "in the margins" definitional or postulatory status to any of the consequences.

Another alternative would be to adopt a formalization like the one described two paragraphs above and declare that this was the formalization of a theory different from that first described. For, we could maintain, Robb was wrong in thinking that his "definitions" of such things as inertial line and congruent interval really were definitions; they could be viewed as empirically testable consequences of the theory, found false by observation.

We could adopt the second kind of formalization discussed above and, again, maintain that it formalized a different theory than that formalized

234

by the maximal Robbian. We could admit that there was no observational means of telling whether the maximal Robbian or this alternative theory was correct, but go on to declare that there could be genuinely incompatible theories that were observationally indistinguishable. We would supplement this approach by offering, one hopes, an account of how a term could have a meaning in a theory without having "directly observational" criteria of applicability, and an account of how one could rationally go about deciding just which of two incompatible but observationally indiscriminable theories was correct. In any case we would either provide this last supplementary account or admit ourselves skeptics or conventionalists with regard to theories.

Finally, we could take the view that all the consequences of the maximal Robbian formalization were correct, but that the maximal Robbian formalization was misleading in that, by incorrectly restricting primitive status to causal connectibility, it misrepresented some of its true observational or factual consequences as having mere definitional status.

V. Alternative Formalizations of Minkowski Space-time

A. The Formalizations

We have subjected Robb's theory to various stresses, and we have seen how it holds up. When the stress is so great that his postulates for causal connectibility break down, all would agree, I believe, that his metric definitions must go by the boards as well. But, as we have seen, a weaker stress, insufficient to contradict any of his axioms for causal connectibility, but sufficient to place stress on the additional postulates connecting the behavior of material objects to the space-time structure, can lead to many options. Some of these are maximal Robbian, retaining his metric definitions intact; but others lead us to reject the Robbian formalization of a space-time theory to a greater or lesser degree. Even when the Robbian theory is subject to no observational stress whatever, there may still be dispute as to whether the Robbian formalization of space-time is correct, and, if "correct," whether it is "best."

Robb's theory is now usually thought of as one way of formalizing special relativity, even if Robb would not quite have viewed it that way. Let us suppose we believe special relativity to be correct. Then we believe that Robb's basic axioms of causal connectibility hold, and that the matter postulates we would ordinarily use to supplement Robb's theory hold as well; i.e., rigid rods correctly measure spatial congruences as defined by Robb, clocks correctly determine Robbian temporal congruences and free particles follow Robbian inertial lines. If we do believe this, are we then precluded from adopting any of the options discussed in the sections above? We shall see that all of these options are still available to us, each option differing from the Robbian axiomatization as before. But some of the options will no longer have the "plausibility" they had before, in the sense that the grounds for adopting them will no longer be available. On the other hand, some options, previously unmotivated and unlikely ever to be espoused, will now become plausible alternatives.

As before, Option A is the Robbian. The only primitive term is "causally connectible," and the axioms for it are Robb's. The metric notions and the notion of inertial line are defined as in Robb. For our complete formalization Robb's theory is supplemented with matter postulates that tell us that rigid rods, clocks, and free particles have the usual special relativistic connection with the underlying space-time features.

In Option B we would still take causal connectibility as our only primitive notion, and the class of consequences of the formalization we take as primitive would remain the same. But we would offer different definitions for the metric notions and for the notion of inertial line. What would this amount to? Just as our maximal Robbian option amounted earlier to adopting a flat space-time formalization of a theory which, in general relativistic terms, we would take to be a theory of a nonflat space-time, adopting Option B would amount to proposing a curved space-time formalization of what we would take, from a general relativistic point of view, to be a theory of a flat space-time.

Earlier we saw that one might choose Option B on the grounds that it was "descriptively simpler" than the maximal Robbian formalization, fitting in with the way we would talk in the general relativistic context, as the maximal Robbian formalization did not. But in this case it is clear that the Robbian correlations of the behavior of matter with the space-time structure are, from the point of view of general relativity, the right correlations; and that if either formalization is descriptively simpler it is the Robbian.

Option C remains quite viable. Here we adopt a formalization that agrees with Robb's in counting only causal connectibility as primitive. It has all the same consequences as Robb's formalization. But it either (a) allocates definitional and postulational status among these consequences in a manner different from Robb's, or (b) eschews the marginal annotation of postulational and definitional status to propositions containing nonprimitive terms altogether.

A proponent of alternative (a) might argue like this: "While all the consequences of the Robbian formalization are true, his allocation of definitional and postulational status to them is untrue to the *meanings* of the 'defined' terms. It is true that free particles travel inertial lines. But that is what we *mean* by 'inertial line.' It is also true that inertial lines have the association with causal connectibility which the Robb formalization postulates. But that is a *fact* about the world, not a proposition true by the meaning of 'inertial line.' So Robb's consequences are all true, but he takes some to be true as a matter of fact that are really true by definition, and *vice versa.*"

The exponent of alternative (a) has a notion of meaning—or at least thinks he does—that allows him to tell what is *really* a definition and what is really a postulate. The exponent of alternative (b) thinks that Robb and the exponent of alternative (a) are equally confused. For, he alleges, there is simply no good reason to label consequences postulational or definitional at all. Remember that he is still allowing that we can determine which consequences of a formalization are "observational," and he is taking difference among these consequences to indicate clear divergence of theory. But beyond that, he believes it is fatuous to pretend that the propositions of the theory involving nonprimitive terms can in any important way be characterized as definitional or postulational. Insofar as the nonprimitive terms are "defined" by the theory, he says, the defining is spread throughout the theory as a whole.

Option D is again one that no one is very likely to espouse. Like Option B it offers a formalization that differs from Robb's in some of the nonprimitive consequences. Thus, like Option B, it amounts to adopting a curved space-time formalization for what we would ordinarily take to be a theory of a flat space-time world. Now whereas Option B took this to be a matter of merely "rewording" Robb's theory, Option D takes the new formalization to be that of a theory incompatible with Robb's. Choosing Option D amounts to deciding that space-time *really* is curved. In one alternative Option D claims that this is an observationally determinable fact; in another, that, while the fact that the space-time is curved rather than flat is immune to observational discovery, it is still a factual and by no means a definitional matter.

Earlier we saw that this option was a natural one to take in the case in which space-time was, from the general relativistic point of view, curved but globally conformal to Minowski space-time. We saw that considerations of the general theoretical context of general relativity might very well lead one to say that the maximal Robbian way of formalizing the theory of space-time was not just misleading with respect to meanings, but simply *wrong*. And if the maximal Robbian would argue that the two formalizations did not really differ in primitive consequences, the exponent of Option D argued that they did; for he argued that the maximal Robbian's version of what was primitive was too narrow.

But now, given that we believe that the world is, from the general relativistic point of view, flat, who would be likely to argue for a curved space-time as a theory incompatible with the flat space-time theory, but correct? If one were to believe that Option D and the Robbian formalization really are formalizations of incompatible theories, he would be more than likely to declare the Robbian alternative correct, and on the same grounds on which the proponent of Option D argued earlier for the correctness of his theory as opposed to that of the maximal Robbian.

The proponent of Option E argues for a formalization that differs from the Robbian in no consequences. But he disagrees with Robb about the scope of primitive concepts. Let us suppose, as we did earlier, that he differs from Robb only in wanting more concepts as primitive, and let us again postpone arguments about the primitiveness of causal connectibility. The proponent of Option E would then argue that, whereas Robb's formalization is based on a correct theory of space-time, it is misleading in that it falsely declares some of its consequences to be nonobservational or nonfactual. He would say: "I, like Robb, prefer the consequences of Option A to those of Option B. But whereas the Robbian and the proponent of Option B agree that their two formalizations are based on the same theory, I take it that they formalize incompatible theories. So the choice of the Robbian formalization over that of Option B is one of fact, not of mere expression. Like the proponent of Option D I consider it a factual matter whether space-time is or is not flat. But whereas he opts for an implausible curved space-time picture, I opt for the flat."

B. Can Definitions Be "Right" or "Wrong"?

Let us look more closely at the first alternative version of Option C and at Option E. Both of them find the Robbian formalization defective. The first version of Option C finds Robb's formalization unjust to the meanings of words. Option E finds Robb's formalization unfair to the question of what is "factual" as opposed to "definitional."

Consider the following argument: "In the Robbian formalization the notion of congruence for spatial and temporal intervals is *defined* by means of causal connectibility. So the connection between the notion of congruence and the causal notion used to define it can hardly be said to be correct or incorrect. These, like all, definitions, in order to be acceptable, must meet certain formal requirements of existence and uniqueness of defined entity. Beyond this, there is no sense in speaking of a definition as 'right' or 'wrong.'"

But the Robbian definitions of spatial and temporal congruence are not the only possibilities for causal definitions of "congruence" relations in his theory. One can, as Malament has shown, causally define, in Robbian fashion, a congruence relation which "renders as congruent timelike and spacelike vectors whose Minkowski lengths are equal in absolute value (although opposite in sign). And one can define a relation that renders timelike and spacelike vectors congruent if the absolute value of the Minkowski length of the former is p/q times that of the Minkowski length of the other, where p/q is any particular positive rational number. Both these definable congruence relations otherwise agree with the standard congruence relation."[18] Of course, if we so define 'is congruent to,' rods and clocks will not, even in Minkowski space-time, determine congruity correctly when used in the usual way. But why should that bother us if the only meaning attribution we have for 'is congruent to' is our causal definition?

But these definitions of the congruence relation would be *wrong*. They would not be causal "definitions" of "congruent," but of some other relationship misleadingly so named. This shows that, in some sense at least, these "definitions" are not really the appropriate definitions.

Here I think the argument could go in one of two directions, one appropriate to the advocate of the first version of Option C and the other appropriate to the advocate of some version of Option E. In the first version it is agreed that 'is congruent to' can only be a defined term in a formalization. But, this version argues, it is wrong to the *meaning* of 'is congruent to' to give it a *causal definition*. For example, one might argue that the association of congruence with rigid rods and clocks really captures the meaning of these terms. One might argue that the usual defini-

239

tion of congruence in terms of rigid rods cannot be "wrong" or "right," for it is this association that really fixes the meaning of 'is congruent to' as we usually use that expression. Congruent intervals are *just* those intervals determined congruent by the usual method of transported rods and clocks. What masquerade as definitions in the Robbian formalization are really postulates that associate certain causally definable structures with the congruence relation. The association will indeed hold if space-time is Minkowskian, but one hardly wants every general truth about the world to count as a "definition."

A similar but more complex tale about "meaning," or perhaps a denial that 'meaning' has any useful meaning and that out with it goes any useful notion of 'definition' in a formalization of this kind, might accompany the view that we should either talk about the meaning of 'is congruent to' as being fixed by its total role in the theory, or not talk about its meaning at all. In any case this position would still reject the claim that the causal definitions correctly "fix the meaning" of 'is congruent to' in any sense beyond correctly designating as congruent those intervals that are in fact congruent when the space-time is Minkowskian.

The second version argues that 'is congruent to' should be taken as part of the primitive vocabulary. This signifies that both the postulates connecting causal structure to the congruence relation, *and* those associating the behavior of rigid rods and clocks with the congruence relation, are empirical postulates that all turn out to be true because of the Minkowskian structure of space-time. As usual one might argue that these postulates are observationally determinable to be true or false, or that they are factual but not directly observational. But, in any case, in this view none of these postulates constitutes a "definition" of the congruence relation, for that is primitive and undefined. From both these general points of view one can see why we would instinctively feel we can judge Robb's definitions to be "right" or "wrong" in a manner entirely inappropriate for definitions.

We noted earlier that in Minkowski space-time (and in a more general class of space-times as well), we could "define" a topology—the Alexandroff topology—that would agree with the usual manifold topology on the space-time. We might note here (and we shall return to this point later) that what we have said about causal definitions of metric notions holds for causal definitions of topological notions as well.

Zeeman, for example, has shown how one can, in Minkowski space-

240

time, causally define (if one allows oneself the full resources of set-theoretic definition) a topology quite different from the Alexandroff topology, and hence from the usual manifold topology. The Alexandroff topology is the coarsest topology in which the interiors of light cones are all open sets. The Zeeman topology is the finest topology on the space-time that induces the usual three-dimensional Euclidean topology on every space slice of the space-time and the usual one-dimensional Euclidean topology on every time axis.

This topology has a number of curous features. For example, the topology induced on light rays by it is discrete. In this topology the paths of freely moving particles are all piecewise linear. Interestingly—and perhaps not too surprisingly, in the light of Zeeman's other work mentioned above—every one-to-one mapping of a Minkowski space-time onto itself which is bicontinuous in this topology is first of all a causal automorphism (i.e., it preserves all causal connectibility and symmetric absolute temporal order relations); and is, second, one of the group of mappings generated by inhomogenous Lorentz transformations and dilations![19]

Thus, just as many "congruence relations" are causally definable, so are many "topologies." Later I shall deal to some extent with the question of how to pick the right topological definition. I believe the discussion above, however, in the light of such noval "causal topologies" as Zeeman's, is enough to show us that one can no more glibly assume there is no question of rightness or wrongness about a causal definition of the topology, than one can assume there are no such questions about the causal definitions of the metric structure. And if there is a question of rightness and wrongness, in what sense can these "definitions" be said to be "merely definitions?"

VI. On Drawing Philosophical Consequences from Formalizations

A. Winnie's Thesis

At long last we can consider one example of an answer to our second overall question. So far we have been asking: what are the possible alternative formalizations of a theory, and how do they differ from one another? We have pursued in some detail our first primary question: what could be the scientific and/or philsophical grounds for preferring one formalization to another? Now we wish to ask: what are the philosophical consequences of adopting a given formalization? Since we see that much philosophical decision-making goes into choosing a formalization, we shall really be exploring the question: how do the philosophical choices one makes in choosing a formalization affect further philosophical consequences?

The philosophical theses Winnie extracts from Robb's formalization of Minkowski space-time provide, as the reader will recall, an example of consequences drawn from formalization. Winnie's argument went like this: Robb has shown us that in the case of Minkowski space-time the metric is reducible to the topology. But some authors have claimed that it is because the metric is undetermined by the topology that the metric is a matter of convention. So, in the case of Minkowski space-time, we see that the metric is not conventional, at least not in this sense.

There are two issues we must pursue:

(1) In what sense, or senses, is it true that in Minkowski space-time the metric is "reducible" to the topology? In what senses is this false?

(2) Is the reducibility, in any sense, of the metric to the topology ever good grounds for declaring the metric nonconventional? Is nonreducibility of the metric to the topology, in any sense of 'reducible,' ever good reason for declaring the metric conventional?

B. On Causal Theories of the Metric

Robb has shown, beyond question, a number of interesting and important things. He has shown that if space-time is Minkowskian, then the fundamental metric relations of congruence, and the property of being an inertial line, are coextensive with relations and properties definable purely in terms of the absolute "after" relation—or, if you will, in terms of causal connectibility. But has he shown that, if space-time is Minkowskian, the metric is "reducible" to the topology?

Using the results noted earlier concerning the causal definability of the Alexandroff topology and its extensional equivalence with the manifold topology when space-time is Minkowskian, together with Robb's results, the following can be derived: if space-time is Minkowskiań, we can determine all the topological properties of regions of the space-time simply by determining which events are and which are not causally connectible. Using the very same determinations, and no others, we can also find all the metric properties of regions of the space-time.

But this hardly means that we can, by using causal connectibility alone, determine *if* space-time is Minkowskian, knowing only that it is pseudo-

Riemannian. We can tell that it is not Minkowskian if one of Robb's postulates fails; but even if none does fail, this is no guarantee that the space-time is flat. And the correctness and importance of Robb's results within his formalization hardly force us to accept the view, even if we take it that space-time is Minkowskian, that he is right in saying that causal connectibility is an immediately determinable fact in a theory-independent way, and hence a proper member of the class of primitive terms. Nor does Robb's formalization mean we must agree that he is right in consigning all the metric notions to the nonprimitive class on the basis of "non-observability." Even if we do partition the terms as he suggests, we need not accept his "definitions" as truly definitional, nor agree that the *meanings* of the metric terms are correctly fixed by their causal "definitions." As we have seen, there are generally good reasons for adopting just such "anti-Robbian" positions. And, I submit, if we consider just how little we must accept, given Robb's results, and just how weak the connection is between causal order and metric structure, compared with what philosophers have usually demanded of an "X theory of Y," we shall probably conclude that Robb's results certainly do not establish a "causal theory of the metric," nor a "reducibility of the metric to causal order" in many of the most important philosophical senses of 'reducible.'[20]

To espouse an "X theory of Y" might be to maintain that X and Y are co-extensive predicates. Now we do not believe that metric features will in fact be coextensive with the Robbian causal correlates, since we believe that space-time is not Minkowskian. But if we believed that space-time were Minkowskian, I think we would have to agree that a causal theory of the metric was correct in this very weak sense. Alternatively, to espouse an "X theory of Y" might be to maintain that X and Y are coextensive as a matter of scientific *law*. Obviously we do not believe that the causal and metric predicates are so lawlike coextensive, since we do not believe they are coextensive at all.

But suppose we believe that space-time is Minkowskian. Believing, as we do, in general relativity, I think we would still deny lawlike coextensiveness to the pair of predicates. For, from the point of view of general relativity, the coextensiveness is a matter not of the fundamental laws of physics, but of the de facto flatness of space-time. Minkowski space-time is just one possible model for the laws of general relativity, but hardly the only "lawlike possible" space-time. But if we were working in 1905, say, and had never even seriously contemplated the possibility of curved space-time, the universal (for all of space-time) coextensiveness of the causal and metric predicates might be taken by us as a fundamental law of nature.

We noted earlier that if we believed in general relativity, observations on causal connectibility alone would not be sufficient to fix for us the curvature of space-time. This illustrated a fundamental fact: the question of what kinds of observations would or would not be sufficient to determine fully all relevant physical features of the world depends upon what possible worlds one is willing to contemplate. And that depends upon the fundamental theory one believes. Thus if we presupposed that space-time was Minkowskian, we would agree that we could map its metric using causal connectibility alone. But if we only presupposed space-time to be pseudo-Riemannian, we could not. Now we see that our presupposition about the possible worlds, framed on the basis of our most fundamental theory, has other consequences. For whether a correlation of features is taken to be lawlike or merely de facto is again a function of our *assumptions* about which possible worlds we will consider.

To espouse an "X theory of Y" might be to maintain that the predicates are coextensive as a matter of their *meaning*; that 'all Y's are X's' is an analytic truth. In the philosophical contexts we are interested in, the argument for this usually takes a standard form: we can tell what an X is without presupposing the theory in question. But we have no independent check on what a Y is; we simply call Y's the things that are X's. For a causal theory of congruence this would amount to claiming that we can, without presupposing the theory in question, determine which events are and which are not causally connectible. But we have no independent check on the metric. What we mean by 'congruent intervals' is some relationship between two pairs of points characterizable in terms of causal connectibility.

Believing in general relativity, we do not accept this argument. For we believe that there are nonisometric space-times with all the same causal connectibility features between the point-events mappable one-to-one on each other. But suppose we thought space-time was Minkowskian? Would we believe in such "analytic coextensiveness" then?

Once again our thinking depends, I believe, on a complex estimate of our theoretical situation, not only on the theory we do believe, but on the theories we did but no longer believe, and on the theories we think we might believe should observation ever lead us to give up our present theory. Our views about the formalization of Minkowski space-time are nice test cases for us. For this theory is the successor to previous space-time theories, and has itself been succeeded by a more general theory. This case will, we shall see, be applicable in interesting ways to the case in which we have no specific "future theory" yet in mind. For example, it will be useful when we come to discuss the "meanings" of terms in formalizations of general relativity.

Which propositions of a theory are "analytic", and which are true "by the meanings of the words alone?" We know we do understand what we are asking here; philosophical positions range from the view that there is something to analyticity but we do not really know what it is, to the view that the whole analytic/synthetic distinction is just a tissue of confusions. But perhaps the following will throw some light on the problem.

The "analytic" propositions of a theory should be at least more or less "fixed points" in theoretical change. For if a proposition is analytic, what on earth could lead us to deny its truth? Obviously the situation is not that simple, for one could always offer a new theory in which the proposition is denied, yet attribute analyticity to it in its old role by arguing that the meanings of the terms have changed in the theoretical transition. After all, we have seen how one can argue for different formalizations of even one and the same theory, which may or may not differ in asserted consequences but which do differ in their "meaning" attributions to defined terms.

Yet I think the following position has some validity: at least a prima facie guide to the analyticity of a proposition in a theory—if there is such a thing—is the stability of that proposition in the transition from an older theory to the one now believed, and from the one believed to one of the ones we presently contemplate as theoretical alternatives to which we would move under the pressure of novel observations.

For example, in Newtonian space-time theory we maintain that rigid rods measure spatial intervals. They do so as well in Minkowski space-time understood from the special relativistic point of view. An in general relativity we still say that rigid rods correctly determine spatial congruence. So it may be plausible to argue that the connection between rigid rods and spatial congruences in Minkowski space-time theory is "analytic," for it is that connection that remained fixed when we moved from the Newtonian to the Minkowskian theory, and remains fixed when we

245

move from Minkowski space-time to general relativity and curved space-times.

On the other hand, Robb's association of spatial congruence with causal connectibility holds only in the Minkowskian theory, and not in the Newtonian or in the general relativistic case. Isn't that some reason for attributing analyticity to the rigid rod-congruence connection in the theory of Minkowski space-time, and for denying analyticity to the Robbian association of congruence with causal connectibility? What other, or better, reason could there be for distributing analyticity or nonanalyticity than this "historical" fixedness or nonfixedness of the connection?

Once again we see that our philosophical attribution will depend on many things. First of all, it will depend on assumptions about the independent determinability of various features of the world. And it will depend not only on the relation of the theory we are interested in to the theory which preceded it, but on its relation to the class of theories that we *assume* to be plausible candidates for a novel theory to adopt under observational pressure which would lead us to relinquish our present views.

To espouse an "X theory of Y" might mean to maintain that the coextensiveness of the predicates was somehow a *necessary* truth, not identifying this notion of necessity with either lawlikeness or analyticity. Despite recent invocations of possible world semantics and doctrines concerning necessary truths about individuals or substances which are neither a priori nor, perhaps, analytic, I cannot understand what such an allegation would amount to. Once again the clearest notion I can think of relies upon putting the theory in question in the framework of antecedent theories and of assumed alternatives. Perhaps it would be plausible to speak of some proposition of a theory as necessary if it held not only in this theory, but in all antecedent theories and in any plausible alternative to which we would turn if we abandoned our present theory in the light of new data.

Under this interpretation the causal theory of the metric in Minkowski space-time again seems false. For although the congruence relations are coextensive with certain causally definable relations in Minkowski space-time, they are not so coextensive in either the theory that antedates Minkowski space-time or in the general relativistic theories to which we would turn, even if we believed in Minkowski space-time, should the data indicate to us that something about the special relativistic picture of the world was wrong in its assumption of flatness of the space-time.

246

The reader may object that my notions of the lawlikness, analyticity, or necessity of the connection between the relations were all defined only relative to a body of assumed theoretical alternatives to the believed theory. I reply that this is the major virtue of my analysis of these notions. Once again the claim is that while *facts* and *conventions* have received their due in the philosophical theory of the epistemology, semantics, and metaphysics of theories, *assumptions*—assumptions about the class of theories we will even consider as possible candidates for belief—have not.

C. Can One Get from Causal Definability to Nonconventionality?

I have argued that a causal theory of congruence is plausible, even if we believe in Minkowski space-time, only in its very weakest sense—the sense of mere coextensiveness in this space-time of some congruence relations and some relations definable by causal connectivity. But what about the claims that are made for what this shows concerning the relation of metric to topological structure in Minkowski space-time? And what about the claim that this "causal theory of the metric" is relevant to showing the metric "nonconventional?"

First we should note that what is offered in the Robbian formalization, even in the weakest sense of an "X theory of Y," is a theory of congruence in terms of causal connectibility. How then is congruence associated with *topology* in this view? The connection is mediated by the allegation that one can offer a theory of the topology in terms of causal connectibility as well. We do not offer topological "definitions" of metric quantities; rather, we attempt to "define" both metric and topological notions by causal ones.

Later I shall argue that the causal "definition" of the topology in Minkowski space-time is a "definition" only in the very weak sense in which we have seen it possible to "define" the metric causally. That is, I shall argue that while it is possible to construct, out of causal connectibility, properties and relations coextensive in Minkowski space-time with all the relevant topological properties and relations, there is good reason to think these coextensivities are neither lawlike, analytic, nor necessary—just as was the case with the causal and metric coextensivities.

So let us speak of a *causal* theory of the metric in Minkowski space-time, rather than of a *topological* theory of the metric. Is there any reason to think that the metric of Minkowski space-time is "nonconventional" because it is *causally* "definable?" I think one could argue from "causal

247

definability" to "nonconventionality" only in the following way: which events are and which are not causally connectible is a fact discernible totally independently of any theoretical assumptions. Hence no elements of "convention" enters into deciding which events are or are not so related. But there is, in addition, no "optionality" or "choice" in going from causally connectible events to the metric of the space-time. Therefore the "nonconventionality" of the metric follows from its reducibility to causal connectibility.

Since other arguments have been used to associate nonconventionality with "reducibility to the topology," we might first say something about them. Riemann suggests, and Grünbaum has held, the view that if the topology of space-time were discrete, we could determine metric relations from topological by "counting numbers of points between events." This is mysterious in the case of multiple dimensional space-times since we do not yet understand, I believe, what such discrete multidimensional cases would amount to. Even in the one-dimensional case of a simple ordered set, it is not clear why we should take the *real* distance between points to a function of the number of points between them. Further, even if the metric were determined by counting points, it is still hard to see the relevance of this to issues of conventionality without some demonstration, never given, as to why counting points is not in any way conventional. In any case, this is hardly relevant with respect to Minkowski space-time, where the topology is admittedly dense and not discrete.[21]

Again, it has been argued that since many metrics are "compatible" with a given topology, this is why the metric is "conventional." But if "compatible" means only that both the metric and the topology are consistently assignable to the underlying point set of events, then, trivially, many topologies are "compatible" with a given metric. Zeeman's results show that there are even many topologies compatible with the Minkowski metric in the further sense that the metric is continuous with respect to each of them. In any case, the fact that many "metrics" in the formal sense are "compatible" with a given topology hardly shows that the real metric—the real measure of intervals between points—is·a matter of "convention" in any interesting sense. And why should we, without further *epistemic* argument, take the topology as somehow "less conventional" than the metric?

We have seen many reasons for thinking that even given a "nonconventional" notion of causal connectibility, we should be dubious about the move from that to a "nonconventional" metric. In many space-times no "definition" of the metric congruence in terms of causal connectibility is even possible. Even when one would be possible (i.e., when the Robbian postulates for causal connectibility hold), we might be reluctant to accept these "definitions" as even extensionally correct (the case of space-times globally conformal to Minkowski space-time but not flat). And even when we admit the co-extensivity to hold, we have seen good reasons for denying that even then the "definitions" of the metric in terms of causal connectibility are really definitions in any deeper sense than that of true statements of coextensiveness. And if that is all the definitions are, it is hard to see how they provide an adequate bridge from the nonconventionality of the causal connectibility relation to that of the metric. Couldn't we still "conventionally choose" a different metric, simply by paying the price of denying the coextensiveness of the metric and causal connectibility relations, on the grounds that they were not, after all, in any sense "necessary" or even "lawlike"?

Even if we could establish the causal connectibility relations so as to preclude any invocation of conventionality in the process, and even if the relations so established allowed for some "definition" of the metric in terms of causal connectibility, what is there to prevent us from asserting that accepting those "definitions" is "just a matter of convention?"

But is even the establishment of the causal connectibility relations a matter totally immune from "conventional" elements? Let us look at two versions of theories of how causal connectibility relations are established, and see how their "nonconventionality" fares under each. One version is that espoused by Robb and, as remarked, interestingly like the position taken by Bergson. Let us do the analysis in terms of Robb's 'after' relation. Why is it not a matter of convention which events are truly after which others? Well, says Robb, because we could imagine a human consciousness moving from one event to another absolutely after it in time. The 'afterness' relation between the events would be an immediate apprehension by this consciousness, and such immediate apprehensions are matters in which no conventionality can rear its ugly head.

Notice, first of all, how we have retreated "behind the veil of perception." Isn't it *subjective afterness* which we are now determining, and not an *objective* physical relationship? The reader should compare this with Reichenbach's remarks, noted above, concerning which notion of coincidence of events should be taken as primitive in formalizing special relativ-

249

ity. In addition, it is hardly alleged that some consciousness does actually record the afterness between any pair of events—only that one could. So the nonconventionality is a matter of *potential* direct apprehendability, rather than *actual* direct apprehendability.

A minor technical difficulty with this Robbian approach also arises. Events that are light-connectible, but not particle-connectible, are still in the absolute 'after' relationship. But what sense does it make to talk of even a possible consciousness experiencing directly the "afterness" between two such light-connectible events, given the embodiment of consciousness in material bodies and the impossibility that such bodies, even in principle, could travel at the velocity of light? Some "limit of possible experience of a single consciousness" might salvage Robb's idea; or he could take as primitive, instead of 'after,' the notion of 'connectible by a genidentical *material* signal.' But I shall not pursue this here. It is interesting to note that in Woodhouse's formalization, which we shall later examine, continuous timelike paths are taken as primitive, but continuous light paths are not.[22]

Another item of philosophical interest arises if we imagine a plurality of consciousnesses, one for each possible timelike world-line, "checking up" on afternesses. Would an actual or potential *social* direct apprehendability do, or must we imagine—contrary not only to fact but to lawlike possiblility—a *single* "direct apprehender" to check up on which events are really, and nonconventionally, after which? Here the reader may want to reflect on the relation of this issue to the so-called problem of indistinguishable space-times. This shows that there can be space-times that no single observer could ever tell apart on the basis of his experience, even though we postulate them to have quite different structures. This is because of the causal limits on the information a single observer can receive from or communicate to other observers. If we allow "potential social awareness," an imagined plurality of observers who cannot, in fact, pool their information, but whose total knowledge *we* can imagine in toto, then such universes are "distinguishable" by the "assembled observational data." The notion of direct apprehendability, then, is far from clear. And even if it were clear, it is hard to see how one could get from the subjective realm of appearances to a *distinct* realm of "nonconventional" objective relationships.

The other approach goes like this: let us now work in terms of causal connectibility instead of in terms of "after." What makes causal connectibility "nonconventional" is the direct empirical test for it in terms of genidentical particles or light waves. Two events are causally connectible if and only if a possible genidentical signal can be sent from one to the other.

One difficulty with this approach is its "scientific unreality." Given that we really believe in quantum mechanics, and not in the classical pseudophysics which usually serves as the arena for foundational discussions in relativity, to what extent do we really want to pin down our foundational expositions to those that rely on such concepts as single genidentical particles and their paths, when, on the basis of our real physics, we are in some doubt about the very existence of such classical objects? But let this be for the moment.

A more direct philosophical criticism is this: we are trying to establish a "nonconventionality" for causal connectibility. To do this we are using the notion of being connectible by the path of a genidentical particle or light wave. But is it obvious that the question of which events are so connectible is a matter totally devoid of "conventional" elements?

Reichenbach, for example, considered the very topology of space-time as conventional as its metric, in the sense that we could choose, "conventionally," different topologies for space-time and still "save the phenomena" as long as we made sufficient changes elsewhere in our physical theory—in particular, changes in what we took to be paths of genidentical particles or light rays. One such choice would depend upon how we identified or "disidentified" events. This is what happens when we "conventionally" choose between a space-time and a different spacetime that is a covering space for it. But, Reichenbach thought, even the *local* topology was conventional, for we could "conventionally" select space-times with different "nearness" or "neighborhood" relations by deciding differentially on which paths were or were not continuous. Now Reichenbach thought that such "conventional" choices were only choices about "manners of speaking" or "descriptive simplicity"; but we have seen that other kinds of conventionalism could make use of the same arguments. For example, we could espouse the conventionalism that tells us that the Reichenbachian topological alternatives are *really* incompatible theories among which we could make no choice on observational grounds.[23]

251

So if we try to argue the nonconventionality of causal connectibility on the grounds of its "empirical testability" by genidentical particles, we must first of all make the usual philosophical moves of allowing possible genidentical particle connectedness to test causal connectibility, and, more important, have some further argument, *contra* Reichenbach, to the effect that connectibility by a genidentical particle was itself "nonconventional."

I shall argue later that there are some grounds for making this move. That is, I shall give some reasons for maintaining that genidentical paths are nonconventional. I shall base this view on arguments for defining genidentical paths in terms of continuous paths, rather than vice versa; and on arguments for taking one kind of topological nearness to be as primitive as directly observationally determinable—in a theory-independent way—as anything can be in a foundational analysis of general relativity. So I am not saying that causal connectibility isn't correctly designated 'primitive.' I am only saying that if one makes the decision to so consider it, one must argue for this decision in a complex philosophical way; one should not just assume it.

My general response to Winnie's argument from Robb's formalization to the nonconventionality of the metric in the theory of Minkowski space-time is similar. One can, perhaps, draw philosophical consequences from the existence of certain formalizations of scientific theories. But one can do so only after one has chosen the formalization. In doing so we are operating on the basis of numerous, complex, and sophisticated *philosophical* presuppositions. Even having chosen a formalization, further *philosophical* assumptions may be necessary to get the philosophical results out of the formalization. We can get philosophical conclusions out if we put philosophical assumptions in. These assumptions should be clearly laid on the table. And once they are, we see that they are rarely obvious or self-evident, but rather the end product of a complicated chain of philosophical reasoning.

In Winnie's particular example, two grand steps must be taken. First we must agree that in the theory of Minkowski space-time, the metric is truly definable, and not just "definable" in terms of causal connectibility. We have seen just how dubious this thesis is. Second, we must agree that causal connectibility itself is a nonconventional matter. While this may be true, it must be argued. And the arguments for it are neither obvious nor simple.

VII. Formalizations and General Relativistic Space-time

A. Introduction

So far we have been studying, primarily, the problem of the formalization of Minkowski space-time from a philosophical point of view. But we know that space-time is not Minkowskian. Why should we have gone to all this effort over a dead theory? This answer is clear: the apparatus we have developed will be of great use in trying to understand, philosophically, the numerous problems that arise when we try to formalize the space-time theory of general relativity, our currently most viable theory of space-time. It will also be helpful, I believe, in the general methodological problem of formalizations, even outside the area of space-time theories.

When we come to examine formalizations in general relativity, we should note initially a couple of interesting features of this case that distinguish it from formalization of Minkowski space-time. First, in formalizing a theory of Minkowski space-time, we are concerned with the theory of a single space-time model. In considering formalizations of the space-time theory of general relativity, we must remember that this one theory allows at least the lawlike possibility, relative to its laws, of many distinct space-time worlds. So when we wanted to put "stress" on the Minkowski space-time theory by looking at alternative space-times, we had to "go beyond the theory whose formalization was in question." But we can make progress in understanding the philosophical presuppositions of formalizations in general relativity, by considering the wide variety of space-times compatible with the theory. This is a variant on the old remark that in special relativity, as in the usual Newtonian theories, space-time is a "passive arena" of material events. But in general relativity, space-time is itself a variable dynamic element in the theory.

Second, when considering the philosophical aspects of formalizations in the theory of Minkowski space-time, we were able to avail ourselves not only of the theories of space-time which preceded the one we were primarily concerned with, but also of its successor theory, general relativity. We saw that our views about the adequacy and desirability of various formalizations could be radically affected by the fact that we had this newer theory of space-time in mind. Essentially we could ask how well a given formalization of Minkowski space-time "fit in" to the future developments in the theory of space-time whose nature we already knew. In examining the formalizations of general relativistic space-time theories, we are not in this enviable position. For while alternative theories to general relativity have certainly been espoused, and while some of these are "generalizations" of the general relativistic theory, in the sense that this theory is a generalization of Minkowski space-time theory, there is no more general theory of space-time beyond general relativity that is universally agreed to have plausibility either as a correct theory of space-time (general relativity being assumed incorrect), or even as the theory of space-time to which we *would* move were we to come to reject the general relativistic theory. We shall see that this is of some importance in our philosophical analysis.

B. From Rods and Clocks to Clocks: Synge's Chronometric Formalization of General Relativistic Space-time

Suppose we believe in general relativity. How should we go about determining just which of the many space-times compatible with the theory is the actual space-time of the world? And how should we determine the actual intervals between events within this space-time world?

The natural suggestion is to use infinitesimal rigid rods ("taut strings," if you wish) to measure spacelike separations in some observer's frame, and to use ideal clocks to measure timelike intervals. From these the general intervals between events can be determined, with enough such measurements performed on a sufficiently fine scale, we can determine the g-function at every point of the manifold and hence both the structure of the space-time and the particular intervals between particular events.

Now many physicists have been reluctant to accept rigid rods, even infinitesimally, as "primitives" in their formalizations of general relativity. Others wish to eschew as well the use of the notion of ideal clocks as a primitive. For the moment I want to avoid discussing their reasons for this self-denial. I shall, however, look closely at this issue in a later section.

Synge is reluctant to use infinitesimal rigid rods, but he is perfectly happy with clocks. And he has shown that using clocks alone as primitive we can formalize the theory.[24] How does he proceed? First, he takes as primitive a class of ideal clocks. Their rates are all linearly related to one another, in the sense that two of these clocks transported together from one event to another along the same timelike world-line will have their number of "ticks" elapsed between the two events in a constant numerical ratio.

Next he makes the following assumption: suppose x and $x + dx$ are two events connectible by a timelike world-line. Let ds be their interval separation. Then $ds = f(x, dx)$, where f is positive homogeneous of degree one in the differentials. Now it has been alleged that he offers no justification for making such an assumption.[25] But the justification which, I presume, he is presupposing is that offered by Riemann for assuming at least this much about the metric form in his original inaugural dissertation on Riemannian geometry.[26] First we must assume that the g-function will vary continuously in the coordinates to show that ds is a function only of x and dx. Next, if we assume that, ignoring magnitudes of second and higher order, the length of a line element remains unchanged when all its elements undergo the same infinitesimal change of place, then the homogeneity of f in the first order in the differentials follows.

But for general relativity we need much more than this. We need the usual *quadratic* form for ds as a function of the dx's. Synge assumes this form for the g-function, and here he does so with no further rationalization of the assumption. Once we have made this assumption, however, Synge shows how, by making a sufficient number of interval measurements between timelike connectible events, using his ideal clocks, we can calculate the g-function at any point. But if we have the g-function, we can determine the lightlike and spacelike separations between all point-events as well. Thus using ideal clocks as our only primitives, assuming them to measure timelike intervals correctly, and assuming, boldly, the full quadratic form of the g-function, we can determine the metric structure of our space-time fully, and we can determine the particular interval separations between all events, even those lightlike or spacelike separated.

C. Causal-Inertial Theories of the Metric

But there are those who will, for reasons we shall examine later, eschew *clocks* as primitives along with rods. What will they take as primitives? In a "rods and clocks" formalization, or in a "clocks alone" formalization, we shall need to supplement our initial postulates about the adequacy of the primitive devices as interval measurers, and our initial assumptions about the form of the g-function, by postulates that relate the behavior of light rays in a vacuum and free particles to the structure of space-time. Usually we shall simply assume that light rays travel null geodesics, and free particles, timelike geodesics. Perhaps we can get a different formalization of space-time theory by taking these entities as the primitives of our theory. Since light rays in a vacuum mark out the boundaries of the sets of causally connectible events, we shall be searching for a formalization in general relativity that is as close as we can come to Robb's formalization in Minkowski space-time theory.

Since we know that nonisometric pseudo-Riemannian space-times are mappable one-to-one, preserving all causal connectibility relations, we shall not attempt a formalization that takes light rays alone as primitive. This was Weyl's mistake, as Lorentz pointed out. But the following fact gives us reason to believe light rays and free particles together will be enough: any two pseudo-Riemannian space-times that are causally isomorphic under an isomorphism which takes timelike geodesics into timelike geodesics will be isometric, up to a constant factor. The fact that, in general relativity, free particles travel timelike geodesics (either because we introduce this as a separate postulate or because we show that they must so travel because of the nonlinearity of the field equations), suggests that light rays, marking out the causally connectible from the noncausally connectible, and free particles, marking out the timelike geodesics, will be enough. Actually, we must be careful here, as we shall later see. For by 'free' particle we mean more than is apparent at first glance. Free, infinitesimal, gravitational monopole would be more correct. But I shall deal with this later.

Several methods for measuring the space-time using light rays and free particles alone ahve been outlined. The interested reader should consult Marzke and Wheeler, and Kundt and Hoffman, for details.[27] The formalization I shall focus on, which is a generalization of the Robbian approach, is the extremely interesting and important one of Ehlers, Pirani, and Schild, which I shall call the EPS formalization.[28]

To understand the EPS formalization we shall first have to look at some of the mathematical aspects of general relativistic space-times. The full pseudo-Riemannian metric manifold of general relativity is a complex structure. From it we can abstract some simpler structures.

First of all, the manifold is a topological space. Second, it is a differential manifold. Next, these space-times have a *conformal structure*.

Infinitesimally this means the possibility of constructing the hyperplanes orthogonal to a given line element. Such a construction results in the infinitesimal null-cone structure, i.e., in the distinctions among timelike, null, and spacelike connected events, or, physically, in the causal connectibility structure on the space-time.

The space-times also have a *projective structure*. Infinitesimally this means that there is a well-defined notion of parallel-transporting, in its own direction, a vector from tangent space to tangent space. Globally this results in a geodesic structure on the space-time. For the geodesics are just those curves whose tangent vectors remain tangent vectors when they are parallel-transported in their own directions.

The space-times have an *affine structure*. Infinitesimally this means that there is a well-defined notion of parallel-transporting, in any direction, a vector from tangent space to tangent space. Globally this results in the definability of an affine parameter along the geodesics. Equidistant points along a geodesic (with respect to the affine parameter) are those points whose "connection vectors" are parallel.

Finally there is the full metric structure, which allows us to compare the intervals between arbitrary pairs of pairs of point events.

EPS assumes the notions of event, light ray, and particle. Particles are meant, intuitively, to be paths of "free" particles. Axioms are introduced concerning the smoothness of the particles and of messages sent between them by light rays. An axiom is introduced which allows us to coordinatize the space by particles and light rays, and it is assumed that any two such coordinatizations are smoothly related to each other. This lets us define a differential topology for the manifold of events. There are some problems here, but I shall discuss them later. Next, axioms on the light rays are introduced that are sufficient to allow us to define the usual light cone structure on the space-time, i.e., to define its conformal structure. Then axioms on the particles are introduced that are sufficient to let us define the projective structure of the space-time.

But we still don't have a full affine structure for the space-time, much less its full metric structure. EPS shows that only one additional axiom is necessary to guarantee the existence of the full affine structure. This is an axiom of compatibility between the conformal and projective structures already introduced. It states that the interiors of the light cones, definable by the conformal structure, are filled with the particles. That is, we can approach the light rays as closely as we like by paths of free particles in the interiors of the light cones. It follows that a full affine structure is definable on the space by the light rays and particles alone.

Finally, we can get the full metric structure in either of two ways. We can demand that if we have two nearby timelike geodesics, and if we mark off equal intervals along them relative to their affine parameters, the events marking off the equal intervals will be simultaneous (to the first order) using the "radar" method for establishing simultaneity familiar from special relativity. Alternatively we can define congruence for vectors that are located at the same point by using the conformal structure. We can define parallel transport of a vector along a curve as well, using the affine structure: if we take a vector and transport it parallel to itself to a new point by two different paths, then, if the two resulting vectors are congruent at this second point, a full pseudo-Riemannian metric structure on the space-time is well defined. Thus if we add either of these last two conditions as an axiom, we can fully define the metric structure of the space-time using light rays (causal connectibility) and particles (timelike geodesics) alone.

Ehlers, Pirani, and Schild claim a number of virtues for their formalization. First of all, one does not take as an axiom the extremely powerful assumption that the metric has its usual pseudo-Riemannian quadratic form. Instead one assumes a group of axioms each of which is, individually, a much weaker assumption. Of course together they are enough to guarantee the existence of the usual metric form. While historically our assumption about the metric form for space-time followed from the Pythagorean results about space established in terms of a metric usually thought of as defined by "rigid rods," these authors argue that the *generation* of this form from a number of far weaker assumptions, each testable locally in a very simple way, is a major advance.

Second, they believe their formalization is ideally suited to considering what theories might be plausible candidates as *generalizations* of general relativity. For by putting on the full metric structure a piece at a time, one can now easily generate theories more general than general relativity by asking what kind of space-time we get as the axioms are progressively rejected, presumably starting from the last imposed and working backwards.

Now let us look at the EPS formalization from a philosophical point of view. Fortunately we can be fairly brief, for we can avail ourselves of much of the apparatus introduced in our earlier discussion of Robb's formalization of Minkowski space-time. As usual there will be two basic areas of philosophical interest—the EPS choice of *primitives* and the philosophical arguments for and against this choice, and the EPS choice of *definitions* and the arguments for and against them.

It is clear that Robb chose the single primitive 'after' because he felt that facts about the after relationship between events were in principle immune to claims of conventionality. EPS makes no such claim. There it is only suggested that one is deriving the full structure of space-time from "some qualitative (incidence and differential-topological) properties of the phenomena of light propagation and free fall that are strongly suggested by experience."[29] But suppose someone were to make the claim that the construction of space-time in the manner of EPS is "convention free." Light rays play the part of boundaries of causally connectible sets of events. And we have already given some discussion to the claim that causal connectibility is "totally nonconventional." I shall continue that discussion later.

When we move on to the role particles play in the EPS formalism, we see that any hope of totally undercutting "conventionality" theses about the metric is unsupported by this formalization. The EPS formalism assumes that we know what a free particle is. And I suppose we do, given our vast array of background theory. But to support a "nonconventionality" thesis we would need reason to believe that we could determine when a particle was free totally independently of our theoretical assumptions. But how could we do this? Certainly not by seeing that it followed timelike geodesics! And if a particle is claimed to be free, we can always "conventionally" deny this by postulating "universal forces." After all, is a particle free or not when it is gravitationally attracted by another particle? If we say it is, and use the EPS construction, we shall get one space-time. But what if we say it is not? Won't we get a whole "conventionally alternative" space-time by using the EPS construction?

Again, even assuming that the particle is "free from forces," we still have difficulties. First of all, we need infinitesimal particles; but that is a standard idealization. Now the choice of free particles to pick out the timelike geodesics in the EPS formalism follows because, according to general relativity, free particles travel timelike geodesics. But not all free particles do—only those that are spherically symmetric and spin-free. For if the particle has a gravitational multipole structure, it will not generally travel the timelike geodesics, even if no forces act on it.[30] And how, without already *knowing* the space-time structure, are we to determine which particles are gravitational monopoles and which are not? There is no need here, I trust, to rehearse all the possible meanings of "conventional" and "conventionality," nor to look at our options once again. Enough has been said to indicate that, whatever its virtues—and they are real indeed— resorting to the EPS method to determine the metric, or to any other method that utilizes free particles as well as causal connectibility, will not release us from any burden of conventionality imposed by the use of transported rods or clocks.

There may be other virtues, however, in choosing light rays and free particles instead of rods and clocks as our primitive instruments. Let us discuss one aspect of this claim here. What about the EPS definitions? Concerning the rods and clocks or chronometric formalisms, EPS argues like this: "If the g-function is defined by means of the chronometric hypothesis, it seems not at all compelling—if we disregard our knowledge of the full theory and try and construct it from scratch—that these chronometric coefficients should determine the behavior of freely falling particles and light rays, too. Thus the geodesic hypotheses, which are introduced as additional axioms in the chronometric approach, are hardly intelligible; they fall from heaven. . . ."[31] But of course the authors are aware that on their construction one must somehow stipulate the connection between intervals as they define them, and the behavior of rods and clocks.

The authors say that, first, one can define a "clock" by means of light rays and particles alone. Thus: "The chronometric axiom then appears either as redundant or, if the term 'clock' is interpreted as 'atomic clock,' as a link between macroscopic gravitation theory and atomic physics: it claims the equality of gravitational and atomic time. It may be better to test this equality experimentally or to derive it eventually from a theory that embraces both gravitational and atomic phenomena, than to postulate it as an axiom."

But we can test experimentally the correlation between atomic clocks and clocks constructed with light rays and particles, whether we choose the chronometric or the EPS formalization. And if our eventual overall theory allows us to derive their synchronization in one direction, it allows us to derive it in the other. Why then is the EPS method better than the chronometric?

Perhaps the authors reason as follows: if the nice agreement between atomic and light ray-particle clocks holds up forever, then it will not really matter too much which formalization we choose. But we must look forward to what we would do if we ever discovered the synchronization to break down. The behavior of light rays and free particles in space-time is an integral part of our theory of space-time. Their behavior follows from the most fundamental principles of our space-time theory. But the explanation of behavior of atomic clocks depends upon theories, like quantum theory, which are not—at least not yet—so intimately related to our theory of space-time. Should the synchronization break down, wouldn't we be more likely to look for an explanation in some aspect of the quantum behavior of atoms, than we would be to drop our fundamental postulates of space-time theory? Wouldn't we, in other words, retain the belief that free particles and light rays travel timelike and null geodesics respectively, and the view that time intervals as measured by them are correctly measured intervals of the space-time, blaming the discrepant results obtained using the atomic clock on some peculiarity of its makeup?

I don't know just what we would do. We might gain some insight by looking at what did happen when a generalization of general relativity was attempted but was rejected on empirical grounds. This was Weyl's affine but non-Riemannian space-time. This case is discussed by Ehlers, Pirani, and Schild themselves; they consider the plausibility of the additional axiom needed to go from affine to Riemannian space-time. The authors invoke Einstein's criticism of Weyl's non-Riemannian theory.[32] The trouble with this theory was that it proposed that a vector could change length upon being transported around a closed curve. *If we assume that atomic clocks measure time intervals*, atoms should radiate at different wavelengths depending upon their histories; but they do not. This fact led physics to reject Weyl's theory. No one to my knowledge tried to save Weyl by arguing that the conformity of atomic clocks to each other was due to the fact that in traversing the closed paths they ceased measuring time properly in just such a way as to be in agreement with each other despite their journey. Nor did anyone propose "checking up" on the atomic clocks by using the pure geodesic method.

I am suggesting that, although our explanation for the periodicity of atoms may be more complex in our ultimate theory than would be our explanation for the periodicity of a particle-light ray clock, what we *mean* by 'time interval' may be more closely associated with such atomic clocks

261

than with "gravitational" clocks. The same argument applies regarding spatial separations and rigid rods as opposed to a "gravitational" measure of spatial interval using only light rays and particles.

Suppose we discovered in the future that rods we took to be rigid, and clocks of the atomic sort we took to be ideal, no longer measured intervals as defined by the particle and light ray methods of EPS, Marzke, and Kundt and Hoffman. What would we say? EPS et al would have us say that we had discovered that "rigid rods" and "ideal clocks" are not very good indicators of spatial and temporal separations. From the alternative point of view, we would say that while we thought light rays and free particles traveled null and timelike geodesics respectively, they do not.

Let us consider the second position first. After all, it might be argued, just look at the history of general relativity. In Newtonian theory it never occurred to anyone, except retrospectively, to say that particles acted upon by gravity traveled geodesics in space-time. They were "forced," and hence deviated from straight-line (geodesic) motion. Now it is true that Einstein saw that one could "geometrize" gravitation because of the principle of equivalence. But wasn't his real reason for going to curved space-time the plausibility arguments, from the red shift owing to gravity, etc., that gravity must affect our interval measurements as normally carried out by rods and clocks? If we see that space-time as measured with rods and clocks does not coincide with space-time as measured by the geodesic hypothesis for light ray and particle motion, why not just drop the geodesic hypothesis? Wouldn't that be either the more plausible scientific decision—if you think there is a real decision to be made here—or, in any case, the better way of talking, if you think that all that is involved is descriptive simplicity?

But I think the reply would go like this: in our best available current theory—general relativity, the theory we are trying to formalize—the geodesic motion of light rays and free particles is a fundamental result of the theory. We cannot even separate out the field equations from the equations of motion of test objects in the field as we could in the Newtonian theory, for the very equations of motion follow from the field equations. On the other hand, the behavior of rods and clocks is a complex matter whose explanation requires not just our gravitational-geometric theory, but a quantum theory of matter as well. Now "every theory should determine its own interpretation," in the sense that the "definitions" connecting theoretical structures to observable elements in formalizations of the theory should be fundamental propositions that follow in the unformalized theory from that theory alone. On this ground, the geodesic "definition" of the metric is preferable. Admittedly one might want to say that in adopting these definitions we have "changed the meaning of the metric terms." For, whereas the previous "fixed" propositions, invariant under theoretical change, were those connecting rods and clocks to intervals, we are now taking the "fixed" propositions to be those connecting the geometry to motions of light rays and free particles.

But we may have good reason for making this "change of meaning," if, indeed, that is what it is. We saw earlier that we could criticize Robb's definitions of the metrical quantities by seeing how the propositions connecting causal connectibility to congruence held up when we moved from Minkowski space-time to the space-time of general relativity. We saw that the Robbian connections were anything but "fixed point" propositions in this theoretical change. Now in critically examining a proposal to take some propositions as "definitional" in a formalization of general relativity, we have no accepted "newer, more general" theory to use as a standard. But perhaps choosing a formalization is a *proposal* about which broader theories to consider. In other words, to adopt the causal-inertial, or geodesic, definitions of the metrical quantities is to suggest that the future theories we should be considering as plausible candidates for replacing general relativity, need we do so, would be ones in which the association of light rays and free particles with geodesics is retained, whereas the association of rods and clocks with the space-time intervals is allowed to differ from the association postulated in general relativity.

We might say, then, that whereas in going from Newtonian theory to general relativity we kept the association of rods and clocks with intervals constant, and changed our views about the association of light rays and "free" (i.e., acted upon only by gravity) particles with geodesics; in our new theoretical shifts we are likely to keep the geodesic motion of light rays and free particles intact and allow changes in the propositions which associate rods and clocks with intervals.

I think these arguments are of great interest. But I also think that they are not overwhelmingly persuasive. For one thing, the allegation that the geodesic motion of light rays and free particles "follows from general relativity alone," but that the behavior of rods and clocks requires quantum theory for its explanation, is a little misleading. To be sure, if we want to account for the fact that a particular atomic clock, say, "ticks" with a particular frequency, or the fact that a particular ensemble of atoms in a crystalline array has a particular length, we must invoke our complex quantum theory of matter. But what we need for the association of clocks and rods with space-time intervals is not this full theory, but only a few fundamental assumptions about how clock rates, for example, *vary* with the varying gravitational potential. And the rationale for these assumptions does not use the complex quantum nature of these material entities at all. After all, the famous Einsteinian rationale for assuming that the gravitational field will have metric effects works by demonstrating that on the basis of very general considerations drawn from the equivalence principle, ideal clocks must respond, in the manner of the gravitational red shift, to changes in gravitational potential. These arguments can be classical in nature, and certainly do not depend on any detailed theory of matter we have in mind. The same holds for the usual arguments given to rationalize the belief that spatial measuring devices must also respond to the gravitational potential.[33] Now these arguments rest upon very fundamental assumptions of general relativity, perhaps even more fundamental than the field equations. So even if a theory should "provide its own interpretation," is it clear that this in any way motivates a preference for the geodesic definitions of the metric over the rod and clock or chronometric definitions?

Of course it remains true that if we go beyond general relativity to some newer theories, say by going to an affine but non-Riemannian space-time, we shall still be able to retain the geodesic hypotheses intact, but shall no longer have any clear place for the chronometric. But since we do not know where we *shall* go from general relativity, is it clear that the EPS type formalisms will provide in their "definitions" the real "fixed point" propositions in our future scientific changes, and that the chronometric formalizations will not? The most we can say is that EPS formalizations are one *proposal* for a formalization best suited for the future evolution of science. If we adopt them we are implicitly assuming that in any future change we shall take the association of light ray and free particle motion with the geodesic structure of space-time to remain correct; or, if we think the geodesic hypotheses are only "matters of convention," that we shall retain these conventional choices or retain this "manner of speaking." But it really is not clear, and cannot be until we have our future science, that we shall not instead hold fast to the propositions associating rods and clocks with intervals as the invariant "truths" or "conventions" or "ways of speaking."

Whatever one thinks of the EPS, as opposed to the chronometric, method of formalizing general relativity, it is still useful to contrast the arguments given above in favor of the EPS formalization with some arguments we gave earlier in favor of Robb's "definitions" of the metric. For the rationales differ in a very important way. The primary argument in favor of Robb's "definition" of the metric—and this was Robb's argument— is that it was made in completely "nonconventional" terms, using, as it did, only the allegedly totally nonconventional notion of causal connectibility as a primitive. Now, as we have seen, the causal-inertial definition of the metric throws away this advantage altogether. The metric so defined is at least as "conventional" as a metric established using transported rods and clocks.

Rather, the argument for the causal-inertial method is a "scientific" one. This definition is more "natural" and better suited for anticipated future theorizing than is the chronometric method. Whether or not we believe there are good "scientific" reasons for preferring a causal-inertial formalization of general relativity to a rods-and-clocks or to a chronometric formalization, we should clearly realize that the grounds for preference being alleged here are of a very different nature from the "philosophical" grounds offered by Robb and by others in favor of his causal theory of the metric.

D. Causal Theories of Topology

I have frequently promised the reader that I would return to some crucial questions about the relationship of causal connectibility and topology. If, as we have seen, a causal theory of the metric is impossible in general relativity, and correct only in a very weak sense given in the case of Minkowski space-time, is it still possible to hold to a causal theory of the topology of space-time in any interesting sense? And if, as we have seen, there is no escaping accusations of conventionality with regard to the metric of space-time, is there any interesting sense in which we can at least claim that the topology of space-time is in no way a matter of convention?

The Robbian association of causally defined features with metric features breaks down once we leave Minkowski space-time. But what about the coextensiveness, in Minkowski space-time, of the causally defined Alexandroff topology with the manifold topology? Does that association hold up in all space-times compatible with general relativity? If so, does this support the plausibility of a causal theory of the topology of space-time?

It is now known that the Alexandroff topology will not, in fact, coincide with the usual manifold topology in all those space-times that satisfy the usual requirements imposed on space-times by general relativity. One can be sure that the two "topologies" will coincide only if the *strong causality condition* holds in the space-time. The strong causality condition holds in a space-time if for every point p of the space-time and for every neighborhood of p, there is a neighborhood of p contained in the given neighborhood which no nonspacelike curve intersects more than once. Strong causality holds if the space-time has no "almost closed" causal (lightlike or timelike) curves. If strong causality is violated, one can construct space-times in which the Alexandroff and the manifold topologies will not be identical.[34]

One interesting question is: should we, as a matter of principle, exclude from the realm of reasonable physical possibilities space-times with almost closed causal curves; or should we, rather, admit them as real physical possibilities? For our purposes the following question is more important: do causally pathological space-times that have almost closed causal curves block the possibility of a philosophically satisfactory causal theory of topology?

An affirmative answer to the second question is argued in this way: if a causal theory of topology is correct, we should be able to determine the topology of the space-time using causal connectibility alone. But if we do this in the plausible way by taking the causally defined topology to be the Alexandroff topology, then if there are almost closed causal curves in the space-time, the topology we attribute to it on the causal basis will not be the usual manifold topology. And isn't this latter what we wished to causally define?

As David Malament has pointed out to me, the existence of closed causal curves, or even of any two distinct events x and y that cause each other, not only forces a discrepancy between the Alexandroff and the manifold topology, but blocks any plausible attempt at "causally defining" the manifold topology. For if the manifold topology is defined by causal relationships it must be preserved under any causal automorphism of the

266

space-time onto itself; i.e., all such causal automorphisms must be homeomorphisms with respect to the manifold topology. But consider the mapping which takes x into y and y into x and leaves all other points the same. By the transitivity of causation it will be a causal automorphism; but it will certainly not be continous in the manifold topology. The hopelessness of a program of causal definability in general is made even more manifest when we reflect on the fact that there are space-times in which every event is causally connected to every other event and to itself.

On the other hand, it is interesting to note that we can, by causal observation alone, determine whether or not the Alexandroff topology does in fact coincide with the manifold topology.

The causal theory of the topology seems to be foundering on the shoals of space-times with almost closed causal curves, just as the causal theory of the metric foundered on the rocks of space-times that were not Minkowskian. Now in our critique of the causal theory of the metric in Minkowski space-times we had to look at non-Minkowskian space-times. But since space-times with almost closed causal curves are themselves compatible with general relativity, we can attack the causal theory of topology within general relativity by using its own space-time models.

In discussing the causal theory of the metric in Minkowski space-time we showed that even if space-time were believed by us to be Minkowskian, we would be skeptical of any causal theory of the metric that went beyond the mere de facto coextensionality of some causally definable notions and the metric notions and asserted some lawlike, analytic, or necessary connection between the notions. A similar line of reasoning here would lead us to assert, I think, that even if space-time had no almost closed causal curves, the coincidence of the causal Alexandroff topology with the topology of the space-time would just be a "matter of fact," and not a lawlike, analytic, or necessary truth.

We can see some of these points "in action" if we look at Woodhouse's interesting program for formalizing the causal and differentiable structure of space-time.[35] He wishes to develop the structure of space-time a bit at a time on the basis of axioms that have "simple and intuitively obvious physical interpretations." Further, he wishes to put the topology and differential structure on first since this will provide us with a method for characterizing singularities in the space-time, the notion of singularity being essentially nonmetric, without presupposing the whole metric structure.

Woodhouse wants to consider only space-times that have no almost closed causal curves. How does he do this? He introduces as primitive the set of events and *particles*. The particles, once again corresponding intuitively to possible paths of free particles, are assumed to be one-dimensional subsets of M, the set of events that have a continuous structure and are homeomorphic to the real line. So closed timelike lines are excluded ab initio. Using the topology of the particles, built into the primitives, it is possible to introduce an axiom equivalent to assuming the space-time to be free of almost closed causal curves. Having done this, one can introduce the Alexandroff topology and take it as the topology of the space-time. The topology so defined induces the already assumed topology on the particles, and as we shall see later, has other "virtues" as well. The important thing to note here is the necessity to first assume some topological property as a primitive in one's formalization before a "causal" introduction of the full topology even becomes possible.

Even if the possibility of space-times with almost closed causal curves had not occurred to us, I think we could still have some reason to be cautious in accepting a causal theory of topology. In considering earlier definitions of the metric we asked the question whether the definitions in a particular formalization could be considered right or wrong. If they could—if there were some criterion of rightness or wrongness for some "definitions"—we saw that one might hesitate to call these "definitions" really definitions. For it seemed that we had some check on the correctness of the correlations asserted by the "definitions," and this suggested either that these "definitions" were *factual* propositions, or at least that we had some other notion of the *meaning* of the defined expression in mind. Thus if we can say that X is or is not *correctly* defined by Y, it would seem that we already had some notion of what X meant.

We also saw that in looking at causal "definitions" of the topology even in the case of Minkowski space-time, more than one such causally definable "topology" would be constructed. It then became a question of which "topology" was *the* topology of the space-time. Once again, what criteria could we use to judge this if our only notion of a topology was that which the causal "definition" introduced?

In the case of causal topologies for Minkowski space-time, for example, Zeeman was careful to indicate that his causal topology, although it differed from the usual Alexandroff—and hence from the usual manifold—topology, did, like the more standard topology, induce the usual three-dimensional Euclidean topology on spacelike hypersurfaces and the *usual one-dimensional real line topology on time axes*. If a causal topology did not do that, would we even consider it as the "right" topology for Minkowski space-time?

Woodhouse also takes as a check on the "correctness" of the Alexandroff topology the fact that it induces back on the particles the initial topology presupposed for them, that of the real line. Further he shows this: if a sequence of events, x_n, converges to some given event, x, in the Alexandroff topology, and if p is a particle through x, then the time interval (on p) needed to travel from p to x_n and then back to p goes to zero as n goes to infinity. So, he says, there is "an obvious physical interpretation of the convergence."

Perhaps the following is suggested by these points: we may not really have any a priori idea of what the full topology of space-time should be. Witness the difficulty of trying to decide whether the correct topology for Minkowski space-time should be the Alexandroff or the Zeeman. But we do believe we have a *primitive* idea of what a continuous particle motion, and hence a continuous timelike line, is. Whatever topology we impose, causally or otherwise, on the space-time, we can judge it to be correct or incorrect depending on whether or not it induces on the particle paths the continuous topology we had in mind in the first place.

This suggests that if anything is a "nonconventional" fact about the space-time, it is whether a given timelike connected set of events is continuous or not. So at least *that* topological notion should be a primitive in any formalization. And, of course, once we have this, the notion of a genidentical particle can be introduced by *defining* a genidentical particle to be a continuous timelike world-line. Thus while a causal theory of topology seems ruled out, at least if we are going to allow ourselves to countenance space-times with almost closed causal curves as *intelligible* (we need not believe them physically possible for the argument against the causal definition of topology to apply as an argument against the causal "definitions" being analytic or necessary, at least), perhaps some version of a topological theory of causal connectedness is in order.

What I am suggesting is that in any full formalization of a space-time theory the notion of a continuous particle path will be assumed as a primitive. Like facts about the incidence or nonincidence of events, facts about the continuity of particle paths will be, in Reichenbach's terminology, "elementary facts." Insofar as we take aspects of our theory to be conventional, in any of the senses of that much abused term, incidence and continuity of particles will be "nonconventional" consequences of a theory.

I do not know what makes a term primitive. I have already expressed my doubts that "direct apprehensibility" will be the criterion, because, again following Reichenbach, this might force us to take our primitives as "items of subjective awareness" and lock us forever behind the veil of perception. All I am suggesting here is that if some notions are primitive, at least one topological notion, that of continuity of a possible particle path, should be one of them.[36]

VIII. Ontological Questions

The questions discussed in this paper have been primarily epistemological and semantic: how do we know what space-time is like? To what extent is this a matter of convention? How do the terms of the theory of space-time get their meanings, and what are these meanings? To what extent can alleged definitions of terms be criticized and judged right or wrong? Let me make a few brief ontological remarks before I conclude.

Suppose we accepted some definition in a space-time formalization, taking it that to be an "*X*" meant to be a "*Y*." What would this say about the reality of "*X*'s" or their "ontological reducibility" to "*Y*'s"? Suppose, for example, we define 'timelike geodesic' as 'path of a free particle.' What does this say about the reality of timelike geodesics, or about the reducibility *ontologically* of space-time entities to material particles? Notice that in most formalizations in which such a definition would appear, it is usually taken for granted that the world is full of particles; essentially we act as though every timelike geodesic was actually traversed by some *actual* free particle.

But, of course, in the real world we do not believe this. What we really believe, if we accept this definition, is that every timelike geodesic is the path of some *possible* free particle. So instead of saying "*X* is a timelike geodesic if it is the set of events occupied by the history of some free particle," we should really say that "*X* is a timelike geodesic if it is the set of space-time locations that *could* be the locations of the events constituting the history of some free particle."

Unless we want a metaphysics replete with "permanent possibilities of location," ungrounded on any actual locations that might be occupied by events but just are not, the natural way to understand this is in a substantivalist, not a relationist, ontology of space-time. Just because we believe that the meaning of 'Y' is definable in terms of 'X,' where Y's are space-time entities and X's material events, and just because we believe that our only epistemic access to Y's is through X's, we are not committing ourselves to relationism and eschewing a realistic and substantivalistic theory of the metaphysics of space-time.

IX. Facts, Conventions, and Assumptions in Theorizing and Formalizing

I suppose all but the most adamant a priorist would admit that we must rely upon *observational facts* in deciding, on a scientific basis, what the space-time of the world is like. Insofar as we at least *seem* to theorize beyond the realm of pure observational testability, we shall be open to the suggestion that we must make conventional assumptions as well, although, of course, dispute will arise over whether or not these conventions are merely "trivial semantic conventions" or whether they are "real choices unconstrained by observational data"; and over whether or not we have grounds beyond the observational data for making the choices in a rational rather than merely conventional way.

Many people would now admit, I believe, that more is needed as well. For, they would agree, before our decision-making apparatus can even begin to make use of the factual and conventional material we feed it, we must add a number of *theoretical assumptions* about just what theoretical options we consider open as real possibilities for being a correct theory of the world.

My aim in this paper has been to argue that the choice of formalization of a theory is replete with scientific and philosophical consequences. For this reason, we must utilize our full resources in determining facts by observing, making conventional choices, and acting within a framework of assumed theoretical options in order to choose a formalization that is rational on scientific and philosophical grounds.

Our characterization of terms as primitive or defined, and our characterization of consequences as definitional or postulational, are decisions that implicitly reveal our beliefs about the limits of observational testability of theories and about the ability of theories to outrun these limits in their content; our ideas about the meanings of theoretical terms, how they are fixed, and how they change; and our views about the place of the theory formalized both in the historical context of the theories which preceded it and from which it evolved and in the assumed future science which, we anticipate, will perhaps evolve from it under pressure of new observation and new theorizing.

Given that so much hinges upon the formalization we do choose, we should not be surprised to find that this choice is no trivial matter, but one that requires the full utilization of our best available scientific and philosophical methodology. If formalizing is to be more than mere "logicifying," it *is* theorizing, and demands, if it is to be done adequately, the full resources needed for theorizing in general.

Notes

1. See, for example, P. Suppes, *Introduction to Logic* (Princeton: van Nostrand, 1957), chap. 8.

2. A. Robb, *A Theory of Time and Space* (Cambridge: Cambridge University Press, 1914). A second edition published under the title *Geometry of Time and Space* appeared in 1936. An abbreviated treatment is A. Robb, *The Absolute Relations of Time and Space* (Cambridge: Cambridge University Press, 1921). An outline of Robb's work is contained in J. Winnie, "Space-time Congruence: the Causal Theory of Space-time," published in this volume.

3. R. Latzer, "Nondirected Light Signals and the Structure of Time," *Synthèse* 24 (1972): 236–280, sections I and II.

4. J. Winnie, "Space-time Congruence."

5. See, for example, S. Hawking and G. Ellis, *The Large Scale Structure of Spacetime* (Cambridge: Cambridge University Press, 1973), pp. 196–197.

6. H. Weyl, *Space, Time and Matter*, 4th ed. (New York: Dover, 1950). See pp. 228–229 and Appendix I, pp. 313–314. See also Hawking and Ellis, *Large Scale Structure*, pp. 56–65.

7. The "up to a constant factor" qualification here is worth a moment's notice. The adequacy of using light rays and free particles alone is usually justified by arguing that to fix the constant factor is, after all, just to choose a scale of measurement. In one sense this is surely right. Even using rods and clocks to determine a metric we need, to get actual interval values, an assignment of a specific length to a particular spatial or temporal interval—and not just congruence by transported rods and clocks. So given our light rays and free particle paths, the full metric, including the actual intervals between events, is fully specified if we once pick two events and designate their interval separation as unit. Sometimes, however, people go beyond this to claim that the notion of two distinct possible worlds which differ only in their relative scale is an absurdity. I do not believe that this follows from anything said above in this note. This is the question not whether it makes sense to talk about the universe "doubling in size overnight," but whether it makes sense to talk about a universe in which all intervals have always been twice what they actually are. I shall not pursue this question here, but only remark that the answer is, I think, nontrivial and depends greatly upon just what possible worlds are in one's metaphysics and semantics.

8. E. Zeeman, "Causality Implies the Lorentz Group," *Journal of Mathematical Physics*, 5 (1964): 490–493.

9. E. Kronheimer and R. Penrose, "On the Structure of Causal Spaces," *Proceedings of the Cambridge Philosophical Society* 63 (1967): 481–501. See p. 483.

10. See J. Ehlers, F. Pirani, and A. Schild, "The Geometry of Free Fall and Light Propagation," in L. O'Raifeartaigh, ed., *General Relativity* (Oxford: Clarendon Press, 1972), pp. 63–84, esp. pp. 68–69. The original result is in H. Weyl, *Mathematische Analyse des Raumproblems*, (Berlin: Springer, 1923), esp. Lec. 3.

11. Robb, *Absolute Relations*, Appendix, pp. 78–80.

12. A. Trautman, "Comparisons of Newtonian and Relativistic Theories of Space-Time," in B. Hoffman, ed., *Perspectives in Geometry and Relativity* (Bloomington: Indiana Univ. Press, 1966), pp. 413–425. See also C. Misner, K. Thorne, and J. Wheeler, *Gravitation* (San Francisco: Freeman, 1973), chap. 12, "Newtonian Gravity in the Language of Curved Spacetime."

13. See H. Reichenbach, *The Philosophy of Space and Time* (New York: Dover, 1958), chap. 1. See also L. Sklar, *Space, Time and Spacetime* (Los Angeles: University of California Press, 1974), chap. 2.

14. Robb, *Geometry*, Introduction. H. Bergson, *Duration and Simultaneity* (Indianapolis: Library of Liberal Arts, 1965), chap. 3.

15. H. Reichenbach, *Axiomatization of the Theory of Relativity* (Berkeley: Univ. of California Press, 1969), Introduction.

16. For further treatment of the matters which follow immediately see Sklar, *Space, Time, and Spacetime*, pp. 119–146.

17. See L. Sklar, "Methodological Conservatism," *Philosophical Review* 84 (1975): 374–400.

18. D. Malament, private communication. These results are proved in Malament's forthcoming doctoral dissertation for the Rockefeller University. Also discussed in this thesis are a number of distinct notions of "definability" slurred over, I hope without philosophical loss, in this paper.

19. E. Zeeman, "The Topology of Minkowski Space," *Topology* 6 (1967): 161–170. I am indebted to David Malament for pointing out to me the relevance of this work in the present context.

20. For a discussion of the philosophical issues to follow in a slightly different scientific context see Sklar, *Space, Time and Spacetime*, pp. 318–343.

21. Sklar, *Space, Time and Spacetime*, pp. 109–112.

22. N. Woodhouse, "The Differentiable and Causal Structures of Space-Time," *Journal of Mathematical Physics* 14 (1973): 495–501.

23. Reichenbach, *Philosophy of Space and Time*, section 12.

24. J. Synge, *Relativity: the General Theory* (Amsterdam: North-Holland, 1960), chap. 3, esp. sections 1 and 2.

25. A. Grünbaum, "Geometrodynamics and Ontology," *Journal of Philosophy* 70 (1973): 775–800. See p. 780. I am indebted to this article by Grünbaum for suggesting to me many of the crucial questions discussed in this paper.

26. B. Riemann, "On the Hypotheses Which Lie at the Foundations of Geometry," in D. Smith, *A Source Book in Mathematics* (New York: McGraw-Hill, 1929), pp. 411–425. The relevant passages are on p. 416.

27. R. Marzke and J. Wheeler, "Gravitation as Geometry—I: The Geometry of Spacetime and the Geometrodynamical Standard Meter," in H. Chiu and W. Hoffman, eds., *Gravitation and Relativity* (New York: W. A. Benjamin, 1964). W. Kundt and B. Hoffman, "Determination of Gravitational Standard Time," in *Recent Developments in General Relativity* (New York: Pergamon, 1962).

28. Ehlers, Pirani, and Schild, "The Geometry of Free Fall," pp. 63–84.

29. Ehlers, Pirani, and Schild, *ibid.*, p. 65.

30. See Fock, *The Theory of Space Time and Gravitation* (New York: Pergamon, 1959), esp. pp. 371–374. See also B. Tulczyjew and W. Tulczyjew, "On Multipole Formalisms in General Relativity," in *Recent Developments in General Relativity*. See also Grünbaum, "Geometrodynamics," pp. 789–791.

31. Ehlers, Pirani, and Schild, "The Geometry of Free Fall," p. 64.

32. *Ibid.*, p. 82.

33. A very good presentation of these arguments is by Schild himself, as a matter of fact. See Schild, "Gravitational Theories of the Whitehead Type," *Proceedings of the School of Physics "Enrico Fermi," Course XX, Evidence for Gravitational Theories* (New York:

273

Academic Press, 1962), pp. 69–115. The relevant pages are 110–115. On the general problem of the alleged superiority of one method of formalizing general relativity over another see the remarks of J. Stachel in his review of the O'Raifeartaigh volume in which the Ehler, Pirani, and Schild article appears—*Science* 180 (1973): 292–293.

34. Hawking and Ellis, *Large Scale Structure*, p. 192, pp. 196–197.

35. Woodhouse, "Differentiable and Causal Structures," pp. 496–497.

36. David Malament has shown (Ph.D. Thesis, Rockefeller University and his forthcoming "The Class of Continuous Timelike Curves Determines the Topology of Spacetime") that we need assume only past- and future-distinguishingness in order to be sure that the manifold topology is at least implicitly causally definable. Of course if the space-time is not strongly causal, the Alexandroff topology and the manifold topology will still not coincide. (For relevant definitions see Hawking and Ellis, *Large Scale Structure*, p. 192.) Malament has also shown that the class of continuous causal curves "defines" the topology in the sense that given two space-times equipped with the usual manifold topologies, if there is a one-to-one function between them taking continuous causal curves into continuous causal curves, this function will also be a homeomorphism. This result holds generally in space-times. Again we see that while causal connectibility will not, in general, "define" the topology in even the weakest possible sense, the notion of continuity along causal paths will come closer to doing the job. But even in this case much caution is called for if one is to decide in just what senses even the class of continuous causal curves does and does not "define" the topology. For more on these matters see L. Sklar, "What Might be Right About the Causal Theory of Time," forthcoming in *Synthèse*.

Geometry and Observables

These remarks are intentionally brief and semioutline in form. The opinions expressed are expected to be controversial, at least in part.

1. The Concept of Observable

The notion of observable is to be understood within both classical (i.e., nonquantum) and quantum physical theories. As long as the dynamical laws obey a strict and formal Cauchy principle—i.e., as long as Cauchy data at one time are sufficient to predict (or retrodict) unequivocally (expectation) values of all dynamical variables—the notions of dynamical variable and observable are coincident. It is assumed, of course, that within the conceptual framework of the theory, predicted quantities are in principle observable as well.

The data given usually provide information concerning not only the overall (hypothesized) physical situation, but also the (Galilean or Lorentz) frame of reference. By implication, instruments used for observation can be placed within this frame. This remark is nontrivial in that it makes sense only for invariance groups that are Lie groups (finite-dimensional groups) or, more generally, groups that permit the fixation of the frame of reference once and for all by means of information provided on a Cauchy surface (at one time). Groups such as the Bondi-Metzner-Sachs group, although not Lie groups, are included.

With respect to so-called gauge groups, the situation is fundamentally different. For the purpose of this discussion I define a gauge group as a group that involves arbitrary functions of the time (ordinarily as well as of the spatial) coordinates, so that no Cauchy data given at one time provide sufficient information to fix the dynamical variables at any other time a finite distance away. For instance, with ordinary gauge transformations of the second kind, the value of an electromagnetic potential off the Cauchy

NOTE: Prepared for the conference "Absolute and Relational Theories of Space and Space-Time," Andover Center of Boston University, June 1974.

hypersurface is in principle unpredictable, although the value of a field strength is not, nor is the value of an integral of the form $\oint \mathbf{A} \cdot \mathbf{dx}$ (Bohm-Aharonov).

If some of the dynamical variables are not predictable from Cauchy data, one might conclude that such a theory is noncausal. This conclusion appears unpalatable because (a) some dynamical variables remain predictable (all those that are gauge-invariant), and (b) the imposition of gauge conditions, which by assumption do not modify the physical characteristics of the theory, render it formally causal with respect to all dynamical variables. A way out, and the one that I have adopted, is to say that in a theory with a gauge group no Cauchy data fix the frame of reference (the gauge frame), but that in those theories that we are concerned with, Cauchy data do fix the physical situation. This formulation implies that only gauge-invariant dynamical variables are physically significant; the inference is that only gauge-invariant quantities are susceptible to observation and measurement by physical instruments. This conclusion is warranted if all physical interactions, including those with physical instruments, are necessarily gauge-invariant. Admittedly, this point is susceptible to further exploration and discussion; it will be adopted for what is to follow.

If the dynamical laws of a gauge-invariant theory are to be obtained from an action principle, then the action itself must be gauge-invariant, at least with respect to gauge transformations confined to the interior of the domain that supports the action functional. If the dynamics is local, i.e., if the action functional is an integral over a local integrand, then the generators of infinitesimal gauge transformations vanish (weakly, not identically, but modulo the dynamical laws) or equal (again weakly) exact divergences. This is a consequence of Noether's theorem, which states that the generator of an infinitesimal invariant canonical transformation is a constant of the motion.

Whether or not the gauge group is Abelian, the generators of its infinitesimal elements must be first-class in the sense of Dirac, just so as to mirror the assumed group properties. That is to say, in its Hamiltonian version a gauge-invariant theory involves first-class constraints, which generate the infinitesimal gauge group. These first-class constraints will appear in the Hamiltonian, with arbitrary coefficients, which formally reflect the lack of uniqueness in the prediction of variables that are not gauge-invariant, in spite of the fact that in their canonical version the dynamical laws relate Cauchy data (the canonical variables) on one Cauchy hypersurface to those on another Cauchy hypersurface once a Hamiltonian has been fixed.

In a quantum theory one may visualize a linear vector space that permits the formulation of state vectors corresponding to both physical and nonphysical states. If the constraints are considered to be operators satisfying commutation relations akin to those of the infinitesimal gauge group, then those state vectors satisfying the constraints form a subspace. Only gauge-invariant variables map the subspace of physical states on itself. If a Hilbert metric, and hence self-adjointness and expectation values, are defined only on that subspace, and for gauge-invariant variables, then the resulting physical implications of that theory do not depend on the accident of a particular choice of (non-gauge-invariant) dynamical variables, nor will the physical statements of that formalism be affected by the adoption of gauge conditions. For all these reasons I reserve the term *observables* for gauge-invariant dynamical variables.

2. Application to Coordinate Invariance

Electrodynamics and Yang-Mills type theories are prime examples of physical theories whose complete invariance group consists of gauge and Poincaré transformations. Uniformly, the gauge group is an invariant subgroup, and the Poincaré group is the corresponding factor group. Hence there exist variables whose transformation properties are a faithful representation of the factor group only: precisely the gauge-invariant variables. If the gauge group is the group of curvilinear coordinate transformations (or of all diffeomorphic mappings of space-time on itself), there exists no such homomorphism, with the exception of theories that admit as physically meaningful solutions only those satisfying stated conditions of asymptotic flatness at infinity. In those latter theories groups resembling the BMS-group assume the role of factor group, with the details depending on the precise statement of the condition of asymptotic flatness—cf. treatments by R. K. Sachs, R. Geroch, R. Penrose, and others.

What makes the group of diffeomorphisms peculiar is that the mapping of one Cauchy hypersurface on another is not separable from the other gauge transformations, and hence the Hamiltonian of any general relativistic theory is necessarily a linear combination of gauge constraints. Dirac's treatment of Einstein's theory in the fifties is but a specific example of a very general state of affairs. If, for instance, one were to cast the

Brans-Dicke theory into canonical form, there would again be four first-class constraints at each point of a spacelike Cauchy hypersurface satisfying commutation relations isomorphic to Dirac's.

If in such a formalism the definition of observable introduced in the preceding section is adopted, then all observables are constants of the motion, i.e., their values are the same on all conceivable Cauchy surfaces. This result has been dubbed "the frozen formalism". Its adoption is unpalatable to many, as it appears to eliminate from the formalism all semblance of dynamical development.

In its defense I would make two points. First, problems in ordinary mechanics can be restated in terms of a frozen formalism. One has only to parametrize the theory, making the time variable the $(n + 1)$st configuration coordinate, with its canonical conjugate, $p_{n + 1}$, being equal to $-H$. $H + p_{n + 1} = 0$ is indeed the generating Hamiltonian constraint, and all the observables are constants of the motion. The transformation to these new canonical coordinates is well known to be generated by any solution of the Hamilton-Jacobi equation, and no one has taken offense at the Hamilton-Jacobi theory as a normal part of classical mechanics.

But this argument brings me to the second point, which is far more subtle. Even in classical mechanics the transformation generated by $S(q, P)$ is generally not global; in fact the requisite number of constants of the motion usually does not exist. By the same token, the observables of a general relativistic theory are probably not defined globally. But they are almost certainly not locally definable variables either, unless they are defined as coincidences, as by Bergmann and Komar, or in terms of asymptotic quantities (Bondi's news functions, Newman-Penrose constants) in theories admitting such quantities. It seems to me that there are conceptually incompletely understood problems here, whose analysis might teach us something fundamental concerning the nature of general relativistic theories.

In recent years B. DeWitt and C. Smith have published attempts to analyze the observability of variables from a nuts-and-bolts point of view, in terms of at least conceptually possible instrumental procedures.

3. Nonmetric Fields

John Stachel has asked me about the notion of observables in a theory with mixed variables, such as occur in Einstein-Maxwell theory. My response is partly implied by the fact that in the preceding section I have not confined myself to any particular kind of dynamical variables, and certainly not to a pure metric.

Suppose we deal with a theory that is orthodox Einstein 1916 but contains a number of additional field variables describing "matter." The Lagrangian of such a theory will consist, additively, of the standard Einstein term and one or several terms introducing the additional variables. The Brans-Dicke theory, including the electromagnetic field, may be put into this form. Presumably the number of independent constraints does not depend on the number of field variables introduced but only on the structure of the gauge group. If the gauge group consists of space-time diffeomorphisms plus electrodynamic gauge transformations, for instance, there will be five constraints, and their commutation relations, or Poisson brackets, will be predictable without reference to the details of the Lagrangian.

Presumably it is possible to construct observables only from the metric, for instance by the Komar-Bergmann method of intrinsic coordinates. Once this has been done, the remaining field variables can be considered observables in that intrinsic coordinate system. This approach is anything but aesthetic, or even practical, and in my opinion will serve at best only as an example guaranteeing existence. At any rate, all fields in addition to the metric will give rise to an equal number of additional observables.

4. Quantization of the Metric Field

This topic too has been suggested by John Stachel. Surely the components of the metric tensor play a dual role. They form the metric backbone required for the introduction of all other fields, but they are dynamical variables in their own right, and hence ought to be quantized in a full quantum theory of the physical universe. I am aware of the fact that C. Møller has pointed out that quantization of the metric field is not required *logically*. He has been supported in this view by the late L. Rosenfeld, who has thought about the quantization of the metric field earlier and longer than anyone else (his first paper on the subject known to me is dated 1930). I assume, however, that there are strong intuitive grounds for attempting its quantization; after all, gravitation is a physical field like all others that we know.

It seems to me that a quantized metric field should not be thought of as a local quantum field defined on a space-time whose world-points possess individual and classical identity. To me the physical universe is a function space defined on a four-dimensional manifold; what has physical significance is the quotient space of the constraint hypersurface within this function space over the mappings associated with the full gauge group of the theory. Even on a three-dimensional Cauchy hypersurface it appears risky to think primarily of a diagonalized configuration space (i.e., of a sharply defined three-dimensional metric g_{mn} on a three-dimensional spacelike hypersurface). Although the constraints restrict to some extent the range of the canonically conjugate variables, their uncertainty is unbounded, sufficiently so that the assumed sharpness of the 3-metric does not propagate at all.

Perhaps it is irrelevant whether we think of well-defined world-points with a fuzzy light cone, or conversely, of a sharp light cone, with considerable uncertainty as to which world-points lie on it. Most likely, both of these viewpoints are too naïve. Suppose we attempt a physical measurement, by means of instruments that had better not intrude too crudely on the physical situation, lest their large masses and stresses (if they are to contain any rigid components) modify the gravitational field far beyond the minimal effects required by the uncertainty relations. In elaborating what such an instrument measures we must discuss in detail not only which components of the fields are to be observed, but also in which space-time region these observations are to take place.

Perhaps it is just as well if I conclude my introductory remarks on this uncertain note, with all the technical and nontechnical connotations of "uncertain" you can imagine. It is this uncertainty that makes the whole field of quantum gravitation attractive to me.

═══════════WESLEY C. SALMON═══════

The Curvature of Physical Space

If one were seriously to entertain, even in a highly programmatic fashion, the thesis "there is nothing in the world except empty curved space. Matter, charge, electromagnetism, and other fields are only manifestations of the bending of space,"[1] it would seem highly germane to examine the nature of this curvature, which is to serve as "a kind of magic building material out of which everything in the physical world is made."[2] Such an examination has been carried out in depth by Adolf Grünbaum in "General Relativity, Geometrodynamics, and Ontology," a chapter that appears for the first time in the new edition of his *Philosophical Problems of Space and Time*.[3] The present discussion is intended primarily as an addendum to that chapter[4]—although, I should hasten to add, not necessarily one that he would endorse.

1. Metrical Amorphousness

The question I shall be addressing can be phrased, "Does physical space possess intrinsic curvature?" This way of putting it is liable to serious misunderstanding on account of the term "intrinsic," for it would be natural to call the Gaussian curvature of a surface "intrinsic" because, as Gauss showed in his *theorema egregium*, it can be defined on the basis of the metric of the surface itself, without reference to any kind of embedding space. The mean curvature, in contrast, is not intrinsic in this sense. It is entirely uncontroversial to state that, *in this sense*, any Riemannian space—not just a two-dimensional surface—possesses an intrinsic curvature (possibly identically zero) which is given by the type (0, 4) covariant

NOTE: The author wishes to express his gratitude to the National Science Foundation for support of research on scientific explanation and related matters, and to his friend and colleague, Dr. Hanno Rund, Head of the Department of Mathematics, University of Arizona, for extremely helpful information and advice. Dr. Rund is, of course, not responsible for any errors that may occur herein. His thanks also go to David Lovelock and Hanno Rund for making available a copy, prior to publication, of the manuscript of their book, *Tensors, Differential Forms, and Variational Principles* (New York: John Wiley & Sons, 1975).

curvature tensor R_{jmhk}. For present purposes I shall, however, deliberately avoid use of the term "intrinsic" in this sense, and shall use the term *internal curvature* to characterize those types of curvature that can be defined in terms of the metric tensor g_{hj}—or more generally, those types of curvature that do not depend upon an embedding space. This use of the word "internal" is nothing more than a terminological stipulation made for purposes of this particular discussion. I hope it will prove convenient and not misleading or confusing.[5]

In line with the foregoing stipulation, when I now ask whether physical space has intrinsic curvature, I am asking a question that is similar and closely related to one discussed by.Grünbaum in various of his works, namely, "Does physical space possess an intrinsic metric?"[6] In the context of Riemannian geometry, this question is motivated by the fact that the salient geometrical properties of a space—such as its Euclidean or non-Euclidean character, or the type of internal curvature it possesses—are determined entirely by congruence relations to which its metric gives rise. The converse is, of course, not true. The internal curvature—e.g., the Gaussian curvature for a two-dimensional surface—does not determine a metric uniquely (even up to a constant factor k). This fact does nothing to undermine Grünbaum's claim about the nonintrinsicality of the Riemannian curvature. Although a given curvature tensor R_{jmhk} of type (0, 4) does not determine a unique metric tensor g_{hj}, it does determine a unique class of metric tensors. Given two metric tensors from two such distinct classes, they do determine distinct curvature tensors.

In his well-known "theory of equivalent descriptions," Reichenbach maintains that, by employing different definitions of congruence, one and the same physical space can be described equivalently by the use of alternative geometries.[7] Different choices of coordinating definitions of congruence lead to different metrics, and these different metrics are such as to lead to different curvatures (or geometries). Reichenbach is not making merely an epistemological claim when he points to the existence of equivalent descriptions. He is arguing instead that since the descriptions are genuinely equivalent, they describe the same aspects of the same reality.[8] Given that the resulting descriptions are equivalent—including the fact that they are either both true or both false—a geometry involved in one but not the other of these descriptions cannot, in itself, represent an intrinsic characteristic of the space. This leads Reichenbach to his thesis of the *relativity of geometry*.[9]

Grünbaum has argued for the extrinsicality of the metric, and in consequence, the curvature, on the basis of what he has called "the metrical amorphousness of space." Appealing to an argument similar to one advanced by Riemann (which he calls RMH for Riemann's metrical hypothesis),[10] he maintains that the congruence or incongruence of two intervals in a continuous homogeneous manifold cannot be an intrinsic property of those intervals. Since I plan to use a similar argument below, it will be well to state it explicitly. I shall not present the argument in the same manner as Grünbaum, but I intend to offer a schematic restatement of his argument, not an alternative argument. Again, *he* might not regard it as a mere reformulation.

Although Riemann was obviously unaware of Cantor's theory of the cardinality of the linear continuum, he did seem to recognize that any closed linear interval is isomorphic to any other closed linear interval. In the jargon of set theory, any simply ordered, dense, denumerable set containing its end points is of the same order type as any other simply ordered, dense, denumerable set with end points. Even though the linear continuum is no longer considered denumerable, Riemann's basic notion is not invalidated, for it is easily proved that any closed, continuous, linear interval is of the same order type as any other closed, continuous, linear interval. Indeed, Cantor's proof that any two line segments ("regardless of length") have equal cardinality proceeds by establishing an *order-preserving* one-to-one correspondence between the points of the two intervals. Hence, if one regards a linear continuum merely as a point set that is ordered by a simple ordering relation, it follows that any closed interval of any linear continuum is isomorphic to any other closed interval of any linear continuum.

In dealing with "Zeno's metrical paradox of extension," Grünbaum takes pains to show how it is possible without contradiction to regard a linear continuum of finite nonzero length as an aggregate of points whose unit sets have "length" or measure zero.[11] This is done, essentially, by assigning coordinate numbers to the points on the line, and identifying the length (measure) of a nondegenerate closed interval with the absolute value of the difference of the coordinates of the end points.[12] Given the arbitrariness of the coordinatization of the line, it is evidently possible to assign coordinates to that line so as to yield *any* desired positive value as the length of *any* segment of that line. Since the coordinate number assigned to a point on the line obviously does not represent any intrinsic property of the point, we may say that the length of a segment or closed linear interval is an extrinsic property. And since the line can also be coordinatized in such a way that *any two* nonoverlapping segments receive the same length, we may add that equality or inequality of length is an extrinsic relation between them.[13]

We could say that two line segments are nonisomorphic only if we could invoke some further property or relation of the segments or of their constituent points such that no property-and-relation-preserving one-to-one correspondence between their members exists when the new property or relation is taken into account. This is Grünbaum's reason for insisting emphatically that continuity by itself is *not* sufficient to guarantee metrical amorphousness. Homogeneity is required in order to exclude possible further properties or relations that would destroy the isomorphism of any closed interval with any other nonoverlapping closed interval.[14] Although it may be hard to certify that, as points on the line, the elements have no property or relation that would render one segment nonisomorphic to another nonoverlapping segment, the obvious arbitrariness of the coordinatization lends strong prima facie plausibility to the supposition that no such properties or relations exist. The absence of any reasonable suggestions as to what properties or relations might render two nonoverlapping segments intrinsically equal or unequal in length lends stronger presumptive evidence to the claim that, as geometrical intervals on a line, any two nonoverlapping segments are isomorphic to each other, and that this isomorphism holds with respect to *all* of the *intrinsic* spatial properties and relations among the elements.[15]

The same considerations apply whether we are dealing with segments of a single one-dimensional continuum, or with all sorts of finite closed intervals on one-dimensional curves (which do not intersect themselves) in a continuous manifold of higher dimension. Given the fact that any segment of any curve is isomorphic to any other nonoverlapping segment of any curve, we have extremely wide latitude in the choice of a congruence relation and a metric. This exhibits a facet of the metric amorphousness of physical space. The full force of this metric amorphousness is revealed by the fact that a Riemannian space is coordinatized by regions, and that any coordinate region of an n-dimensional Riemannian space is isomorphic to any other coordinate region of any n-dimensional Riemannian space. The argument to support this latter claim is essentially the same as that designed to show the isomorphism of nonoverlapping segments in the one-dimensional line. This isomorphism among coordinate neighborhoods of equal dimension will play a major role in the subsequent discussion.

If physical space is, in fact, metrically amorphous, then obviously it can, with equal legitimacy, be endowed with metrics that differ from each other nontrivially, even to the extent of giving rise to different curvatures. In that case, neither the metric nor the curvature based thereon can be held to represent intrinsic properties of the space. It is tempting to argue for the converse proposition: if a given manifold can, *with equal legitimacy*, be metrized by means of two different metric tensors, g_{hj} and $\tilde{g}_{hj}$, which are associated, respectively, with two distinct curvature tensors, R_{jmhk} and $\tilde{R}_{jmhk}$, then the curvature tensor R_{jmhk} does not reflect an intrinsic characteristic of that manifold. Appealing as this principle is, it must be treated with caution, as is shown by Grünbaum's discussion of the logical relations between alternative metrizability and metrical amorphousness.[16] However, for spaces composed of homogeneous elements—the most likely case for physical space—van Fraassen has offered a plausible account of alternative metrizability and metrical amorphousness that equates the two concepts.[17] The success of his program hinges upon our ability to recognize the difference between trivial variants of the same metric and pairs of metrics that differ significantly from each other. I have offered a different account of alternative metrizability which is also designed to exclude all cases in which the alternative metrizability rests solely upon the existence of trivial variants of a single metric.[18]

Even if we refuse to admit that, with suitable explicitly stated caveats, alternative metrizability of a space *entails* metrical amorphousness, it still seems reasonable to construe alternative metrizability as usually or frequently symptomatic of metrical amorphousness. Since I am not attempting to repeat in full detail the arguments for the metrical amorphousness of physical space, and the consequent extrinsicality of the Riemannian curvature given by the type (0, 4) curvature tensor, I shall simply accept the conclusion that alternative metrizability is, in this case, indicative of the relevant sort of amorphousness. I am thus inclined to agree with Grünbaum et al. that such extrinsic curvature does not seem to constitute a fundamental property of empty space that qualifies it as a "magic building material" from which everything else in the physical world is to be constructed. The main purpose of this discussion is to show that considerations of the same type furnish equally strong grounds for claiming that another type of curvature is likewise extrinsic.

2. The Mixed Curvature Tensor

Clark Glymour has quite properly pointed out that there is a type (1, 3) mixed curvature tensor $K_h{}^j{}_{mk}$ that can be defined on a differentiable manifold endowed with an affine connection, even if it does not possess a metric.[19] The existence of such a curvature tensor is not controversial. This shows that there is a type of curvature that does not depend upon a metric; consequently, it does not follow immediately from the thesis of the *metrical* amorphousness of space that space lacks intrinsic curvature of *this* type. At the same time, to show that curvature represented by the type (1, 3) tensor may exist independently of a metric does not show that this type of curvature is indeed an intrinsic property of space. Glymour's consideration shows simply that the intrinsicality of this type of curvature is still an open question.

Glymour's argument does nothing to vindicate the original geometrodynamic program of constructing everything in the physical world out of curved empty space; at best, it provides a temporary reprieve. For the question now becomes: is the curvature represented by the mixed tensor a genuinely intrinsic property of space, or is it extrinsic in precisely the same sense as the metric is extrinsic to the Riemannian manifold? The question can be rephrased: is space as amorphous with respect to the curvature tensor furnished by the affine connection as it is with respect to the metric tensor? This is the crucial question, but neither Glymour nor Grünbaum has addressed it.[20]

The answer to this question could be furnished, I believe, by means of the following consideration. We asked above whether a Riemannian manifold, endowed with a particular metric, could with equal legitimacy be described by means of a different metric—one that leads to a different curvature and a different geometry. In a completely parallel fashion, we can now ask whether a differentiable manifold, endowed with a particular affine connection yielding a particular mixed curvature tensor $K_h{}^j{}_{mk}$, could with equal legitimacy be endowed with a different affine connection that would yield a different curvature tensor $\bar{K}_h{}^j{}_{mk}$. If so, we can provide a pair of equivalent descriptions of the same manifold embodying different curvatures. We could then argue, along the same lines as Reichenbach did with respect to the metric, that this type of curvature is also nonintrinsic, for it can vary nontrivially among equally legitimate descriptions of one and the same manifold. This amounts to an argument, similar to that based on alternative metrizability, which might be said to rest upon alternative connectability.

3. Affine Amorphousness

I shall now attempt to make a case for the view that the curvature associated with the type (1, 3) mixed tensor is extrinsic—in other words, that differentiable manifolds are amorphous, not only metrically, but also with respect to their affine connections. Let an n-dimensional differentiable manifold X_n be given. This manifold can, by definition, be covered by a finite number of overlapping coordinate neighborhoods; in each of these neighborhoods, every point can be assigned coordinates by means of a biunique continuous mapping of the points of the neighborhood onto n-tuples of real numbers. These n-tuples constitute an open subset of the n-dimensional space of real numbers R_n. Since R_n is obviously isomorphic to the Euclidean n-space E_n, the coordinatization of any coordinate neighborhood of our differentiable manifold X_n establishes an ismorphism between that coordinate neighborhood and an open n-dimensional region of E_n. This is, of course, a local isomorphism between a region of X_n and a region of E_n; it is not possible in general to extend this isomorphism to the entire manifold X_n, for it is not possible in general to cover the entire manifold with any single system of coordinates.

The question I am raising is, however, a local question. I am attempting to clarify the relationship between the curvature associated with the type (0, 4) covariant curvature tensor R_{jmhk} and that associated with the type (1, 3) mixed curvature tensor $K_m{}^j{}_{hk}$. The two types of tensors are defined at each point of their respective manifolds. The question of the metrical amorphousness of space is a local matter; it does not involve the global topological characteristics of the space. In raising the question of the intrinsicality of the curvature represented by the mixed curvature tensor, I am putting aside the global considerations in precisely the same fashion. In dealing with the nature of the affine connection and the curvature based thereon, it will therefore be sufficient to restrict attention to one coordinate neighborhood.

To state this point explicitly is to give the whole show away. As already remarked, the coordinate neighborhood of X_n is isomorphic to a region of

Euclidean n-space. If I am correct in saying that the fundamental basis for claiming that Riemannian space is metrically amorphous is the isomorphism of any finite closed interval to any other, then it would seem plausible to maintain that the isomorphism of any coordinate region of X_n to some region of Euclidean n-space has a fundamental bearing upon the question of whether the differentiable manifold is amorphous with respect to the curvature based upon the affine connection. It shows that any differentiable manifold that can be endowed with any affine connection whatever may be endowed locally with a connection whose components are identically zero in some given coordinate system. Obviously, the curvature tensor based upon this connection will also have components that vanish identically, and this property holds in *all* admissible coordinate systems.

The affine connection is not a tensor; under special circumstances it may therefore vanish with respect to some sets of coordinates but not with respect to others. For instance, while it vanishes identically for a Euclidean space with Cartesian coordinates, it does not vanish for the same space referred to curvilinear coordinates. But the related curvature is tensorial, and if it vanishes in one coordinate system it will vanish in all. This means that any coordinate region of any differentiable manifold may legitimately be provided with a set of coordinates and an affine connection such that the type $(1, 3)$ mixed curvature tensor vanishes. It is easy to see an important analogy here between the two types of curvature. Given even a non-Euclidean space, such as the surface of a sphere or the surface of a torus, it is possible, on account of metrical amorphousness, to remetrize an arbitrary region (provided it is not too large) in such a way that the region becomes Euclidean and its Riemannian curvature vanishes throughout that region. In a completely analogous way, any coordinate region of a differentiable manifold can be coordinatized and endowed with an affine connection such that its mixed curvature tensor vanishes throughout that region. This shows, I believe, that the *presence* of non-vanishing curvature of the type indicated by the type $(1, 3)$ mixed curvature tensor cannot be an intrinsic local property of a coordinate neighborhood of a differentiable manifold.

One further point should be added. Because of the transitivity of the isomorphism relation, the foregoing argument shows that any coordinate neighborhood of a differentiable manifold X_n is isomorphic to any other coordinate neighborhood of any differentiable manifold of the same dimension n. Thus, if any coordinate neighborhood of such a manifold can be endowed with an affine connection which gives rise to a nonvanishing curvature, then any other coordinate neighborhood of equal dimensionality can be endowed with the same connection and the same nonvanishing curvature. Given the obvious fact that some spaces are *so endowed with a metric or affine connection* that they are flat,[21] we may thus conclude that *absence* of nonvanishing curvature (i.e., the presence of zero curvature) of the sort associated with the type (1, 3), mixed curvature tensor is not an intrinsic local property of a differentiable manifold. In other words, a coordinate neighborhood of a differentiable manifold is neither intrinsically flat nor intrinsically nonflat.

4. Parallelism

I do not wish to rest the argument there, however, for I believe it can be made more compelling by considering the nature and function of the affine connection. Let us, therefore, look at the grounds for introducing such connections. In order to deal with certain kinds of physical and geometrical problems, we introduce tensors of various types, including vectors as special cases. At each point of our differentiable manifold X_n we construct a series of vector spaces—type (r, s) tensors at a given point constituting the members of an n^{r+s}-dimensional tangent vector space. The vectors or tensors that are elements of these tangent vector spaces are *not* elements of the differentiable manifold X_n; they are members of abstract vector spaces associated with the points p of X_n.[22] Such algebraic operations as addition, multiplication, and contraction are performed on the elements of the vector spaces associated with one and the same point p of X_n. At this stage of the analysis, the vectors that are elements of these various vector spaces have no physical or metrical significance; they are simply elements of an abstract mathematical structure.

There are many circumstances in which we must deal with relationships among vectors or tensors located at different points of X_n. For this purpose, we introduce vector or tensor fields. For the present discussion it will be sufficient to confine attention to contravariant vectors, i.e., type (1, 0) tensors. A contravariant vector field is simply a collection of uniquely specified contravariant vectors, each of which is associated with a distinct point of a region of X_n. The vector field thus consists of members of the various tangent spaces of contravariant vectors associated with the different points of the region of X_n over which the vector field is defined, one member being selected from each such tangent vector space. Of particular importance is the so-called *parallel vector field*.

We may, indeed, define the concept of parallel displacement. The general idea is this. We say loosely that a vector can be moved about in a Euclidean space, and as long as it retains the same length and direction it is still the same vector.[23] In the very special case of Euclidean space and Cartesian coordinates, this condition is equivalent to saying that its components are always the same, regardless of its point of application. A better way to say the same thing is to say that we can define a parallel vector field consisting of one vector at each point of the space, and that relative to the Cartesian coordinates, each of these vectors has precisely the same components. If this same Euclidean space is recoordinatized with curvilinear coordinates, the *same* parallel vector field is defined in a different way. Given a contravariant vector X^j at a point p with coordinates x^k, a vector $X^j + dX^j$ at a neighboring point q with coordinates $x^k + dx^k$ results from parallel displacement or is a member of the same parallel vector field as X^j if its components satisfy the equations

$$dX^j + \{{_h}^j{_k}\} X^h \, dx^k = 0 \tag{1}$$

where $\{{_h}^j{_k}\}$ is a Christoffel symbol of the second kind. Vectors related in this way are said to have the same magnitude and direction. It is of crucial importance to be clear about what is going on here. Given a particular vector at p, which is a member of a tangent space at p, we associate with it a unique member of the corresponding tangent space at q. Indeed, by means of the Christoffel symbol we establish an isomorphism between the members of the tangent space at p and those of the tangent space at the neighboring point q. It is in this sense that we specify vectors at q which uniquely correspond with vectors at p.

An analogous procedure can be carried out, not only in Riemannian spaces in general, but also in differentiable manifolds that do not possess a metric, provided they are endowed with an affine connection. To handle this more general situation, one defines parallel displacement of a vector from point p to point q along a curve C (defined by a parameter t) by the condition

$$X^j(t) = X^j{_{(0)}} - \int_{C}{_{t_0}}^{t} \Gamma \, {_h}^j{_k} \, X^h \, \frac{dx^k}{dt} \, dt, \tag{2}$$

where $\Gamma_{h}{}^{j}{}_{k}$ is the affine connection, and t_0 corresponds to the point p. This equation serves to define the differential dX, enabling us to write (2) in differential form:

$$\frac{dX^j}{dt} + \Gamma_{h}{}^{j}{}_{k} X^h \frac{dx^k}{dt} = 0 \qquad \text{along } C(t) \tag{3}$$

For arbitrary points p and q, the specification of a vector at q as parallel to a given vector at p may not be unique, but may depend upon the choice of the curve $C(t)$. For a point q in the neighborhood of p, however, a unique vector $X^j + dX^j$ corresponding to the vector X^j at p is given by the condition

$$dX^j = -\Gamma_{h}{}^{j}{}_{k} X^h dx^k \tag{4}$$

Thus $\Gamma_{h}{}^{j}{}_{k}$ provides a unique local mapping of the tangent space $T_n(p)$ onto $T_n(q)$.

Since the function of the affine connections is to establish isomorphisms between the members of vector spaces located at neighboring points of our differentiable manifold, it is natural to ask what restrictions are to be imposed. Given any two vector spaces of equal dimension, there is obviously a vast array of possible isomorphisms to choose from. In a Euclidean space (where the metric is given) we want all of the members of a parallel vector field to have the same magnitude and direction. In the absence of a metric, however, no sense can be attached to the question: are the members of the two spaces that are correlated by a given isomorphism *really* parallel (equal) to each other? As entities that are correlated with one another by the isomorphism of (4), they are parallel *by definition*, regardless of the particular isomorphism involved.

As equation (4) shows, the parallel vector field is defined in terms of relations between the respective components of the vectors at the two neighborhing points p and q. Given the continuity conditions imposed in characterizing a differentiable manifold, it is natural to impose continuity requirements upon the components of the vector under parallel displacement. This is, of course, built into the definition of the affine connection. In addition, the connection must have certain properties with respect to coordinate transformations. This is obvious from the following consideration. If a particular vector X at a point p is chosen, it will have certain components X^j relative to a system of coordinates x^j. If a coordinate transformation to a system $\bar{x}^k$ is performed, the *same* point p will have a different set of coordinates, and the *same* vector X will have components $\overline{X}^k$ that are different functions of its new coordinates. The same is true of the vector $X + dX$ located at the point q, which is said to belong to the same parallel vector field as X. The condition that must be satisfied by the affine connection is that parallelism is invariant under coordinate transformations. Two vectors which are parallel with respect to one set of coordinates must also be parallel with respect to any other set of coordinates. This fact is exhibited clearly when we state the defining relation in terms of the absolute differential

$$DX^j = 0 \qquad \text{(along } C\text{)} \tag{5}$$

which is equivalent to equation (3). This property of the affine connection is secured by demanding that it obey the transformation law

$$\overline{\Gamma}_{m\ p}^{\ j} = \frac{\partial \overline{x}^j}{\partial x^n} \frac{\partial x^h}{\partial \overline{x}^m} \frac{\partial x^k}{\partial \overline{x}^p} \Gamma_{h\ k}^{\ n} - \frac{\partial^2 \overline{x}^j}{\partial x^h \partial x^k} \frac{\partial x^h}{\partial \overline{x}^m} \frac{\partial x^k}{\partial \overline{x}^p} \tag{6}$$

Moreover, *any* set of 3-index symbols that satisfy this law qualify as an affine connection. This law, which embodies the necessary invariance conditions, constitutes the only restriction on the affine connection. Looking at this transformation law, one would hardly be tempted to suppose that it determines a unique affine connection. This impression is correct.

5. A Simple Example

Let us consider the situation concretely by reference to a simple example. Suppose we have two two-dimensional differentiable manifolds X_2 and Y_2; we confine attention to a coordinate neighborhood of each. Thus, in the region of X_2 which we are examining, we have a coordinate system x^1, $\overline{x}^2$, and in our region of Y_2 we have a coordinate system y^1, y^2. We endow X_2 with an affine connection whose components are, relative to the coordinate system $(\overline{x}^1, \overline{x}^2)$, identically zero. That this is a legitimate set of connection components is evident from the fact that it is the appropriate connection to use in Euclidean 2-space referred to Cartesian coordinates. We may introduce a new set of coordinates (x^1, x^2) according to the transformation

$$\overline{x}^1 = x^1 \cos x^2 \tag{7}$$
$$\overline{x}^2 = x^1 \sin x^2$$

which is recognized immediately as a transformation from Cartesian to polar coordinates. (We assume that the pole is not in the coordinate neighborhood we are discussing.) With this new system of coordinates is associated a new affine connection, which is identical with the Christoffel symbol of the second kind in this simple example:

$$\Gamma_2{}^1{}_2 = -x^1 \tag{8}$$
$$\Gamma_2{}^2{}_1 = 1/x^1 = \Gamma_1{}^2{}_2$$

all other components being zero. Routine calculation shows that the curvature tensor $K_h{}^j{}_{mk}$ based upon this connection, like that based on the connection all of whose components vanish, is identically zero.

We may note at the same time that the affine connection is sufficiently lacking in intrinsicality that it would have been entirely possible to assign the identically vanishing affine connection to our unbarred system of coordinates (x^1, x^2), or the nonvanishing affine connection to the barred coordinates $(\bar{x}^1, \bar{x}^2)$.[24] The differentiable manifold X_2 has no intrinsic characteristics that render one set of coordinates intrinsically Cartesian and another intrinsically polar.[25] Of course, this example, though it has some heuristic value, does not make the real point I am driving at. For although we have seen that the choice between the two affine connections does not rest upon any intrinsic property of the manifold X_2, the two connections yield the same curvature—namely, identically zero—and hence the space is flat under either description. This feature is, however, peculiar to this particularly simple example; it does not arise from any intrinsic characteristic of our differentiable manifold which demands an affine connection that will render it flat.

Let us therefore look at our other differentiable manifold Y_2. In the neighborhood under consideration, we supply coordinates (y^1, y^2). We then endow it with an affine connection whose components are

$$\Gamma_2{}^1{}_2 = -\sin x^1 \cos x^1 \tag{9}$$
$$\Gamma_1{}^2{}_2 = \Gamma_2{}^2{}_1 = \text{ctn } x^1,$$

all others being zero, which are nothing but the Christoffel symbols of the second kind associated with the usual metric $ds^2 = d\theta^2 + \sin^2\theta \, d\phi^2$ for the surface of a sphere of unit radius. When we calculate the components of the curvature tensor $K_h{}^j{}_{mk}$ we find that they do not vanish identically. In particular,[26]

$$K_1{}^2{}_{12} = 1 \tag{10}$$

Our neighborhood of Y_2 is not flat.

Since, however, our coordinate neighborhood of Y_2 is isomorphic to our coordinate neighborhood of X_2, there is no reason why we could not endow X_2 with the same affine connection we imposed upon Y_2, and conversely. Although the simple examples of affine connections I have discussed are all Christoffel symbols associated with familiar metrics, it is clear that the situation would be no different in principle if we were to deal with nonsymmetric connections that are not related to any metric at all. One could, in fact, write down an arbitrary set of differentiable functions as connection components, attach them to some definite coordinate system x^j, and *stipulate* that the components of that connection for any other coordinate system $\bar{x}^k$ be given by equation (6). The connection is *that* arbitrary, and *that* insensitive to selection of an original coordinate system to which it is to be attached.

6. Physical Manifolds

There may be some feeling that this discussion, thus far, has rested too heavily upon mathematical considerations, to the neglect of relevant physical factors. I have, it is true, emphasized such aspects as the isomorphisms among coordinate neighborhoods of equal dimension, and the alternative metrics and affine connections that are abstractly possible in such regions of differentiable manifolds. This emphasis is *not* the result of a mistaken notion that the problem under discussion is one of pure mathematics; rather, it comes from an attempt to compare the extrinsicality of the curvature exhibited by the type (0, 4) curvature tensor with that of the type (1, 3) curvature tensor. In constructing this comparison, I have tried to focus upon those features of physical space that underly the potent arguments of Reichenbach and Grünbaum in support of the thesis of extrinsicality of the metric (and associated curvature). The crux of the argument seems to me to hinge upon the kinds of isomorphisms I have mentioned. If such arguments for extrinsicality are to be defeated, it is necessary to show what intrinsic properties of the physical manifold, over and above its structure as a differentiable manifold of given dimension, can be invoked to undermine the alleged isomorphisms of segment with segment, or neighborhood with neighborhood. With regard to the metric and the curvature associated with it, I have explicitly stated my agreement with Reichenbach, Grünbaum, et al. The only question remaining is whether a continuous physical manifold (of space or space-time) might be said to possess an intrinsic affine structure that determines a unique affine connection, and consequently, an intrinsic curvature of the sort represented by the type (1, 3) tensor.

If one were to accept the Reichenbach-Grünbaum argument, then, it seems to me he would be hard put to imagine what sort of intrinsic structure of space (or space-time) could support the claim that the affine connection is uniquely determined. We recall that the function of the affine connection is to establish a biunique correspondence between vectors in a tangent space at point p and those in another tangent space at the neighboring point q. It is difficult to see what intrinsic property of the underlying manifold could determine which isomorphism is the correct one to represent parallel displacement of vectors. What conceivable intrinsic property could it be, and in what way could it compel the choice of an isomorphism? At this stage of the discussion, it seems to me that the burden of proof (or the burden of suggestion, at least) shifts to the proponent of intrinsicality.

One way to use the concept of parallel displacement is in the definition of an *autoparallel* curve as a curve whose tangent vectors are parallel to one another. Such autoparallel curves, or *paths* as they are sometimes called, bear striking resemblance to the stright lines of Euclidean spaces and the geodesic curves of Riemannian spaces. In a physical theory that employs an affinely connected differentiable manifold, the autoparallel curves may be interpreted as trajectories of gravitational test particles. At this juncture it is essential to remember that gravitational test particles are as extrinsic to a spatial or spatio-temporal manifold as Einstein's rods and clocks.[27] Just as one can argue for the metrical amorphousness or alternative metrizability of physical space on the basis of alternative admissible coordinating definitions of congruence, so also can one argue for "affine amorphousness" or "alternative connectability" on grounds of the possibility of alternative affine connections.[28]

When we endow our physical manifold with one affine connection, we may notice that gravitational test particles have autoparallel curves as trajectories, while with another affine connection we find that the trajectories of these test particles are not autoparallel. Such an observation would be entirely analogous to the commonplace that certain kinds of solid bodies remain self-congruent wherever they are located if our space is endowed with one metric, while under a different metric these same solid objects change their size as they are transported from place to place.

And just as we can ask (1) whether physical space has some intrinsic structure that determines whether the amount of space occupied by the measuring rod in one place is equal to the amount of space occupied by the same rod in another place, so also must we ask (2) whether physical space has an intrinsic structure that determines whether the tangent vectors of the particle trajectory are *really* parallel or not.[29] If, on the one hand, one accepts the Reichenbach-Grünbaum negative answer to question (1), it is difficult to see how he could then go on to answer question (2) affirmatively. At this point, it seems to me, the burden of proof becomes acute. If, on the other hand, one wants to avoid the embarrassment of trying to give an affirmative answer to one of these questions while giving a negative answer to the other, it becomes necessary either to refute the powerful arguments that have been advanced for the negative answer to (1), or else admit that the affine structure of physical space is no more intrinsic than its metric structure.

7. Conclusions

The considerations advanced in this discussion seem to me to lend strong support—albeit inductive support—to the view that curvature is not an intrinsic local property of physical space, whether that curvature be of the kind associated with the type (0, 4) tensor or that associated with the type (1, 3) tensor. Thus Glymour's observation that there exists a curvature tensor which is independent of any metric does nothing to show that there is a kind of curvature which is intrinsic, and which could therefore be employed by geometrodynamicists as a "magic building material" from which to construct "the furniture of the world."[30]

Appendix
Alternative Metrizability of Space

The concept of metrical amorphousness, and the related concepts of alternative metrizability and intrinsicality of metrics, are difficult to define in the generality needed to cover wide varieties of *mathematically* interesting spaces, as is attested by the volume and complexity of recent literature on the subject.[31] As Gerald J. Massey has written, it would have been reasonable to interpret some of Grünbaum's earlier writings as maintaining that the concepts of metrical amorphousness and alternative metrizability are interchangeable.[32] In his recent detailed investigation of

296

intrinsic metrics, however, Grünbaum explicitly rejects this equivalence by exhibiting a manifold which possesses a nontrivial intrinsic metric—thus disqualifying it from metrical amorphousness—but which possesses at least two nontrivially distinct intrinsic metrics.[33] At the same time, van Fraassen has offered an account of alternative metrizability *for a restricted class of spaces* which seems to allow for the equivalence of metrical amorphousness and alternative metrizability within the range of his discussion.[34]

In this note, I shall not attempt to lay down necessary and sufficient conditions for the general applicability of the concepts mentioned above; rather, I shall attempt to enunciate a sufficient condition which seems to me to capture their import as applied to the sorts of manifolds that have figured prominently in discussions of the structure of physical space, physical time, and physical space-time. For this purpose, it seems to me, sufficient generality is achieved if we can explain the applicability of these concepts to differentiable manifolds—roughly, spaces of dimension one or greater that satisfy certain continuity requirements. The foundation for the whole development will be the concept of *unconditional alternative metrizability*, which I shall proceed to define. To what extent this concept can be usefully applied to spaces that are not differentiable manifolds is not clear to me at present.

We begin by regarding a differentiable manifold X_n of finite dimension n (≥ 1) as a space of dimension n that can be covered by a finite number of coordinate neighborhoods, each of which can be provided with a coordinate system. A coordinate system for a coordinate neighborhood is an assignment of n-tuples of real numbers to the points of that neighborhood; specifically, it is a bicontinuous one-to-one mapping of a region of the n-dimensional space R_n of real numbers onto the points of the coordinate neighborhood of X_n. Since R_n is isomorphic to the n-dimensional Euclidean space E_n, we can equally well conceive the coordinatization of a neighborhood of X_n as a one-to-one bicontinuous mapping of that neighborhood onto the points of a region of E_n which has already been endowed with Cartesian coordinates. The coordinates of the points of our neighborhood of X_n are simply the coordinates of their image points in E_n under that mapping. Since the supply of one-to-one bicontinuous mappings is very large, a given coordinate neighborhood of X_n may be outfitted with a wide variety of distinct systems of coordinates. It should be explicitly stated, moreover, that the region of E_n involved in the coordinatization of a given neighborhood of X_n is by no means uniquely determined by the choice of the neighborhood of X_n.

With the wide latitude of choice of coordinate systems available, it is useful to consider tranformations from one system of coordinates to another. If we begin with a set of coordinates x^i ($i = 1, \ldots, n$), we can transform to a new set of coordinates $\bar{x}^j$ ($j = 1, \ldots, n$), where

$$\bar{x}^j = f^j(x^1, \ldots, x^n), \tag{11}$$

and the functions f^j are continuous, possess continuous partial derivatives to some specified order, and possess inverses. For present purposes, let us assume that the functions f^j are of a class C^∞. Any system of coordinates that results from an admissible coordinate system by such a transformation is considered an admissible coordinate system.

Let us now endow our coordinate neighborhood of X_n with a metric g_{hk}, which is a field of type $(0, 2)$ symmetric tensors. At each point P of our neighborhood we have n^2 quantities that are given as functions of the coordinates of the point. To keep the functional dependency of the components of g_{hk} on the coordinates x^i clearly in mind, let us write explicitly

$$g_{hk} = g_{hk}(x^i) = g_{hk}(x^1, \ldots, x^n). \tag{12}$$

Each member of our tensor field is attached to a point of our coordinate neighborhood, and the point is identified by its coordinates x^i.

Let us now recoordinatize our neighborhood by performing a coordinate transformation to a new set of coordinates $\bar{x}^j$. Let us further define a *new* tensor field on the same coordinate neighborhood by the rule,

$$g_{hk}(x^i) = \tilde{g}_{hk}(\bar{x}^i). \tag{13}$$

This simply has the effect of carrying a tensor attached to a point p with coordinates x^i in the old coordinate system to a *different* point $\bar{p}$ whose coordinates $\bar{x}^j$ relative to the new coordinate system are equal to those of p relative to the old system, i.e.,

$$x^i(p) = \bar{x}^i(\bar{p}). \tag{14}$$

This *new* tensor field $\tilde{g}_{hk}$ is a *different* metric for our coordinate neighborhood.

We may now ask whether the two metrics g_{hk} and $\tilde{g}_{hk}$ are *equally legitimate* metrizations of our coordinate neighborhood. This is a difficult

298

question, to which such authors as Poincaré, Reichenbach, and Grünbaum have devoted considerable attention.[35] It seems to me that they have effectively established a general affirmative answer to the question of equal legitimacy; for example, this would seem to be the import of Reichenbach's theory of equivalent descriptions of physical space based upon the admissibility of alternative coordinating definitions of congruence. It is not necessary to try to argue the case again here; for purposes of the present discussion, I shall simply assume that we know how to answer the question at issue. I now propose the following definition:

Definition 1. A coordinate neighborhood of X_n is *unconditionally alternatively metrizable* if and only if every remetrization of the type defined in (13) with respect to *every admissible coordinate* system for that neighborhood is an equally legitimate metrization of that neighborhood.

In other words, suppose we begin with a legitimate metrization of our neighborhood, then perform a coordinate transformation as given in (11), and finally introduce a new metric according to (13). If this *always* results in an equally legitimate metrization of the neighborhood, the neighborhood is unconditionally alternatively metrizable.

The idea behind this definition is straightforward. If the legitimacy of a metrization of a space (or a coordinate neighborhood thereof) is totally insensitive to coordinate changes, then, in view of the evident arbitrariness of the assignment of coordinates, the metric cannot represent any intrinsic property of that space. Consider a simple example. Suppose we begin with a two-dimensional space which we recognize intuitively as a Euclidean plane, and suppose further that it is provided with a coordinate system which we recognize intuitively as Cartesian. The metric is given by

$$ds^2 = (dx^1)^2 + (dx^2)^2; \ g_{hk} = \delta\,^h_k \tag{15}$$

We now transform to a coordinate system we recognize intuitively as polar (with the pole falling outside our coordinate neighborhood).[36] Nevertheless, we do not change the *form* of the metric, writing

$$d\bar{s}^2 = (d\bar{x}^1)^2 + (d\bar{x}^2)^2; \ \bar{g}_{hk} = \delta^h_k \tag{16}$$

This constitutes a drastic change of metric, in the sense that the distance between two points p and q will in general be changed, and intervals that were congruent under the old metric will in general not be congruent under the new metric. Consider any two pairs of points (p_1, q_1) and (p_2, q_2) that represent equal intervals in the old metric, but unequal intervals in the new metric. If there is no intrinsic property of these two intervals that determines that they are either intrinsically congruent or that they are intrinsically incongruent, then there is no valid basis for maintaining that one (at most) of these metrizations is legitimate—that at least one of them must be illegitimate. If it can be argued validly that there are no intrinsic properties of this space that show that the congruences delivered by one of these metrics are correct while those produced by the other are incorrect, then that argument for the legitimacy of this kind of remetrization is an argument for the metrical amorphousness of the space.

As I said at the outset, I do not claim to have provided necessary and sufficient conditions for the applicability of such concepts as alternative metrizability and metric amorphousness in all generality.[37] I am prepared, however, to offer the following sufficient conditions:

Condition 1. A coordinate neighborhood of X_n is *alternatively metrizable* if it is unconditionally alternatively metrizable.

Condition 2. A coordinate neighborhood of X_n is *metrically amorphous* if it is unconditionally alternatively metrizable.

Condition 3. A coordinate neighborhood of X_n *has no intrinsic metric* if it is unconditionally alternatively metrizable.[38]

Unconditional alternative metrizability may be a fairly strong condition; nevertheless, it seems to exhibit a type of alternative metrizability from which metrical amorphousness can reasonably be inferred. It is strong enough to rule out those cases of alternative metrizability in which all of the alternative metrics can be considered, in any sense, trivial variants of one another. Moreover, I am convinced that the arguments of Poincaré, Reichenbach, Grünbaum, et al. are sufficient to qualify such continuous *physical* manifolds as physical space, physical time, and physical space-time for unconditional alternative metrizability. Conditions 1–3 thus seem to apply to the physical manifolds of interest.

Notes

1. J. A. Wheeler, *Geometrodynamics* (New York: Academic Press, 1962), p. 225.

2. J. A. Wheeler, "Curved Empty Space-Time as the Building Material of the Physical World," in E. Nagel, P. Suppes, and A. Tarski, eds., *Logic, Methodology and Philosophy of Science* (Stanford, Calif.: Stanford University Press, 1962), p. 361.

3. Adolf Grünbaum, *Philosophical Problems of Space and Time*, 2nd ed. (Dordrecht/ Boston: Reidel, 1973), chap. 22. In this chapter Grünbaum carefully documents Wheeler's *recent* renunciation of his earlier pure geometrodynamic thesis.

4. In particular, the present discussion is a supplement to section 3b—an extended footnote to p. 788, it might be said.

5. Grünbaum, *Philosophical Problems*, p. 501, carefully distinguishes between "intrinsic" (German: *inner*) in the sense applicable to Gaussian curvature, and "intrinsic" or "implicit" (German: *schon enthalten*) in the sense in which Riemann denied it with respect to the metric of continuous space. It is in this latter sense that I am using the term "intrinsic" in the present discussion. For the former concept, I am using the term "internal."

6. See chiefly Grünbaum, *Philosophical Problems*, chap. 16, "Space, Time and Falsifiability," which also appeared in *Philosophy of Science* 37 (1970): pp. 469–588.

7. Hans Reichenbach, *Philosophy of Space and Time* (New York: Dover Publications, 1958). This point had been argued with great cogency by Poincaré much earlier.

8. Reichenbach does, it is true, adopt a verifiability criterion of equivalence, but he construes this as a criterion of what can be meaningfully said about physical reality. Thus the claim that physical space actually has one of these geometries and not the other is without any possible justification, and would constitute a totally unwarranted claim *about reality* (not merely about our knowledge).

9. Some authors, including Grünbaum, prefer the term "conventionality" to "relativity" in such contexts. Reichenbach may have eschewed the former term to avoid any suggestion that he was adopting a *thoroughgoing* conventionalism of the sort be found in Poincaré. The term "relativity" emphasizes the fact that, in a given physical space, the geometry is relative to a choice of congruence standard—i.e., given a coordinating definition of congruence, the geometry of the space is a matter of empirical fact. In the presence of suitable coordinating definitions, there is nothing conventional about the geometry.

10. Grünbaum, *Philosophical Problems*, pp. 495–99, 527–32, has offered a detailed statement and elaboration of RMH.

11. Grünbaum, *Philosophical Problems*, chap. 6, especially pp. 170–72.

12. By a closed interval [*a, b*] on a line I shall mean the set consisting of points *a* and *b* and all points lying between them. The interval is degenerate only if *a* = *b*. In this discussion I shall use the term "interval" to refer only to nondegenerate intervals. The term "segment" will be used as a synonym for "nondegenerate closed interval."

13. The continuity of the line, which *is* an intrinsic structural feature, is reflected in the fact that the coordinatization of the line is a continuous one-one relation between the points of the line and the continuum of real numbers. We specify that the intervals be nonoverlapping, for the inclusion of one interval entirely within another is an intrinsic relation between them, and this relationship should be reflected in the metric.

14. This point is brought out clearly in Grünbaum's comparison of the geometrical continuum of points with the arithmetical continuum of real numbers: *Philosophical Problems*, pp. 512–14, 526–31.

15. Grünbaum, *Philosophical Problems*, pp. 498–501, 529–31 explicitly acknowledges that RMH is only inductively confirmed, not a demonstrated truth. The strength of support for RMH would seem at least adequate to shift the burden of the argument to its opponents.

16. Grünbaum, *Philosophical Problems*, pp. 547–56.

17. Bas C. van Fraassen, "On Massey's Explication of Grünbaum's Conception of Metric" *Philosophy of Science* 36, no. 4, December 1969.

18. In the Appendix "Alternative Metrizability of Physical Space," I have defined *unconditional alternative metrizability* as a type of alternative metrizability that excludes cases in which the alternative metrics are simply trivial variants of one another. Unconditional alternative metrizability does, I claim, entail metrical amorphousness.

19. Clark Glymour, "Physics by Convention," *Philosophy of Science* 39, no. 3, September 1972.

20. Grünbaum's answer to Glymour's attempt to impugn Grünbaum's philosophical attack

upon the geometrodynamic program follows an entirely different tack; see Grünbaum, *Philosophical Problems*, pp. 773–88.

21. When I use the term "flat", it is to be construed in the technical sense of the vanishing of the type (1, 3) curvature tensor.

22. This is true of the metric tensor g_{hk} as well as any others—a fact which has not, to my knowledge, been mentioned explicitly in discussions of intrinsicality of metrics.

23. Because of its linearity, Euclidean space may be identified with a tangent space.

24. Note that this is precisely the sort of insensitivity to coordinate changes on the part of the affine connection as I discussed with respect to the metric in the Appendix "Alternative Metrizability of Physical Space." It is as indicative of "affine amorphousness" in this case as it was indicative of metrical amorphousness in the other case.

25. Remember that the origin is not within our coordinate neighborhood.

26. In 2-space the curvature tensor has only one *independent* component.

27. Grünbaum has argued in detail that Einstein's rods-and-clocks method, Synge's chronometric method, and the geodesic method of Weyl et al. depend equally upon extrinsic standards for the determination of the space-time metric. *Philosophical Problems*, chap. 22, section 2.

28. Just as one definition of congruence might be preferable to another on grounds of *descriptive* simplicity, so also might one affine connection be preferable to another on the same grounds. Lack of descriptive simplicity does not render either a metric or an affine connection inadmissible as factually incorrect.

29. In discussion, Howard Stein made reference to a particle "sniffing out" a path through space. If we construe the term "path" as autoparallel, then the question becomes: on the basis of what "olfactory" characteristics of physical space itself can an autoparallel be detected? To "sniff out" a path requires discriminable odors, as any bird dog knows, but in a space of *homogeneous* elements, such differences in odor cannot possibly be intrinsic characteristics.

30. Whether topological structure could provide the requisite curvature is, of course, an entirely different question.

31. Especially: Gerald J. Massey, "Toward a Clarification of Grünbaum's Concept of Intrinsic Metric," *Philosophy of Science* 36, no. 4, December 1969; van Fraassen, "On Massey's Explication"; Adolf Grünbaum, "Space, Time and Falsifiability, Introduction and Part A," *Philosophy of Science* 37, no. 4, December 1970, reprinted in Adolf Grünbaum, *Philosophical Problems*, chap. 16.

32. Massey, "Toward a Clarification," p. 332. In the end Massey seems to despair of finding a reasonable explication of "alternative metrizability." See p. 345.

33. Grünbaum, *Space, Time, and Falsifiability*, part A, section 3.

34. Van Fraassen, "On Massey's Explication."

35. Henri Poincaré, *Science and Hypothesis* (New York: Dover Publications, 1952); Hans Reichenbach, *The Philosophy of Space and Time* (New York: Dover Publications, 1958); Adolf Grünbaum, *Philosophical Problems of Space and Time*, 1st ed. (New York: Alfred A. Knopf, 1963).

36. It is, of course, strictly nonsense to identify a coordinate system *as such* (without specifying a metric) as Cartesian, polar, etc. I am nevertheless using these intuitive notions simply to try to give some feeling for the definition of unconditional alternative metrizability.

37. Unlike van Fraassen's discussion, the present one applies (negatively) to various inhomogeneous spaces, such as the arithmetical spaces of real numbers. I am not sure whether it can be extended to apply to discrete spaces.

38. In this context I am, of course, referring only to nontrivial intrinsic metrics; see Grünbaum, *Space, Time, and Falsifiability* part A, section 2b, for an explication of triviality.

Absolute and Relational Theories of Space and Space-Time

I wished to show that space-time is not necessarily something to which one can ascribe a separate existence, independently of the actual objects of physical reality. Physical objects are not in *space*, but these objects are *spatially extended*. In this way the concept "empty space" loses its meaning.

June 9, 1952   A. Einstein

(Preface to the fifteenth edition of *Relativity: the Special and the General Theory*. New York: Crown Publishers, 1961, p. vi.)

1. Introduction

Has the issue of ontological autonomy between Newton's absolutism and Leibniz's relationalism become otiose, defunct, and perhaps even spurious in the context of present-day theories of space-time structure? Or is the dichotomy between absolutistic and relational ontologies as originally understood by Newton and Leibniz illuminatingly germane to space-time as much as it was to pre-Einsteinian space and time? Two

NOTE: This paper is based on a much shorter version delivered at a conference of the same title, held on June 3–5, 1974 in Andover, Mass., under the auspices of the Boston University Institute of Relativity Studies, directed by John Stachel.

Apart from specific other acknowledgments made within the text, I am very grateful to the various Andover Conference participants whose ideas are discussed or mentioned by name within the text, especially Howard Stein. In addition, I wish to thank John Stachel, the organizer of the conference, for the opportunity to present the first draft of this paper there and for valuable private discussions of some of its content. And I am greatly indebted to Allen Janis for having read both the first and final drafts: my treatment of the debate between the relationalist and the "semi-absolutist" had the benefit of some incisive questions from him. Both John Winnie and Alberto Coffa stimulated me to deal with some issues in section 4 (2) which I would otherwise have neglected. Jeffrey Winicour made me aware of some pitfalls in the use of the term "graviton." I also wish to thank the National Science Foundation for the support of research.

statements made in 1974 by space-time physicists will illustrate that the central problem posed in the Newton-Leibniz debate has by no means become ill-conceived; instead it has since become more sophisticated and has ramified.

In his illuminating paper "The Rise and Fall of Geometrodynamics," John Stachel wrote as follows:

Would it not seem more sensible to talk about absolute and relational aspects of the geometrical structure within a given physical theory; and to recognize that generally a theory will contain both aspects? For example, the manifold structure of space-time seems to be an absolute geometrical aspect of all existing physical theories. While geometrodynamics, as originally formulated, envisioned the metric as absolute, from Sakharov's viewpoint, as well as from at least some other interpretations of the formalism of general relativity, the metric is to be viewed as a relational aspect of the geometrical structure. It is not too clear from the comments of Sakharov and Wheeler, but at least one tenable interpretation of this program is that the manifold structure will still be taken as absolute, in the presumed derivation of the metric from the properties of various quantized matter fields. . . . it seems possible to envisage a somewhat different program in which even the manifold structure would be derived—in some approximation, at least—from something more fundamental. Indeed, this seems to be part of what Wheeler has in mind in his discussion of pregeometry. . . .

Thus, the nature of an element of the geometrical structure as absolute or relational may be a changing one, depending on the level of theoretical depth at which the question is being considered. Penrose and others have also speculated on the possibility for deriving various geometrical structures of present-day theories from some more basic structures, which if we can reach a satisfactory definition of 'geometrical,' we might agree to call non-geometrical.[1]

I take it that in this last sentence, Stachel is alluding, for example, to Penrose's scheme of deriving the Euclidean spatial geometry of *angles* from the intrinsic angular momentum of particles. In the concluding paragraph of Penrose's presentation of this scheme, he says: "One has to define the things with which one builds up geometry (e.g., points, angles, etc.) in *terms* of the physical objects under consideration."[2] Leibniz must be smiling with joy in his grave. But Leibniz must also be frowning in the face of Charles Misner's quite contrary aspiration, which Misner formulates as follows:

The immediate philosophical problems which the current tentative supporting evidence for physical geometrodynamics poses, and which its possible definitive establishment would make imperative, are the following. One needs to clarify and expound the idea of self-existing empty space as a reality in the actual physical world. To a considerable extent, this is merely an extension of the demands posed by Maxwell's electrodynamic theory in its special relativistic interpretation, where again abstract mathematical fields occur as the fundamental classical entities. Nor is it entirely different from the problems of describing 'what matter is', on the basis of quantum field theories. An aspect of this would seem to be to take a particular stand on the question of the nature of space. One is not able to interpret space as a relationship between material objects; rather, one sees matter and material objects arising as structures of mathematical relationships constructed in and out of space itself.[3]

My concern with the conflict between absolutism and relationalism will focus largely on some central problems that underlie or grow out of the themes struck in the citations from Stachel and Misner. But I shall begin and end by asking questions without championing theses that purport to furnish viable answers. The latter questions will range from a proposed clarification of Newton's absolutistic conception of metrical interval ratios, to the physical and epistemological status of mathematical devices for turning a non-time-orientable relativistic space-time into the corresponding time-orientable double covering space.

In the course of our inquiry into some facets of the present-day relevance of the Newton-Leibniz controversy, I shall distinguish several significantly different senses in which the terms "absolute" and "relational" and some of their cognates have come to be construed, beginning with Riemann and especially since the advent of the theory of relativity.

2. Newton

It will set the stage for a foundational question posed by Newton's 1693 letter to Bentley, if we begin by considering an intriguing though unsuccessful thirteenth- and fourteenth-century attempt to give an absolutistic account of the metric structure of Euclidean physical space. As we shall see presently, this account clashes head-on with the modern Cantorean view that all intervals of that space other than singletons have the same cardinality. But it embryonically anticipates Cantor by envisioning a *hierarchy* of transfinite cardinalities without, however, associating that hierarchy with the modern concept of superdenumerability.

In her book on the fourteenth-century precursors of Galileo, Anneliese Maier[4] discusses the assumption made by *some* of these precursors that line segments and higher-dimensional regions of Euclidean 3-space are literally aggregates of points, a conception to which she refers as "*compositio ex punctis.*"[5] Referring to the *compositio* doctrine, she writes: "It is tacitly understood by both exponents and opponents of this doctrine, that if a given continuum does consist of points, then it must consist of a definite number of points such that a larger line or a larger surface contains more points than a smaller one."[6] Speaking of such definite numbers of points, Maier says: "the continua consist of (actually) infinitely many points; . . . one *numerus infinitus* can be greater or smaller than another."[7] As shown by examples given in the literature of the time,[8] *differences* in transfinite cardinality were attributed to *metrically unequal* intervals or infinite point sets, be they disjoint, overlapping, or such that one is a *proper* subset of the other. Thus it was held that the metrical ratio between the metrically unequal *peripheries* of two concentric circles is given by the unequal transfinite cardinalities of the relevant point sets, and so is the identical ratio between the respective radii. Even the metric ratio between the diagonal and any side of a square, namely $\sqrt{2}$, is presumed to be given by correspondingly different infinite cardinalities.

Champions of this *compositio* thesis include the thirteenth-century writers Robert Grosseteste and William of Auvergne (or of Paris), as well as the fourteenth-century proponent of actually infinite sets, Henry of Harclay.[9] Let me cite Grosseteste's articulation of this conception by reference to both time and space. He writes: "The time that measures one revolution of the heaven is measured by the maker of time by the infinite number of indivisible instants that are in that time, and a double time by double the infinite number of instants, and a half by half, and a time incommensurable to it by an infinite number incommensurable to that number."[10] After saying that to God himself, any cardinal number which is infinite from man's point of view is in some sense finite, Grosseteste speaks of how God endowed space with an absolute metric, writing:

With some infinite number (determinate and finite to him) he [God] measures a line of one cubit[11] (*lineam cubitabilem*); and with an infinite number double it, a line of two cubits; and with an infinite number half it, a line of half a cubit. And there is one infinite number of points on all lines of one cubit. . . .

In what way then is the first line measured and numbered in the first place? I hold that it is by the infinite number of points of that line (but finite to the measurer), which number of points is not in any greater or lesser line. But in each greater there is a greater infinite number of points and in each lesser a lesser. . . .[12]

Five centuries before Cantor, this version of an absolutistic conception of the metric was shown to be untenable within the framework of Euclidean geometry by means of the concept of one-to-one correspondence.[13] Other fourteenth-century thinkers soon pointed out that the radial lines emanating from the common center of two concentric circles provide a mapping which shows that the points on the metrically larger periphery have the *same* infinite cardinality as those on the smaller one.[14] By the same token, the critics of *compositio ex punctis* had no trouble exhibiting a biunique mapping of the points on the diagonal of a square with those on any side.[15]

This brings me to an intriguing question posed by Newton's 1693 letter to Bentley. In my quotation from the pertinent part of that letter, there will be two instances in which I shall cite in square brackets the wording given in the Turnbull edition, because that wording differs importantly from the text of the Thayer edition, from which my quotation will be taken. Newton writes:

But you argue, in the next paragraph of your letter, that every particle of matter in an infinite space has an infinite quantity of matter on all sides and, by consequence, an infinite attraction every way, and therefore must rest *in equilibrio*, because all infinites are equal. Yet you suspect a paralogism in this argument, and I conceive the paralogism lies in the position that all infinites are equal. The generality of mankind consider infinites no other ways than indefinitely [definitely]; and in this sense they say all infinites are equal, though they would speak more truly if they should say they are neither equal nor unequal, nor have any certain difference or proportion one to another. In this sense, therefore, no conclusions can be drawn from them about the equality, proportions, or differences of things; and they that attempt to do it usually fall into paralogisms. So when men argue against the infinite divisibility of magnitude by saying that if an inch may be divided into an infinite number of parts the sum of those parts will be an inch; and if a foot may be divided into an infinite number of parts the sum of those parts must be a foot; and therefore, since all infinites are equal, those sums must be equal, that is, an inch equal to a foot.

The falseness of the conclusion shows an error in the premises, and the error lies in the position that all infinites are equal. There is, therefore, another way of considering infinites used by mathematicians, and that is, under certain definite restrictions and limitations, whereby infinites are determined to have certain differences or proportions to one another. Thus Dr. Wallis considers them in his *Arithmetica Infinitorum*, where, by the various proportions of infinite sums, he gathers the various proportions of infinite magnitudes, which way of arguing is generally allowed by

307

mathematicians and yet would not be good were all infinites equal. According to the same way of considering infinites, a mathematician would tell you that, though there be an infinite number of infinite [infinitely] little parts in an inch, yet there is twelve times that number of such parts in a foot; that is, the infinite number of those parts in a foot is not equal to but twelve times bigger than the infinite number of them in an inch. And so a mathematician will tell you that if a body stood *in equilibrio* between any two equal and contrary attracting infinite forces, and if to either of these forces you add any new finite attracting force, that new force, howsoever little, will destroy their equilibrium, and put the body into the same motion into which it would put it were those two contrary equal forces but finite or even none at all; so that in this case the two equal infinites, by the addition of a finite to either of them, become unequal in our ways of reckoning; and after these ways we must reckon, if from the considerations of infinites we would always draw true conclusions.

. . . I fear what I have said of infinites will seem obscure to you; but it is enough if you understand that infinites, when considered absolutely without any restriction or limitation, are neither equal nor unequal, nor have any certain proportion one to another, and therefore the principle that all infinites are equal is a precarious one.[16]

This statement raises the following questions: was Newton here in effect espousing the actually infinite *differing* cardinalities of the scholastic *compositio* doctrine for metrically unequal intervals in order to provide a deductive justification or deeper *explanatory* basis for his absolutistic conception of metric ratios among space-intervals? Is it plausible that Newton could have resuscitated the *compositio* doctrine without being aware of the devastating proofs of the equi-cardinality of all nondegenerate intervals which had already discredited its account of metric interval-ratios in Euclidean space? Or are we to understand his statement about the metrical 12:1 ratio of a foot to an inch very differently as referring to the merely *potentially* infinite number of parts involved in taking limits of ratios of *indefinitely* increasing finite numbers of ever smaller parts of a foot and of an inch respectively? On this latter reading, the 12:1 metrical ratio is an unexplained *brute fact* of absolute space rather than explained by a deeper underlying *punctal* structure of intervals, and this brute fact ratio is then merely made manifest by the limit of an appropriate sequence of finite ratios.

These questions about Newton's intention are, of course, primarily *historical*. But I shall document later that the unavailability of a viable deeper explanatory basis for metric absolutism was relevant to the intellectual dissatisfactions that drove thinkers like Riemann and Weyl to advocate a specifiably *relational* conception of the metric structure of an *n*-dimensional (i.e., continuous) *physical* manifold. Hence let me develop my conceptual motivations for my historical questions.

In purely kinematic contexts no less than in specifically dynamical ones, Newton speaks of certain attributes or entities as "absolute" and tells us repeatedly that these are *ontologically autonomous* in the sense of existing "without relation to anything external." Thus, when he outlines the kinematic substratum of his dynamics in the well-known scholium of the *Principia*, he speaks of *metric ratios among time-intervals* as absolute, saying: "I. Absolute, true, and mathematical time, of itself, and from its own nature, flows equably without relation to anything external, and by another name is called duration: relative, apparent, and common time, is some sensible and external (whether accurate or unequable) measure of duration by the means of motion, which is commonly used instead of true time; such as an hour, a day, a month, a year."[17] He goes on immediately to stress the like ontological autonomy of the spatial "figure and magnitude" of any portion of empty absolute space, writing:

II. Absolute space, in its own nature, without relation to anything external, remains always similar and immovable. Relative space is some movable dimension or measure of the absolute spaces; which our senses determine by its position to bodies; and which is commonly taken for immovable space; such is the dimension of a subterraneous, an aerial, or celestial space, determined by its position in respect of the earth. Absolute and relative space are the same in figure and magnitude; but they do not remain always numerically the same. For if the earth, for instance, moves, a space of our air, which relatively and in respect of the earth remains always the same, will at one time be one part of the absolute space into which the air passes; at another time it will be another part of the same, and so, absolutely understood, it will be continually changed.[18]

Clearly Newton asserts that there are *unique* metric interval-ratios, which exist "without relation to anything external" like a physical clock or spatial yardstick.[19] Thus unique metric ratios among intervals of empty container space, on the one hand, and of receptacle time on the other, are ontologically quite independent of measuring standards, in the sense that these ratios are *not* at all first generated or *ontologically* constituted by relations to "sensible and external" physical devices or processes: even when a clock yields an "accurate . . . measure of duration" for "equable" lapses of time, and when a motion is uniform, any and all external physical entities are *ontologically adventitious* for the existence of the *unique*

309

metric ratios of time-intervals and/or of space-intervals. But such *external* physical agencies can then serve as *epistemological* vehicles for human discovery of the uniquely true ratios, if suitable theoretical care is exercised in the interpretation of the ensuing results of measurements. In short, in Newton's view, the role or status of *external* metric standards *as such* is *at best epistemological*, since container space and receptacle time are each endowed with metric properties while containing any such external entities *only fortuitously* or *contingently*. In particular, the metric structure implicit in container 3-space and independent of anything external is Euclidean.

As we shall see, when applied specifically to *metric ratios* among intervals of space or time, the stated Newtonian distinction between absolute or nonexternal and relative or external was destined to play a *twofold* fundamental role in Riemann's Inaugural Dissertation. There Riemann *contrasts* "implicit" metrics with metrics imposed externally. But, as we shall see, Riemann and then Weyl employed this distinction while advocating a *doubly relational* instead of an absolutistic ontology of the metric structure of an *n*-dimensional manifold of space or space-time.

There is a distinctly *post*-Newtonian usage of "absolute" and "relative" that has gained currency since the advent of relativity theory: "absolute" is *synonymous* with "*invariant* with respect to a specified class *CS* of coordinate systems," while "relative" is synonymous with "*non*invariant with respect to *CS*." In the Newtonian context, absoluteness is *intended* to be a property whose ascription is predicated on having singled out a *preferred* frame, even though Newton did not *succeed* in thus singling out his absolute rest frame.[20] In contrast, in the stated post-Newtonian usage, the ascription of absoluteness in the sense of invariance, far from being predicated on the existence of a preferred frame, is intended to stress independence of any such frame. Indeed the post-Newtonian twentieth century usage is not even extensionally equivalent to the Newtonian meanings of "absolute" and "relative," as we shall now illustrate.

(1) Two temporally *successive* events which occur at the *same* place in Newton's preferred absolute rest system Σ (primary inertial frame) and thus have the same *absolute position* are certainly *not* invariantly isotopic with respect to *all* inertial frames, which include an infinitude of secondary inertial frames. Nor is the absolute spatial distance between two nonsimultaneous events that are nonisotopic in the absolute frame Σ numerically invariant with respect to *all* inertial frames. When speaking here

in the singular of Newton's primary inertial system Σ, I ignore the unnecessary complication that, strictly speaking, this is a special class of inertial systems, differing from each other only with regard to the orientation of axes and the choice of spatial and temporal origin.

(2) Newton speaks of moving frames as "relative" spaces while asserting *invariance* in the sense that "absolute and relative space are the same in figure and magnitude."[21]

(3) The *velocity* of a particle which Newton would regard as "absolute" in the sense that it is referred to his presumed ontologically autonomous, *preferred* inertial system is clearly *not invariant* with respect to *all* inertial systems but is, in general, different in a "moving" inertial frame. Needless to say, despite *this* frame-dependence of velocity, the term "absolute velocity" is *not* a *contradictio in adjecto* in Newton's theory: although the particle to which an absolute velocity is ascribed is "external" to Newton's absolute space and time, and thus the velocity attribute is relational, the absoluteness of space and time as metrically structured systems of points and instants is held to derive from their ontological autonomy. And the velocity can be "absolute" in the sense of being the velocity with respect to the (elusive!) absolute inertial frame.

There are *some* Newtonian attributes that qualify as absolute in the latter sense but *also* happen to be invariant with respect to all inertial frames under the Galilean transformations of Newton's theory: the magnitude of the linear *acceleration* of a particle, the simultaneity of pairs of events, the spatial distance ratio among pairs of simultaneous events, and the metric ratios of time intervals are examples. But the existence of such examples should not tempt us to treat the Newtonian distinction between absolute and relative as synonymous with the antithesis between being *CS*-invariant and its negation.

Despite the post-Newtonian use of the terms "absolute" and "relative" to render the latter antithesis, the *Newtonian* distinction is still employed by contemporary writers even in the context of general relativity theory, for example, to characterize *rotation* as "absolute." Thus, J. L. Anderson writes that the matter in a Gödel universe "can be said to be in a state of absolute rotation."[22]

So much for a sketch of some salient features of Newton's absolutistic conception of space and time structure as *not* requiring the existence of external entities causally, let alone logically. This sketch will be useful for our inquiry into the *conceptual significance* of the quite interesting but

311

risky recent historical conjecture by K. C. Clatterbaugh:[23] in the 1693 letter to Bentley, Newton deductively justified his absolutistic conception of metrical interval ratios in space by the postulate that these ratios *derive* from appropriately unequal or equal transfinite cardinalities of the actually infinite sets constituting the intervals. As applied to the length of a particular interval and the lesser lengths of its proper subintervals, this postulate entails that Euclid's axiom, "The whole is greater than any of its parts," holds for the corresponding cardinalities and thus is, of course, strikingly incompatible with the Cantorean characterization of an infinite set as having the *same* cardinality as at least one of its *proper* subsets. For the sake of brevity, I shall refer to this anti-Cantorean postulate as "ACP": thus ACP asserts that all metrical interval ratios among intervals of positive length *derive* from the transfinite cardinalities of the intervals in question. Incidentally, in his proposed exegesis of Newton's 1693 letter, Clatterbaugh makes no mention of the late medieval ancestry of ACP in the *compositio* doctrine, when he represents Newton as an exponent of ACP.

Let us pause to comment on the exegetical problem of interpreting Newton's letter as championing ACP before trying to state the conceptual significance of that interpretation. The exegetical questions I am about to ask are not discussed by Clatterbaugh but have had the benefit of Howard Stein's privately communicated arguments against Clatterbaugh's interpretation.

Quite apart from the text of his letter to Bentley, there is documentary evidence that Newton did believe in the existence of actually infinite sets, as distinct from merely potentially infinite *finite* sets.[24] But Howard Stein has argued that in the work on infinite series by Wallis which Newton mentions to Bentley, one is *not* dealing with any ratios of infinite numbers as such, but rather with the computation of limits of ratios of *finite* numbers, when the numerator and denominator are only potentially infinite by merely increasing boundlessly. Why is Newton urging Bentley to renounce the kind of uncritical treatment of infinites that regards all infinites as equal without restriction? Clearly Newton's *initial* motivation for making this point arose from Bentley's question about gravitational disequilibrium in infinite space. And Newton explains that the uncritical treatment of all infinites as equal leads to error in this gravitational case no less than it would if it were invoked in dealing with two further cases: (1) "the various proportions of infinite magnitudes" which were analyzed carefully by Wallis, and (2) "the infinite number of infinite [infinitely] little parts of an inch" vis-à-vis "the infinite number of those parts in a foot."

Precisely how did Newton construe what is *common* to all three of the cases mentioned by him? It is clear enough that each of these cases involves a ratio *different from 1*: (1) the ratio of infinite gravitational forces that produce disequilibrium, (2) Wallis's infinite magnitudes whose ratio differs from 1, and (3) the 1:12 ratio of the infinite number of infinitely little parts in an inch to the infinite number of those parts in a foot. Newton is surely telling Bentley that each of these three cases furnishes a warning against uncritically equating all infinites. But does Newton intend to say *furthermore* that *in all three cases alike* the failure of the relevant ratio to be unity is to be understood in terms of the merely *potentially* infinite numerators and denominators that Stein sees in Wallis's treatment of infinite series? Or does Newton mean, as Clatterbaugh thinks, that the 1:12 metrical ratio of an inch to a foot is to be *explained* on the basis of the ACP assumption of appropriately different transfinite cardinalities of the points composing intervals that differ in length? In opposition to Clatterbaugh's reading, Howard Stein doubts very much that Newton could have overlooked the *equi*-cardinality of metrically unequal intervals, an oversight which we recall to have been the major error of the medieval advocates of the *compositio* doctrine. Thus, in regard to the relevant notion of infinity, Stein regards Newton's conception of the metrical ratio of a foot to an inch as precisely analogous to Stein's reading of Wallis's ideas on proportions.

I take no stand on the historical dispute over the proper exegesis of Newton's 1693 letter. Perhaps further evidence on the matter will be found in, say, the forthcoming volumes of Derek Whiteside's new edition of Newton's Mathematical Papers (Cambridge University Press), or in the still unpublished *Quaestiones*, which are being edited by James E. McGuire.

But from a conceptual rather than historical perspective, I now venture to speculate on the significance of Newton's advocacy of absolute metric interval-ratios for the subsequent *logical* development of the theory of space. Suppose Newton did actually espouse ACP because he somehow overlooked its untenability. Then he would have thought himself to be in possession of an outline of a *deductive theoretical explanation* of the absoluteness of metrical interval-ratios, rather than of merely some kind of overall *inductive confirmation* of the theory asserting their absoluteness.

313

And this explanation-sketch would have had at least the promise of *specifically* justifying, via a deeper structure, the claim that the metric interval-ratios are absolute: the justification would be given in terms of more fundamental presumed structural features that are absolute, to wit, the presumed infinite cardinalities of intervals.

Whatever the merits of the speculation that Newton espoused ACP, the legacy Newton bequeathed to Riemann surely did *not* include any deeper, *mathematically viable* explanation of the purportedly absolute ratios, but only an inductive confirmation of Newton's theory as a whole. And in the absence of a viable deeper justification of the absoluteness of metric interval-ratios, the stage was set logically for a momentous event at a time when Newton's physics was hardly in jeopardy from adverse experimental findings: as early as 1854, at the *beginning* of his Inaugural Dissertation (Section I, 1), Riemann explicitly *denied* the then received Newtonian thesis that unique metric interval-ratios exist in continuous physical space and time "without relation to anything external": instead he enunciated the opposing relational thesis of external imposition of metric interval-ratios by transported metric standards. And Riemann did so in Section I, 1 long before saying even a single word to claim *further* that the metric structure is *also* relational in the sense that the metric geometry is *dynamically*-dependent on the matter distribution. Indeed, not until the very end of his Inaugural Dissertation did Riemann couple his predynamical thesis of relationality with the further dynamical one to offer their *conjunction* as grounds for denying that the empirical triumph of Newtonian physics must be held to confer credibility on Newton's metric absolutism, and for expecting the Newtonian geometrical structure to require gradual modification "under the compulsion of facts which it cannot explain." [25]

Weyl's exegesis of Riemann's ontology of metric structure in space and time took explicit cognizance of its *doubly* relational character. For, as we shall see, Weyl perceived clearly that, *in Riemann's view*, the dynamicity of the metric geometry is effected by the matter distribution *only* through the (actual or potential) causal *mediation* of external, transported, metric standards which first generate the metric interval ratios as such. It behooves me to try to articulate this exegesis of Riemann's postulational reasoning a good deal more precisely than I have done heretofore. [26] For despite Weyl's contrary statement, Riemann is misportrayed in some quarters—by retrospection from a *particular* interpretation of general

314

relativity theory!—as having *confined* his relationalism to the dynamicity thesis to the exclusion of the following claim of his: ontologically—and *not* just epistemically—some external transported metric standards or other (e.g., rigid rods) function indispensably as *mediating* causal agents, *rather than* as mere manifesting devices, in the dynamical dependence of the *geometry as such* on the matter distribution. In other words, this causal mediation claim asserts that a physical field produced by the matter distribution *acquires geometrical significance* only via its action on particular kinds of external transported physical standards, which, to begin with, generate metric interval ratios in the continuous spatial manifold altogether.

As we shall see in due course, it is one thing to avow that the specific articulation of Riemann's quite sketchy dynamicity thesis in standard general relativity theory ("GTR") does not assert this causal mediation claim and may even be incompatible with the latter claim; or that a suitable physical, second-rank tensor field "carried" by the differentiable space-time manifold does in and of itself endow this manifold ontologically with a metrical structure, without reliance on any external metric standards as such (e.g. atomic clocks, freely falling particles), but that such standards will serve to *manifest* this otherwise autonomously geometric structure, and hence may be *epistemically* indispensable for human discovery of this structure; or yet that the causal mediation claim is at best contrived and thus much less fruitful or exciting for physics than the dynamicity thesis.

But it is quite another thing to foist the stated presumed import of GTR onto *Riemann* by pretending that *he* did not espouse the causal mediation claim, and thereby to obfuscate the difference between his ontology of a dynamical geometric structure in a differentiable spatial manifold, on the one hand, and the avowed ontology of the *metrical field* which is asserted to have the sanction of GTR, on the other.

We now turn to the more precise articulation of the exegesis of Riemann's philosophy of geometry, partly because the logical distinctions that will emerge shall serve our subsequent philosophic concern with the current status of the rivalry between absolutistic and relational ontologies of space-time.

3. Riemann

As will soon become evident, Riemann contrasts discrete space with continuous space by claiming that "the principle of metric relations" is "implicit" in the notion of the former, while it must come from elsewhere in the case of the latter. Hence the following nearly prescient anticipation of Riemann by Thomas Aquinas, which Gerald J. Massey has pointed out to me, deserves mention here: Aquinas regarded discrete space as endowed with a *"mensura intrinsica"* ("intrinsic metric"), while deeming continuous space or time—qua duration of things changeable in their very substance—to have interval congruences only with respect to a *"mensura extrinsica"* ("extrinsic metric").[27] And, as emerges from the context, Aquinas holds (*ibid.*, chap. 5, "What is the subject of Aevum?") that there is latitude for "decree" in the choice of a "mensura extrinsica" not only in regard to the unit but also as between the different congruences generated by alternative choices of an extrinsic metric standard.

Turning to the text of Riemann's 1854 Dissertation, note that he devoted section I to "The Concept of an *n*-ply Extended Manifold." And in that section I, subsection 1, he wrote:

Determinate parts of a manifold, distinguished by a mark or by a boundary, are called quanta. Their comparison as to quantity comes in discrete magnitudes by counting, in continuous magnitude by measurement. Measuring consists in superposition of the magnitudes to be compared; for measurement there is requisite some means of carrying forward one magnitude as a measure for the other. In default of this, one can compare two magnitudes only when the one is a part of the other, and even then one can only decide upon the question of more and less, not upon the question of how many [much].[28]

Note several points at once:

(1) Riemann treats it as *given* whether the spatial "manifold" is "continuous" or "discrete" and addresses his foundational search only to the basis for the metric. In short, he does ask where the metric comes from in each of the two cases considered by him, but he does not ask where the underlying continuity or discreteness comes from.

(2) If physical space is a continuous manifold, he tells us that—ontologically and hence epistemically—"for measurement there is requisite some means of carrying forward one magnitude as a measure for the other." And "in default of this" *external* transported metric standard, metric comparability of spatial intervals degenerates into a system of merely *qualitative* relations of proper inclusion. This is clearly a *relational* conception of metric interval-ratios in the nontrivial sense that determinate metric interval-ratios are held to depend ontologically on an *external*

transported standard in addition to being trivially relational qua being ratios. And yet the former *non*trivial relationality is surely logically *pre-dynamical*, since nothing is said as yet concerning any dynamical dependence of metric ratios on the distribution of matter! I emphatically submit that it is *wrong* to interpret the role here assigned by Riemann to external transported standards as being *purely manifestational and epistemic*. For, as we shall see below, such an interpretation is quite at odds with the manner of his foundational invocation of transported external metric standards in the rationale he offers for his dynamicity thesis in his section III, 3.

Riemann's concluding section III, which is entitled "Application to Space," begins with these words: "Following these investigations concerning the mode of fixing metric relations in an *n*-fold extended magnitude, the conditions can now be stated which are sufficient and necessary for determining metric relations in space." [29] And in the last subsection of that concluding section III, Riemann first recalls from section I, 1 his thesis that in continuous physical space, metrical interval-ratios are imposed by external transported metric standards, writing:

. . . in the question concerning the ultimate basis of relations of size in space . . . the above remark is applicable, namely that while in a discrete manifold the principle of metric relations is implicit in the notion of this manifold, it must come from somewhere else in the case of a continuous manifold. Either then the actual things forming the groundwork of a space must constitute a discrete manifold, or else the basis of metric relations must be sought for outside that actuality, in colligating forces that operate upon it.

A decision upon these questions can be found only by starting from the structure of phenomena that has been approved in experience hitherto, for which Newton laid the foundation, and by modifying this structure gradually under the compulsion of facts which it cannot explain. [30]

Let me call attention to two sets of features of Riemann's reasoning here in order to comment on their cogency thereafter:

1. Note that when Riemann turns to the characterization of "the ultimate basis of relations of size in space" in section III, 3, he *first* stresses the applicability of his "above [section I, 1] remark . . . that while in a discrete manifold the principle of metric relations is implicit in the notion of this manifold, it must come from somewhere else in the case of a continuous manifold." Specifically, as shown by his very wording in this

recapitulation, he had told us in section I, 1 that in continuous physical space, the *non*implicit "principle of metric relations" is *constituted* by external, transported metric standards. Metric interval ratios, *qua ratios*, are, of course, *trivially* relational in discrete space no less than in continuous space. But Riemann's point here is that in a metrically structured, continuous, physical space, "the principle of metric relations" is *itself* relational in the sense of being externally constituted, instead of being absolute by being "implicit." Observe that Riemann uses the strong term "must" ("muss" in the German original) in the phrase "must come from somewhere else," when recalling the contention of his section I, 1. His use of the term "must" strongly suggests that he regards this *predynamical* relationality of the ontologically constitutive standard or criterion of metric interval-ratios to be *deducible* from the presumed continuity of physical space, as I stressed in *PPST*, pp. 9 and 16.

Several of my critics have overlooked the fact that in contrast to Riemann, I have *explicitly* advocated (since at least 1970) that this predynamical thesis of externality or relationality is an ontological claim having the character of an *additional* physical hypothesis. Thus I have maintained in my *PPST* (pp. 498, 527–528, and 753–755) that this thesis is not certifiable a priori for an assumedly continuous and metrically structured physical space, but is an *empirical hypothesis* concerning "the ultimate basis of relations of size" in that space. For example, I characterized as "empirical" (italics in original) the assumption that RMH [Riemann's predynamical thesis of the externality of the ontologically constitutive criterion of metric interval-ratios] is *true*.[31] I said that RMH is empirical, if only because the mere *logical* possibility of the truth of Wheeler's pre-1972, thoroughly absolutistic geometrodynamics suffices to show that RMH is indeed an additional physical hypothesis, *not* deducible from the presumed continuity of space-time.[32]

Thus I use the label "RMH" (short for "Riemann's Metrical Hypothesis") to identify a hypothesis which I construe to be contingently true—if true at all—and certifiable only empirically (a posteriori), whereas I declared earlier that Riemann had apparently meant to enunciate it as necessarily true and certifiable a priori. In the face of this discrepancy, it might be asked why I use Riemann's name to identify the hypothesis in my divergent construal. To this I say that I do not invoke Riemann's authority in order to lend *credence* to RMH, all the less so since I am concerned to stress the consequences that follow, *if* RMH is empirically true, rather than to assert categorically that it is empirically very highly confirmed. My linking of Riemann's name to RMH seems to me justified, even though I do not intend it to lend credence to RMH: the justification derives from the fact that, for whatever reasons, Riemann did espouse the assertion made by RMH, and that he then proceeded to invoke it at the end of his Inaugural Dissertation in order to justify his *dynamical* conception of physical geometry.

Early in my *PPST* (pp. 9 and 16), I had pointed out that Riemann had held the following: continuity furnishes "a sufficient condition for the intrinsic metric amorphousness of any manifold *independently of the character of its elements*." (*PPST*, p. 16, italics in original.) But I neglected to reiterate explicitly that Riemann conceived of his RMH as a necessary, a priori truth, when I set forth my own divergent construal of it as an *empirical*, a posteriori claim in the later chapters (16 and 22) of that book. Perhaps it is therefore understandable that Paul Horwich's criticism of my prior exegesis of Riemann turned out to be at cross purposes with my concern with RMH.[33]

2. Despite the conciseness of Riemann's wording, it is abundantly clear that he is championing *two* kinds of externality in his ontology of the metrical structure of continuous physical space: (1) the inevitable externality of the ontologically constitutive *"principle"* or criterion which first generates determinate metric ratios at all under transport ("das Princip der Massverhältnisse" in the German original); hence he says that the constitutive criterion "must come from somewhere else ["anders woher hinzukommen muss"] in the case of a continuous [spatial!] manifold," and (2) the externality of the *causal basis* of metric ratios ("der Grund der Massverhältnisse"), a "basis" which "must be sought for outside" of space, "in colligating forces that operate upon it," *because* these outside forces determine what particular metric interval-ratios will be *generated* by the constitutive external standards under transport. Clearly, with respect to metric properties, Riemann assigns an essential ontological role both to "das Princip" and to "der Grund": without external, transported metric standards, there would be no metric ratios at all in continuous space, but only qualitative relations of inclusion; but the role of the outside colligating forces ("der Grund") is that their action on the constitutive standards determines the particular metric ratios which the latter generate under transport.

Since the *first* of these two theses of externality is an assertion of *on-*

tological indispensability, I shall hereafter denote this first thesis of externality by the acronym "OI" (for "ontological indispensability"). And when applied to a type of space *other than* continuous space, I shall speak of the claim of ontological indispensability as "the counterpart of OI" for the different kind of space in question. Taken by itself, *without* OI, Riemann's *second* thesis of externality asserts that the metric ratios are *dynamically dependent* on the external forces presumably emanating from the matter distribution. Note that without OI, this second thesis is noncommittal in regard to "the principle" of the metric ratios, and thus does *not* require the *causal mediation* of the dynamical dependence of the geometry by transported metric standards. I shall use "DH" to refer to the mere *dynamical hypothesis* affirmed by Riemann's second thesis of externality.

Having avowed OI, Riemann seems to think that he can then *deduce* DH. Why do I interpret Riemann as thinking that DH thus follows from OI which, in turn, he seems to regard as deducible from the presumed continuity of space? I maintain this interpretation because he employs the *two* words "then" and "must"—"Es muss also" in the German original—in the *second* sentence of our section III quotation from him: there he tells us that if space is continuous, "then . . . the basis of metric relations must be sought for outside that actuality, in colligating forces that operate upon it."

But the logical situation is different from what seems to be Riemann's account of it here. As we saw, OI and DH assert respectively the externality of the constitutive *criterion* and the externality of the *causal basis* of metric ratios. Thus it is quite true that both OI and DH affirm the *external* dependence of the very determinateness of metric ratios. Nonetheless we can now show that OI *neither* entails *nor* is deductively entailed by DH.

That OI does not entail DH is evident from the logical possibility of a world in which OI is true, while DH is false: relationality (externality) to the extent asserted by OI does not logically require, although it allows and even *inductively suggests,* the kind of relationality (externality) inherent in dynamicity. For example, if rigid rods play the role required by OI, then it does not follow from OI alone that the spatial geometry generated alike by all kinds of rigid rods of whatever constitution *must* be dynamical. But OI *heuristically begets* and even strongly suggests DH, so that the latter is a very reasonable speculative or postulational outgrowth of OI. Indeed, it seems very doubtful that Riemann would ever have envisioned DH without this inspiration from OI! Since Einstein's field equations

320

were, of course, unavailable to Riemann, just what *other than* OI could even have *suggested* to Riemann to postulate DH?

Does DH unilaterally entail OI? I failed to acknowledge explicitly in my earlier writings on Riemann's views that DH itself no more logically entails OI than the latter entails the former. Yet it is important to see that DH does not entail OI: the kind of externality inherent in the causal dependence of metric structure on the distribution of the matter immersed in space does *not itself* require logically that this causal dependence *also* be *mediated* constitutively—as distinct from merely *manifested*—by an external, transported metric standard of one kind or another. Instead, taken by itself, DH *allows* that the dynamically dependent metric structure be merely *manifested* rather than first generated by such standards. For DH allows that a physical field produced by the matter distribution does not first have to *acquire geometrical significance* from its action on those transported bodies that count as metric standards.

On the other hand, once it is assumed with Riemann that OI is true, then, of course, the matter distribution can affect the metric geometry dynamically *only* through the then essential causal *mediation* of external metric standards. For if determinate metric ratios are first *generated* ontologically in continuous space by external, transported standards, then it is only by affecting the metric behavior of these standards that the matter distribution can influence the geometry causally. In brief, Riemann holds that without an external standard there can be neither a nondynamical nor a dynamical geometry in continuous physical space, so that any dynamicity of the geometry must be mediated by the transported standard. But as we saw, Riemann *seems* to have gone further by viewing DH as an entailment of OI. Nevertheless, I prefer to assume more conservatively that he regarded the bearing of OI on DH to be that of the heuristic ancestor of DH. This more conservative assumption was my basis for having said earlier that Riemann *"coupled"* DH with OI and offered their *"conjunction"* as ground for impugning the continued empirical success of Newton's metric absolutism.

In contrast to those who would like to believe that Riemann's metric relationalism was *confined* to DH, Hermann Weyl is cognizant of the role of OI.[34] Weyl makes this clear in his account of the "full purport" of Riemann's rejection of Newton's metric absolutism in the 1854 Inaugural Dissertation. Speaking of Riemann's concluding remarks in that Disserta-

321

tion, Weyl tells us that "to make them quite clear," he "must begin" by pointing out that, according to Riemann, in *discrete* space the counterpart of OI is false. On the other hand, when characterizing Riemann's view of the ontological basis of metric ratios in *continuous* physical space, Weyl writes in italics: "*He* [Riemann] *asserts . . . that space in itself is nothing more than a three-dimensional* [continuous] *manifold devoid of all* [metrical] *form; it acquires a definite form only through the advent of the material content filling it and determining its metric relations.*"[35] Thus it would appear that Weyl reads Riemann as having asserted both OI and DH in regard to continuous physical space. Weyl's emphatic statement is preceded by two other statements, which he intersperses with a full citation of the last three paragraphs of Riemann's section III, 3. In these two statements, from which I have previously cited only the opening few words, Weyl says:

To make them quite clear I must begin by remarking that Riemann contrasts *discrete* manifolds, i.e. those composed of single isolated elements, with *continuous* manifolds. The measure of every part of such a discrete manifold is determined by the *number* of elements belonging to it. Hence, as Riemann expresses it, a discrete manifold has the principle of its metrical relations in itself, *a priori*, as a consequence of the concept of number. . . .[36]

If we discard the first possibility, that the reality which underlies space forms a discrete manifold—although we do not by this in any way mean to deny finally, particularly nowadays in view of the results of the quantum-theory, that the ultimate solution of the problem of space may after all be found in just this possibility—we see that Riemann rejects the opinion that had prevailed up to his own time, namely, that the metrical structure of space is fixed and inherently independent of the physical phenomena for which it serves as a background, and that the real content takes possession of it as of residential flats.[37]

These citations from Weyl prompt me to make several further interpretive comments.

1. Riemann tells us that in granular, atomic space, the *cardinalities* of the parts of that space endow it with at least one set of nontrivial metric ratios which are implicit, rather than externally imposed, and which are thus independent of contingent external dynamical factors. Now suppose that—contrary to fact!—the assertions made by the aforementioned fourteenth-century ACP of the *compositio* doctrine about the cardinalities of intervals of *continuous* space *had* been tenable. Then Riemann and Weyl would presumably have deemed this a *sufficient* though *not* a

necessary condition for regarding continuous n-dimensional space as endowed with a system of absolute metric ratios. And the latter would then, of course, be *non*dynamical no less than those they attributed to discrete space.

2. As we saw, Weyl's gloss on Riemann's Inaugural Dissertation accords with my contention that it is wrong to ignore the OI thesis in Riemann's relationalism in an exclusive preoccupation with his championship of DH. Oddly enough, while some have made this error, others have unwittingly assimilated OI into the folklore of mathematical physics even to the extent of egregiously imputing its espousal to *Newton* while erroneously *confining* Newton's absolutism to his denial of DH. Yet Newton's absolutism denied OI and DH alike by *denying each* of the *two* kinds of externality respectively asserted by them. Thus it is not enough for Newton if the geometry is held to be "given once and for all," because he rejects even a *non*dynamic conception that makes the fixed geometric structure ontologically dependent on a relation to the behavior of external metric standards. The stated erroneously one-sided readings of Riemann and Newton have given currency—each in its own way—to an inaccurate portrayal of Riemann as having clashed with Newton's absolutism *only* in regard to the truth of DH, rather than in regard to both OI and DH!

Unless I misread J. L. Anderson's account, he wrongly depicts Newton as having espoused the OI thesis. For note that Anderson has Newton conceiving of space as "a three-dimensional manifold upon which was superimposed a . . . metric geometry." Anderson writes as follows:

For Newton, space and time were conceived to be physical entities. Furthermore, they were absolute entities, unchanging and unchanged by the presence of other physical entities. Hence the definitions [footnote reference omitted here]:
1. Absolute, true, and mathematical time, of itself, and from its own nature, flows equably without relation to anything external.
2. Absolute space, in its own nature, without relation to anything external, remains always similar and immovable.

It is clear that in speaking of space and time in this way, Newton was referring to their geometrical properties. Furthermore the concept of a straight or right line was fundamental to the whole Newtonian picture, as was the concept of angles and distances between points. Since no other types of geometries were known at the time, there can be no doubt that when Newton spoke of space he meant the space of Euclid, that is, a three-dimensional manifold upon which was superimposed a flat, metric geometry. And by "absolute space" it is clear that he considered this geometry to be unaffected by any physical disturbances in this space. Likewise, Newtonian time must be understood as a one-dimensional manifold with a "distance" defined on it that is also unaffected by such physical disturbances.[38]

When the logical existence of non-Euclidean geometries became known long after Newton's death, it did become illuminating to analyze any one metrical structure, including the Euclidean one, into a three-dimensional manifold, logically coupled with the appropriate particular metric properties. One could then distinguish the bare manifold from the metric features by speaking of the metric structure as *conceptually* "superimposed" on the manifold structure. In particular, one could, of course, thus analyze the Euclidean structure which Newton attributed to physical space.

But in the very sentence in which Anderson characterizes *Newton's* conception of Euclidean physical space in terms of a "superimposed" flat metric geometry, Anderson rightly emphasizes that Euclid's geometry was the only type of metric geometry known to Newton. *Unless* Anderson imputes OI to Newton, it is precisely Newton's belief in the a priori monopoly of Euclidean geometry that would make it pointless for Anderson to characterize *Newton's own conception* of the metric Euclideanism of his physical space in terms of superimposition. By speaking of "superimposition" presumably in the sense of OI when characterizing Newton's view of Euclidean physical geometry, Anderson does violence to Newton's crucial and repeated phrase "without relation to anything external." For, as we saw, Newton employed this phrase to mean that the OI conception of "superimposition" of a metric structure is no less alien to his absolutism than the dynamicity of the geometry: the nondynamicity of the geometry is only a necessary and *not* also a sufficient condition for the Newtonian absoluteness of the geometry. Howard Stein has remarked to me that it *may* perhaps not have been Anderson's intent to impute OI to Newton after all. If indeed it was not, then my criticism of Anderson's apparent portrayal of Newton should be understood to apply to the at least very plausible reading of Anderson's wording which I have stated.

In any case, those who wrongly *confine* Newton's absolutism to the anti-dynamical thesis that the geometry is, as Anderson puts it later, *"fixed once and for all,"*[39] naturally think they are employing the *Newtonian* concept of absoluteness when they claim that the nondynamicity of the geometry is not merely necessary but also sufficient for the absoluteness of the geometry.

3. The preceding considerations now allow us to comment on the relatively weak sense in which Wheeler's pre-1972 geometrodynamics ("GMD") conceives of the geometry as "dynamical." If we compare the role of space-time in GMD with that of Newton's absolute 3-space, we can say that GMD goes Newton's absolutism one better in that it postulates not only an ontologically autonomous geometric structure but also the reduction to that structure of the material content, which Newton regarded as only ontologically adventitious with respect to the geometry. Thus GMD not only denies OI but views the matter distribution itself as a mere manifestation of the geometry rather than as an ontologically independent agency that causally influences the geometry. Hence GMD cannot view the space-time geometry as dynamical in the material agency sense of standard 1915 GTR, which Weyl articulated above: the latter sense of "dynamical" assigns a *relational* status to the geometry by according to immersed matter a certain kind of *geometrogenic* role, whereas GMD deems the geometry to be autogenic, as it were. Thus, insofar as the space-time geometry of Wheeler's unreservedly absolutistic GMD is indeed "dynamical," it is so in an ontologically different or weaker sense. What then is that other sense?

The word "dynamics" in GMD's name "geometro*dynamics*" serves the distinct purpose of rendering the assertion of the *time*-dependence of specified geometric entities. Although there is no dynamicity in Wheeler's GMD in the relational material agency sense of standard GTR, there is "dynamicity" in the distinct sense of *time-dependence*. Thus in space-times devoid of a time-like Killing vector field, GMD countenances a time-dependent 4-metric in *every* (admissible) coordinate system.[40] And even in space-times endowed with a time-like Killing vector field (stationary and static cases), GMD provides scope for a time-dependence of the 4-metric in suitable coordinate systems. Hence there is scope for the temporal evolution of specified 3-geometries. In *this* sense, GMD can be both "dynamical" and unreservedly absolutistic.

4. Bertrand Russell has emphasized that in talking about continuous manifolds, "Riemann had [physical] space in mind from the start, and many of the properties which he enunciates as belonging to all manifolds, belong, as a matter of fact, only to [physical] space."[41] Generalizing this

point, one can say that a four-dimensional, real number space R^4 and a four-dimensional physical space-time both have the same formal structure qua being four-dimensional manifolds, but that they differ materially, as it were, if one may allude to Aristotle's distinction between form and matter in this context. The distinction between formal and material seems to bear on the elucidation of Riemann's concept of implicit metric to the following extent: as both a formal *and* material entity, a four-dimensional, physical space-time manifold may well *not* implicitly possess some properties that are, however, implicit to the real number manifold R^4 qua real number manifold, and conversely. [42]

4. Is It Ill-Conceived to Inquire into the "Implicitness" versus Externality of the Ontologically Constitutive "Principle" of *Metricality* of the Physical *Space-Time* Manifold?

Earlier in this essay, I have endeavored to document that in the *pre-dynamical* section I, 1 of his Inaugural Dissertation, Riemann posed and answered essentially the following question: if continuous physical 3-space is indeed metric at all, whence does its very metricality—as distinct from the *particular* metric structure possessed by it—derive ontologically? More specifically, given that physical 3-space is both continuous *and* metric, can the very "principle" (ontologically constitutive criterion) of metricality be held to be *implicit* in that metric physical 3-space? Or does that very "principle" derive ontologically from entities outside that physical metric space, qua being *physically* metric?

Weyl has summarized Riemann's answer very concisely in the following statement, which I did not cite above: "According to Riemann, the conception 'congruence' leads to no metrical system at all, not even to the general metrical system of Riemann, which is governed by a quadratic differential form." [43] And Weyl clarifies this statement on the next page by saying that ". . . metric relations are not the outcome of space being a form of phenomena [as for Kant], but of the physical behaviour of measuring rods and light rays. . . ." Note that this statement contains the assertion that "metric relations are . . . the outcome of . . . the physical behaviour of measuring rods and light rays." Hence I take the word "outcome" here to be intended by Weyl to allude to Riemann's and perhaps Weyl's own claim OI concerning the ontologically constitutive generating role of external metric standards *of one kind or another*. But the latter may well *not* include measuring rods. [44] Lest it be thought that the wider context in which Weyl made this latter statement will not bear this reading, I shall presently cite this fuller context. We shall see that there Weyl first focuses on Einstein's implementation of Riemann's mainly programmatic DH and then concludes by enunciating both OI and DH. Weyl writes:

Einstein helped to lead Riemann's ideas to victory (although he was not directly influenced by Riemann). Looking back from the stage to which Einstein has brought us, we now recognize that these ideas could give rise to a valid theory only after *time* had been added as a fourth dimension to the three-space dimensions. . . . In Einstein's theory (Chapter IV) the co-efficients g_{ik} of the metrical groundform play the same part as does gravitational potential in Newton's theory of gravitation. The laws according to which space-filling matter determines the metrical structure are the laws of gravitation. The gravitational field affects light rays and "rigid" bodies used as measuring rods in such a way that when we use these rods and rays in the usual manner to take measurements of objects, a geometry of measurement is found to hold which deviates very little from that of Euclid in the regions accessible to observation. These metric relations are not the outcome of space being a form of phenomena, but of the physical behaviour of measuring rods and light rays as determined by the gravitational field. [45]

The reader can judge whether my exegesis makes me liable to the charge, leveled in Laurence Sterne's *Tristram Shandy*, that "he would move both heaven and earth, and twist and torture everything . . . to support his hypothesis." [46]

But apart from my interpretation of Riemann and Weyl, we need to consider a challenge to the effect that the question which I believe Riemann to have asked in his section I, 1 is *ill*-posed. On several occasions, Howard Stein has challenged the very notion that, with respect to the physical *space-time* manifold PST, there is a well-conceived ontological issue at all between the denial that PST is implicitly metric and the corresponding affirmation. In other words, he regards the question, "Whence does the very metricality of PST derive ontologically?" to be a pseudo-question *ab initio*, if it is not to be an altogether trivial question. I am very grateful to my valued friend Stein for this challenge and hope that my reply to it further on in this essay will be genuinely clarifying. Let me try to state Stein's doubts as plausibly and clearly as I can in a detailed series of propositions. [47] Only after this rather lengthy statement shall I offer the rebuttal that I now consider appropriate at this writing.

(1) When confronted with the question of whether PST implicitly or

intrinsically possesses properties of a metric character, Stein finds himself stymied from the outset. For he feels a crucial need for identification of the object about which the question is being asked. And he notes that independently of *theoretical* identification, he has no access to space-time. Hence if the question is, "Is metric structure intrinsic to space-time qua space-time?" then he finds that he knows of no way to indicate the referent of the term *without begging the question*. He contrasts space-time with, say, the case of a globe sitting on one's desk. He claims to know in some sense "by acquaintance" what object is meant when one speaks of the globe on the desk, and he asks the following question: does the globe intrinsically possess the metrical property of having a surface which is a Riemannian 2-manifold of constant positive curvature?

(2) Of course, Stein is not interested, nor am I, in a *trivialized* reformulation of the question concerning the ontological status of metricality of PST whose soundness is at issue. Thus it is agreed on all sides that *on the purely deductive mathematico-logical level*—as distinct from the level of *physical ontology*—both of the following are the case: (a) If PST is characterized mathematically as possessing a certain metrical structure, then it follows trivially that PST, thus characterized, is endowed with metricality: (b) If, on the other hand, PST is characterized conceptually *only* as a differentiable manifold, then its metricality is obviously *not* vouchsafed deductively. With respect to PST, Stein contrasts (1) the clarity of the notion of what is or is not *logically* (deductively) implicit in the *concept* of a specified structure, with (2) the problematic character of the notion of what properties are or are not *ontologically* implicit in the *object* possessing that structure. Specifically, when commenting on the distinction between intrinsicness to a *concept*, and being intrinsic to an *object*, Stein has said:

The former notion is an exceedingly clear one—despite the attacks that have been made against the concept of "analytic truth." It extends to quite trivial cases, that is, immediate consequences of a definition (e.g., "the notion of distance is intrinsic to the structure of a metric space"); and to quite non-trivial ones (e.g., "A Riemannian structure is intrinsic to a smooth manifold with free mobility," or "a metric structure in the sense of Weyl is intrinsic to a manifold with compatible conformal and projective structures").

The notion of "intrinsicness to an object" is much harder to treat with precision. Adolf Grünbaum, in his discussion of Riemann's views on measure-relations in continuous manifolds, has undertaken this difficult task; but the explications he has advanced seem to me not really clear. What seems to me a great—and perhaps the primary—difficulty here is that of giving an appropriate and adequate characterization of the "object" in question itself.

For instance, if we ask (assuming Newtonian physics) whether "equality of time-intervals" is a relation *intrinsic to the space-time manifold*, and if this is construed (roughly) to mean "whether that relation is involved in the structure of *the space-time manifold itself, considered apart from all other entities*," the quesion at once arises of how to explicate the notion of "the space-time manifold itself," and of the *conceptual* line between it and "all other entities." I see no way to confront the former question independently of the latter; and yet the converse may also seem to hold: that we cannot give a conceptual explication of "the space-time manifold" without begging the question of its intrinsic properties. In a situation like this, I think it's healthy to remember the hackneyed old positivist rubric of the "pseudo-question": the conceptual tangle, or circle, seems to me merely a sign that the question itself, as I have formulated it and in the context in which I have so far put it, isn't a clear question at all.

I am, however, able to attach a meaning to the claim that a relation having the formal characteristics of that of equality of time-intervals *is an objective structural characteristic of the natural world*. To say this is to say that a theory that we have reasonable grounds to accept posits a structure to which such a relation is "intrinsic" in the first—conceptual or "logical"—sense. ("Reasonable grounds to accept" the theory should be understood to imply grounds also for the belief that the structure it posits does not involve "redundant" or "eliminable" components—as, e.g., "absolute position" or "absolute rest," *unlike* absolute simultaneity or equality of time-intervals, are demonstrably eliminable from Newton's theory.)

And now I can state what I have meant in speaking of "ratio of time-intervals," etc., as "intrinsic" to space-time for Newton: I meant that, in Newton's theory, these relations (or functions) are objective characteristics of the world; and that within that theory (or a certain reformulation of it which is not very far from Newton's formulation) a certain aspect of the whole (dynamical and physical) structure is conceptually separated out as "space-time," in such a way that the relations in question do (in the logical sense) belong ("intrinsically") to the structure of space-time.

I have addressed myself to the stated platitudes that trivialize *ontological* implicitness vs. relationality of the metricality of PST into *deductive* implicitness vs. nonimplicitness in a concept. And I have done so with a view to eliciting clarification from Stein by saying the following:

I have no doubt that there are difficulties which beset my own detailed attempt elsewhere to explicate more clearly Riemann's intuitive and merely illustrated distinction between implicit and nonimplicit metrics. [48]

But I submit that whatever follies I have previously committed in my effort to provide such an explication, these follies do not include the attempt to make philosophical capital from tacitly trading on the stated platitudes. Instead, I construed the issue posed by Riemann roughly as follows: if an n-dimensional physical space is a specified kind of metric space, where do its particular metric interval ratios come from ontologically or physically? Am I asking an ill-conceived question?

I continue *not* to underestimate the ramified difficulties of giving a clear explication of Riemann's foundational distinction between implicit and nonimplicit, or of Newton's quite cognate distinction between absolute and external, for that matter. But whatever the deficiencies of my proposed articulation of the distinction, one thing seems to be importantly the case: the stated *trivial* construal of how Riemann and Weyl conceived of spatio-temporal metrics simultaneously trivializes much of the intellectual history from the thirteenth century through Wheeler and Sakharov of the issue between absolutistic and relational conceptions of *metric* structure.

Howard Stein is certainly one of the writers who have *not* trivialized the issue between Newton and Leibniz in regard to the ontological autonomy vs. nonautonomy of space. But Stein has also been skeptical concerning the feasibility of a nontrivial and noncircular formulation of questions as to the existence of implicit or internal *metric* ratios in specified kinds of physical spaces. I hope it will not be taken amiss if I venture to say that although he is a doubting Thomas, Stein seems to appeal to something akin to Riemann's distinction between implicit and external relations in Stein's own account of Newton and Leibniz. To be specific, Stein gives a statement of the difference between Newton and Leibniz in regard to the *character* of the relations which they regarded as individuating the points of space via the principle of the identity of indiscernibles. And Stein does so by saying:

> . . . the relations that constitute space and give its parts their individuality are according to Newton *internal* relations; that is to say, he is content to postulate the entire structure of space, without attempting to derive it from or ground it in the relations of non-spatial entities.[49]

It is true that unlike Riemann, Stein does *not* single out *metrical* spatial relations from other kinds of spatial relations. Nonetheless, Stein's statement seems to be predicated on a conceptual delineation of the difference between the *internality* of spatial relations, on the one hand, and their being grounded "in the relations of non-spatial entities," on the other. Why then need it be conceptually unsound for Riemann to single out metrical relations and assert of them that these particular relations, as distinct from at least some others, are ontologically grounded in the relations to *non*spatial entities? Does internality vs. externality of relations have to be an all-or-nothing affair? If so, why? Or is the difference between internality and externality of relations conceptually unsound even

330

if it is applied to the totality of *all* relations? If so, were Newton and Leibniz debating a pseudo-issue? Stein tells us in his article on Newtonian space-time [50] that he does find intelligible Newton's distinction between the gravitational mass of a body as a property of the body *itself*, on the one hand, and the weight of that body as a property whose possession by the body involves the interaction of the body with the earth's gravitational field, on the other. Why, I ask, is this so very different in perspicuity from Riemann's distinction between a complex property implicit in or internal to the manifold of physical 3-space, and a property which is not implicit in it, the latter property of 3-space being its metric structure? I ask these questions *not* in the spirit of offering a *tu quoque* argument against Howard Stein's skepticism, but rather to make a plea for help from others in the work of more adequately explicating Riemann's concept of implicit metric.

Stein's replies to the questions addressed to him in my preceding statement begin with statements of my questions and are as follows:

(a) *"Why then need it be conceptually unsound for Riemann to single out metrical relations and assert of them that these particular relations, as distinct from at least some others, are ontologically grounded in the relations of non-spatial entities?"*

Answer: It *need not* be conceptually unsound. In fact, I have myself suggested in my own paper [51] that Riemann had excellent reasons for suggesting that *the metrical structure of space* has in a certain sense its roots—ontological, if you will—in something deeper and more comprehensive, namely *the dynamical structure of the world.* (Of course, in general relativity even the notion of "metrical structure *of space*" loses definite meaning; but "dynamical structure of the world" does not—and one, at least, of its aspects is the metrical structure of space-time.)

On the other hand, it is far from clear—to me at least—that there is a conceptual necessity, or even plausibility, in the view that it is legitimate to postulate—and in this sense to hold as "intrinsic," and not "grounded in" anything prior—a manifold topology and a differentiable structure upon the set of all world-points, and not equally legitimate (from a general methodological point of view) to postulate that this differentiable manifold carries a physical field with the characteristics of the Einstein metric field. I think I am only paraphrasing Newton—with whom I agree—when I say the following: If a theory that postulates such a structure is in good agreement with the evidence, it is a good theory. Whether we can then do better—whether we can ground the postulated structures in yet deeper ones, for which we can find good evidence—is always again a legitimate matter for inquiry; but that we *must* do better, and do so in a specified direction, is *never* a legitimate demand.

(b) *"Were Newton and Leibniz debating a pseudo-issue?"*

Answer: No. They were debating a quite real, and rather complex issue and, in my opinion, there was some merit on each side of the debate—in particular, some of Leibniz's views were suggestive and important. But if we simplify, and take the central issue to have been this: *are spatial and temporal relations physically grounded in some more fundamental structure?*—then the answer, it seems to me, is this (in three parts):

(1) In the time of Newton and Leibniz, no such deeper structure was known; and Leibniz's program to found such relations upon *bodies* failed decisively.

(2) To the best of our knowledge today, spatial and temporal relations in the old fashioned sense are partial manifestations of something more fundamental: the Riemann-Minkowski structure of space-time.

(3) No one today has any idea how to ground the Riemann-Minkowski structure of space-time in any more fundamental reality—i.e., any more fundamental theory. In this respect, the Riemannian metric is as fundamental for us today as the structure of absolute time and inertial motion was for Newton. What the future will bring is the task of the future (starting, of course, now).

A question addressed to Adolf Grünbaum: . . . he suggests (although conditionally) that "one can hardly refrain" from asking . . . on Riemann's behalf: "Whence does the space-time manifold . . . get its particular metric structure?" I should like to ask him why it is easier to refrain from inquiring whence the set of points of space-time gets its structure as a differentiable manifold? A topological structure on a set may be thought to be a fairly simple thing (although that does not clearly make it "intrinsic"). A differentiable structure on a set is a thing of formidable complexity; by what principle does it escape the burden of justification that the metric field allegedly has to bear?

I shall now proceed to offer such replies to Stein's doubts as I consider pertinent in the light of these various prior exchanges.

(1) *Can the question "Whence does the metricality of space-time derive ontologically?" be construed such that it does not at the outset either beg the question by presupposing the answer, or become trivial?*

To lay the groundwork for this inquiry, let me first defer the case of space-time and discuss some examples whose features will be very helpful in dealing with space-time.

Consider a certain man named "Jack," who is an uncle in the biological sense rather than in the metaphorical sense of "avuncular." In saying that Jack is an uncle, one superficially treats Jack's unclehood as a monadic property. But this is just elliptical for the fact that ontologically, Jack possesses this property only because he is the uncle *of* someone else who is *external* to himself. Thus, from the two-placed uncle predicate "$U(x,$

y),” which depicts linguistically the fact that being an uncle OF is a dyadic property, one generates the one-placed predicate “$U(x)$”—“x is an uncle”—by binding the initially free variable y via existential quantification: $U(x) \underset{\text{Def}}{=} (Ey)U(x, y)$. And then the free variable “x” can take Jack as one of its values. Thus, as an ellipsis, “Jack is an uncle” is on a par with “London is north.”

On the other hand, in the case of the fact that Jack is a biped, his bipedality as such is *not* ontologically dependent on a relation to a person or other entity external to his own body.[52]

Assume that Jack is not only an uncle but is in possession of two feet instead of being, say, an amputee. Clearly unclehood is trivially *logically implicit* in the *concept* of “Uncle Jack,” since the judgment “Uncle Jack is an uncle” is a tautology. But this *conceptual* and *trivial* kind of implicitness in no way militates against the following *non*intrinsicality or externality: ontologically, the *object* Jack *qua uncle* qualifies as such only in *relation* to at least one other entity (person) *external* to Jack, rather than intrinsically or monadically! The latter ontological relationality holds regardless of whether the individual Jack in question is *identified* by anyone as “Uncle Jack” and thereby also explicitly characterized as an uncle. Yet no corresponding external relation is required ontologically for Jack's bipedality as such. Hence even after Jack has been explicitly characterized as an uncle upon having been *identified* as “Uncle Jack,” such identification does *not at all beg the question* whether his unclehood is ontologically intrinsic to him in the manner of his genuinely monadic property of bipedality! Nor does the use of the identifying description “Uncle Jack” trivialize the question of ontological monadicity vs. relationality into “Is Uncle Jack an uncle?” Despite the stated, purely logical implicitness of unclehood in the *concept* of Uncle Jack, and despite the feasibility of the linguistic ellipsis that superficially depicts unclehood as an ontologically monadic property, unclehood is ontologically *not* a monadic property, i.e., is *not* intrinsic to a person possessing this property.

Turning to Newtonian physics, Stein himself has rightly remarked that Newton's conception of gravitational mass, on the one hand, and *weight*, on the other, can be contrasted as follows:

(1) The gravitational mass $M_G(A)$ of a body A is a well-defined property when A is specified, because it is unaffected by anything external to the body and constant over time. Thus for Newton gravitational mass is pos-

sessed by a body "in itself" ("*in se*"), or *absolutely*, just as the metricality of physical time is absolute, because "true" time—in Newton's words— "of itself, and from its own nature, flows equably without relation to anything external." Hence the linguistic representation of the quantitative property of gravitational mass by a one-placed functor is not ontologically superficial or misleading, because this quantity is rendered—with respect to a given physical unit—by the numerical values of a function whose argument ranges over *single* bodies.

(2) In contrast, despite the ordinary usage of the term "weight" as a one-placed functor taking the names of single bodies as values of the free variable, the ontological status of weight in Newton's theory is very different from that of gravitational mass with regard to monadicity vs. polyadicity: whereas the property of gravitational mass is nonrelational or *absolute*, that of weight is "*relative*" (relational). For a body's weight depends on the body's situation, because it is the *net* gravitational force on the body exerted by various other bodies at any given time t. Thus we should speak of "the weight of body A toward body B at time t" ("$W(A, B, t)$"). In the case of "particles" ("point-masses") A and B, this quantity does not depend explicitly on the time but rather on the distance between A and B. But if we restrict ourselves to a fixed, small, terrestrial region, then the *monadic* function $W(A)$ can be used elliptically no less intelligibly than the above bona fide monadic function $M_G(A)$.

Let us now go beyond our example of Uncle Jack to consider whether the drawing of a conceptual line between two different entities (e.g., persons) requires that the properties invoked in the respective identifying descriptions are *eo ipso* or automatically intrinsic (monadic). Does the conceptual delineation of two entities as different require a prior sequestration of their monadic (intrinsic) properties from their polyadic (relational) ones for the purpose of using only the former in identifying descriptions? In other words, does the conceptual delineation of two entities prejudge in viciously circular fashion that the properties invoked in their identifying descriptions will willy-nilly turn out to count as respectively intrinsic subsequent to the respective identifications?

Observe that one person (entity) can be identified as the referent of some statement by characterizing him as the president of the U.S.A. in 1975, and *another* person can be identified as a referent by characterizing the latter as the president of France in 1975. Yet the properties ingredient in the respective characterizations that effect these identifications are each relational (polyadic) rather than intrinsic or monadic. Moreover, regardless of which particular properties effect identifications of the two individuals, one can say that for each of the two men, their presidenthood is ontologically a polyadic property, their bipedality is ontologically a monadic (intrinsic) property, while their respective Newtonian terrestrial or lunar weights and even their time-dependent Newtonian gravitational masses are relational. However identified, the difference in identity between the two presidents turns on there being some difference between their properties, *not* on whether the properties with respect to which they do differ are intrinsic (monadic) or relational. I am assuming that "monads" do have "windows"! *By invoking a particular set of properties initially in a definite description to identify these (human) objects, one does* not *thereby prejudge or determine the degree (monadicity vs. polyadicity) which these and other properties of theirs will turn out to have subsequent to the identification.* Thus, by drawing the conceptual line between President Ford himself and President Giscard d'Estaing himself, and indeed by identifying each of them uniquely as himself, one does *not* automatically sequester the identifying properties as intrinsic (monadic) rather than relational properties. And therefore the reliance on properties in effecting the unique identification is *not* such that one will have begged the subsequent question of which properties of the identified entities are intrinsic and which are relational.

Yet in regard to space-time, Stein has suggested the likelihood that "we cannot give a conceptual explication of 'the space-time manifold' without begging the question of its intrinsic properties." Specifically, we recall that when he considers the question whether in a particular theory, physical space-time intrinsically possesses metric properties, he is stymied from the outset, since (a) he does not know how to identify the referent (object) of the question other than theoretically and (b) he does not see how such a theoretical identification can be given and/or the line drawn conceptually between space-time and all *other* entities, without either begging or trivializing the question concerning the intrinsicality vs. relationality of metric properties.

The analysis I gave of the examples of Uncle Jack and of Presidents Ford and Giscard d'Estaing carries over, *mutatis mutandis*, to sustain the following claim: if, say, the attribute of having an infinite volume (in the metrical sense) is a logical part of an *identifying* characterization of the total 3-spaces of Newtonian physics, the mere fact that metrical infinitude is thus invoked identifyingly does *not* prejudge but *leaves entirely open* the question whether the metrical attribute in question or metricality of 3-space generally is (1) ontologically monadic (intrinsic, absolute)—by analogy to Jack's bipedality—by obtaining "without relation to anything external," as Newton maintained; *or* (2) ontologically relational in the sense of OI—by analogy to Jack's unclehood—as claimed by Riemann. Even though Newtonian 3-space is here avowedly characterized as metrical at the outset, this in no way begs the question of whether (1) the metricality of metrically-structured Newtonian 3-space is a bona fide monadic property, ontologically intrinsic to the object possessing it, so that external mensurational devices (e.g., rigid rods) play no ontologically constitutive role in that property but, at best, only a *manifestational* or *probative* (epistemic) role; *or* (2) qua metrically-structured object, the 3-space of Newtonian physics qualifies as so structured only in relation to such external entities as rigid rods, so that these external devices do play an ontologically *constitutive* role in the very metricality of 3-space.

Just as in the examples of bipedality and unclehood, this issue of ontological monadicity vs. polyadicity of the metric structure of 3-space is not settled by the linguistic feasibility of speaking of 3-space as metric in terms of a one-placed predicate. For that linguistic feasibility obtains in the case of ontological dyadicity no less than in the case of ontological monadicity. Thus, in either case, one can say by means of one-placed predicates that Newtonian 3-space is metrically-structured or that it is infinite in size. Nor is it at all material to the ontological degree of the property of metricality of 3-space that metricality is logically implicit in the *concept* of a metrically-structured 3-space, and indeed trivially so. For as shown by the example of Uncle Jack's unclehood, this trivial logical implicitness holds even if metricality is ontologically relational.

I submit that Newton's identification of the 3-spaces of his physics was no less theoretical than the identification of the Newtonian space-time manifold by reference to which Stein stated his argument concerning vicious logical circularity. Similarly for the identification of Einsteinian space-time. And therefore I cannot see that the question of the intrinsicness (absoluteness) of the metricality of space-time is begged at all by including metrical structure in the very identification of physical space-time, any more than this question is begged in the case of the corresponding identification of Newtonian 3-space. It is true, of course, that there are important differences in relative ease of observational or epistemic accessibility between the presumed metrical features of space-time on the one hand, and the *coarser* features of a household globe or of Uncle Jack on the other. And it will be recalled that Stein endeavored to contrast space-time with the household globe in regard to the circularity of the question of their respective intrinsic properties. When claiming that there is such a contrast, he deemed it relevant to emphasize the following: he knows ("by acquaintance," as it were) the globe's identity, whereas the identification of the physical entity called "space-time" is solely theoretical.[53] But, as I have argued, the avoidance of the vicious circle alleged by Stein in the case of space-time hardly requires at all that, *independently* of the identification of physical space-time which is supplied by *theoretical* description, we also have ostensive, direct access to the *referent* of that description: there is no such vicious circle in the case of space-time, any more than in the case of the household globe, and in neither case does its avoidance turn on the availability of some sort of Russellian "knowledge by acquaintance" of the identity of the object in addition to its theoretical identification. It is therefore immaterial here that there may be some sense in which Stein is entitled to claim possession of "knowledge by acquaintance" as to what object is meant when one speaks of the globe on one's desk before asking whether *its* metrical structure is intrinsic to it. The theoretical identification of space-time no more viciously prejudges the answer to the question of *its* intrinsic properties than the theoretical-cum-ostensive identification of the household globe begs the question of the latter's monadic properties.

More generally, there may indeed be *epistemically* significant differences in ostensive access between space-time on the one hand, and other entities (e.g., Uncle Jack, Presidents Ford and Giscard d'Estaing, my household globe) on the other. But whatever the epistemic significance of these differences, they cannot serve to sustain Stein's charge that owing to the unavailability of an independent ostensive identification, any *theoretical* identification of physical space-time ineluctably begs the question concerning the ontological intrinsicality (monadicity, absoluteness) of its identifying properties. Thus I believe my rebuttal to Stein's charge of vicious circularity cannot be gainsaid by the following epistemological differences, which I myself would otherwise wish to emphasize: it is presumably very much *harder* to *discover* whether the metricality of space-time is ontologically absolute or relational than to discover whether a butcher is mistaken if he intuitively regards the weight of a chicken in the supermarket as a property intrinsic to the chicken (at least to within a numerical factor depending on the unit of weight). Indeed, epistemologically the achievement of even a tentative resolution of the issue of intrinsicality vs. relationality of the metric structure of space-time may be even considerably more difficult than some of the known discoveries of physics that had significant philosophic import.

In the case of the metric structure of space-time, securing a tentative answer to the question of absoluteness vs. relationality may be a very tall order indeed; it might take nothing less than the success of, say, Wheeler's pre-1972 GMD version of general relativity to furnish good grounds for claiming to have discovered the falsity of the relationalist ontology of the space-time metric. I shall elaborate on this point below, when I address myself to the ontological status of the g_{ik}-field of Einstein's field equations in standard (1915) general relativity, and then endeavor to deal with the further challenging queries that Stein posed for me. But first I need to offer some concluding observations on Stein's criticism, which I believe to have answered above.

Recall that when Stein discussed Newton's ontology of physical space in regard to its divergence from the one espoused by Leibniz, Stein characterized Newton's view as eschewing the attempt to ground the structure of space "in the relations of non-spatial entities."[54] Thus, in his own formulation of Newton's position, Stein invoked the distinction between space, on the one hand, and "non-spatial entities" or (as I shall say) entities *external* to space, on the other. By the same token, in his challenge to me, Stein did *not* impugn the intelligibility of the very distinction between space-time and entities external to it, a distinction on which the ontological issue of absoluteness vs. relationality of metric structure is indeed predicated. For he did not deny that in the ontology of a particular physical theory, space-time *can* be identified theoretically as such, and thereby can be conceptually demarcated from entities external to space-time, e.g., measuring devices. Thus Stein did *not* challenge me to show that any given physical theory can draw this line conceptually as such; instead, as I understand him, his objection was that precisely because a theoretical identification does effect a demarcation, the latter line—however drawn—must automatically beg the question of which properties are intrinsic to space-time. And hence the challenge to which I have been addressing myself was to show that the objection of vicious circularity is a *nonsequitur*. And in the case of Newtonian 3-space, I have argued that such an objection would be false no less than in the earlier commonsense examples of Uncle Jack, etc.

It is instructive to note in this connection that various twentieth-century physical ontologies differ in important ways as to whether they do draw a line ontologically between certain prima facie or phenomenologically different entities, and, if so, just how they do so. Thus, in regard to the at least prima facie demarcation between space-time and entities external to it, Wheeler asked which of the following two rival ontologies is true of our world: "Is space-time only an arena within which fields and particles move about as 'physical' and 'foreign' [i.e., external] entities? Or is the four-dimensional continuum all there is? Is curved empty geometry a kind of magic building material out of which everything in the physical world is made . . . ? Are fields and particles foreign [i.e., external] entities immersed *in* geometry, or are they nothing *but* geometry?"[55]

As a further example of an issue of ontological demarcation, it has been asked: is there a bona fide ontological line between particles and fields in the sense that particles exist independently of fields? In the view held by Einstein and Rosen in 1935, an atomistic theory of matter and electricity is constructible such that "while excluding singularities in the field, [the theory] makes use of no other field variables than those of the gravitational field $(g_{\mu\nu})$ and those of the electromagnetic field in the sense of Maxwell (vector potentials, ϕ_μ)."[56] According to this conception, "A complete field theory knows only fields and not the concepts of particle and motion. For these must not exist independently of the field but are to be treated as part of it."[57] Einstein and Rosen elaborated this daring ontological idea using the "bridge" representation of the particle in the Schwarzschild gravitational field. And they envisioned a multiplicity of such bridges in the respective gravitational fields, which they hoped to be able to obtain later from Einstein's equations for the case of many-particle systems.

In the same vein, Misner espouses the reduction of particles to fields in his paper cited above, while rebutting an unwarranted objection of circularity to such reduction. In that paper, Misner understands geometrodynamics to comprise only that part of physics which can be obtained from gravity and electromagnetism.[58] And contrary to his earlier resurrection in 1956 of Rainich's "Already Unified Field Theory,"[59] Misner now seems to agree with Pauli's quip that man should not try to join together what God has put asunder.[60] Misner's reason for rejecting the Rainich theory in favor of the coupled Einstein-Maxwell theory is that the former does not allow for a well-posed initial value problem, as pointed out by Roger Penrose.

Misner discusses the physical interpretation to be given to this coupled Einstein-Maxwell field theory. And he illuminatingly cautions against an *unfounded* accusation of logical circularity which has been leveled against that theory's physical interpretation. This accusation results from the confusion of two quite different roles played by charged and neutral test particles: (1) their highly derived, nonprimitive, *ontological* status as mere excrescences of the fields in the reduction claimed by the theory, and (2) their fundamental *epistemological* function as means of *testing* the field theory in question. Misner describes as follows this confusion between the order of existential fundamentality, in which the fields come first, and the order of our coming to know the fields, in which the ontologically derivative particles come first:

The interpretative circle begins in a textbook sort of definition of electromagnetic and gravitational fields. The electromagnetic field at a space-time point is a measure of forces which small charged particles of various velocities would experience if they found themselves at the appropriate space-time event. Similarly the gravitational or metric field defines the inertial properties of test particles, and defines, for instance, a class of space-time curves, any one of which some sufficiently small neutral particle would have followed it it had been at the right point with the right velocity.

Of course, the charged and the neutral test particles which I have used in these sentences to tell you what electromagnetic fields are, are not among the fundamental axiomatically given concepts of the theory; only the fields are fundamental. So one begins with a well defined mathematical object, interpreted in terms of things that are not present—and even then, as a potentiality, as a measure of what would happen if you did something else! That is, if you provided the test particle and put it at the point in question. The differential equations of the theory show that one can then form mathematically identifiable and interesting structures out of these fields. Structures such as wave packets and black holes. The black holes can be made to play the role of particles.

In particular, in the limiting case of vanishing charge and mass on black holes, they play the role of test particles perfectly well. Therefore, the individual space-time as a solution of the field equations has an interpretable existence only to the extent it is conceived as embedded in the possibility of a variety of perturbed parallels of that space-time, parallels in which it is modified by the addition of one or more further very small black holes. The points of the manifold can, finally, be interpreted as all possible limiting cases of the intersections of the world lines which exist as the zero mass limit of the small black hole perturbations to the central manifold under discussion. For these concepts of points and test particles, which we finally produced out of the field itself, then, one proves that the interpretation of the fields postulated at the beginning is in fact satisfied, and therefore you have completed the circle of forming an interpretative tool.

Now, as you can see from what I have just said, the particle is a derived concept in the theory and the fields are fundamental. It may be essential that the concept of a particle can be derived, so that you can provide some exposition of the sense in which the theory allows itself to be interpreted, without requiring that 'point' and 'field' have interpretations provided by some additional, external and deeper physical theory. You are to discover those concepts by self-consistency within the theory itself.[61]

(2) *Must the physical* $g_{\mu\nu}$ *field of standard (1915) general relativity be interpreted as generating a corresponding pseudo-Riemannian metric geometry throughout space-time* without *the ontological mediation of any other entities external to metrically-structured space-time?*

Consider the following two rival ontologies of space-time geometry: (a) when present at all, atomic clocks, free massive particles, photons (light rays), "infinitesimal" solid rods, and other such devices "external" to metrically-structured space-time play only a *manifestational* and *probative* (epistemic) role in its metricality—as opposed to playing an ontologically *constitutive* role!—so that the *physical* $g_{\mu\nu}$*-field* alone suffices ontologically to endow space-time in its entirety with metric structure even if it is *empty* $(T_{\mu\nu} = 0)$ everywhere. (b) In a *non*empty, metrically structured space-time, metric standards external to this space-time do play an ontologically constitutive role in its very metricality or possession of any metric structure at all, and the gravitational fields of *empty* space-times are devoid of *geometrical* physical significance, regardless of the presence or absence of gravitational radiation. As we shall see below, the specific mention of gravitational radiation in the latter thesis (b) is due to the following reason: those empty space-times that do contain gravitational radiation *may* pose a genuine difficulty for what I shall call the "relationalist" ontology of metric structure by impugning thesis (b).

Does the physical content of *standard* (1915) GTR give an unambiguous verdict in favor of the *first* of these two ontological theses, as presumably claimed by Stein? And in the context of standard (1915) GTR, does the existence of well-defined *empty*-space solutions $g_{\mu\nu}$ of Einstein's field equations ("*vacuum* field solutions") refute the predynamical kind of *relationality* of the space-time metric asserted by OI? John Earman has contended, in effect, that the vacuum field solutions are thus the Achilles heel of the relational ontology of the metricality of space-time.[62]

Let us therefore inquire now from what basis the physical $g_{\mu\nu}$ field of Einstein's equations derives the ontological status of being *the metric field* in the sense of endowing space-time with the latter's particular metric structure. In other words, why is the particular physical field tensor $g_{\mu\nu}$ also *canonical* for the metric tensor of the space-time manifold?

Of course, *if* it has already been *granted* that a certain symmetric, covariant, second-rank tensor does indeed have the physical significance of being the metric tensor of space-time, then the scalar ds^2 which it generates by contraction with the contravariant second-rank tensor $dx^\mu dx^\nu$ will have the physical significance of being the space-time metric squared. By the same token, if ds does denote the infinitesimal measure for intervals of space-time, then a symmetric covariant second-rank tensor which generates it (by contraction with $dx^\mu dx^\nu$) has the physical significance of being the metric tensor of the space-time manifold. But if a physical field $f_{\mu\nu}$ is a symmetric, second-rank, covariant tensor field over space-time, this fact alone surely does *not* guarantee that $f_{\mu\nu}$ *must* have the physical significance of being the metric tensor of space-time: after all, qua physical tensor field, the stress-energy tensor field $T_{\mu\nu}$, for example, has the same formal properties as $f_{\mu\nu}$ while certainly *not* being the metric tensor of space-time. It is quite true, of course, that the scalar $|f_{\mu\nu}\, dx^\mu\, dx^\nu|^{1/2}$ has the *formal mathematical* properties of a Riemannian or pseudo-Riemannian metric; but this fact alone does not automatically endow such a scalar with the physical significance of being the *metric ds* of space-time.

Hence we ask: if $g_{\mu\nu}$ is a solution set of Einstein's equations under specified physical conditions, and thus represents some kind of symmetric, covariant, second-rank tensor field over space-time, what warrants the *particular interpretation* that the scalar $(g_{\mu\nu}\, dx^\mu\, dx^\nu)^{1/2}$ *also* has the specific physical significance of being the infinitesimal measure for intervals of the physical space-time manifold? Thus the bona fide physical fieldhood of $g_{\mu\nu}$ is *not* in question, only the reason for its canonicity for the metric of space-time. To deal with this question in the context of standard GTR, we must be mindful of the conjunction of the following two facts: (1) According to the theory, the behavior of measuring rods, clocks, atoms, and photons (light rays) *"adjusts* itself," as Weyl put it,[63] to the physical $g_{\mu\nu}$ field in the sense that, at least approximately and ideally,[64] these entities *physically realize* the scalar $(g_{\mu\nu} \, dx^\mu \, dx^\nu)^{1/2}$; and (2) J. Ehlers and W. Kundt have pointed out that, besides Minkowski spacetime, there are an infinitude of known solutions $g_{\mu\nu}$ of Einstein's *vacuum* field equations such that if these solutions also have the significance of being the *metric* tensors of their respective space-times, then they do *not* admit the same metrical partition of space-time intervals into equivalence classes.[65] And they show that "there exist complete solutions of Einstein's vacuum field equations; that is, complete solutions free of sources (singularities), proving that it is permissible to think of a graviton field independent of any matter by which it be generated. This corresponds to the existence of source-free photon fields in electrodynamics."[66] Within the genus of *empty* space-times ($T_{\mu\nu} = 0$ everywhere) we shall need to distinguish between two different species for which I shall use special labels as follows: (1) empty space-times in which there is some gravitational radiation, propagation of gravitational waves or gravitons—I shall call such space-times *"weakly* empty"; and (2) empty space-times in which there is not even any gravitational radiation anywhere—these I shall call *"purely* empty." The latter species includes globally Minkowskian spacetime as well as others that are only locally though not merely infinitesimally Minkowskian. Moreover, in order to have a name for those *non*empty GTR space-times that are, however, *devoid* of gravitational radiation, I shall call these "nonempty but gravitationally quiescent" or "NE-GQ space-times." On the other hand, I shall refer to space-times that are nonempty and do contain gravitational radiation as "nonempty and gravitationally effervescent" or "NE-GE" space-times.

Our question now is: what qualifies the tensor $g_{\mu\nu}$ qua solution of Einstein's field equations—to do *double duty* physically by representing not only a physical field but also the metric tensor of space-time? Let me try to state, in turn, the answers to this question given by two philosophers whom I shall label respectively "the relationalist" and the *"semi-*absolutist": the relationalist espouses both predynamical and dynamical relationality of metric structure by asserting both OI and—for the nonvacuum case in which matter *is* present—DH; whereas the *semi-*absolutist rejects the relationalist thesis OI while affirming the dynamicity thesis DH for the nonvacuum case. When formulating the reasonings of the two disputing philosophers, I shall also try to show how the relationalist can accommodate the purely empty space-times and under what conditions he can handle the weakly empty ones as well.

My affirmation of the following twofold contention will show why I diverge from Stein and Earman: (i) the physical content of standard (1915) GTR does *not* give an unambiguous verdict at all in favor of the *first* of the two rival ontologies which I stated as (a) and (b) at the beginning of the present reply (2), so that (ii) the answer to the question stated in the *title* of the present reply (2) is *negative*, where it is to be noted that this titular question begins with the strong word "Must." But I am less concerned to argue that the relationalist's interpretation is superior to the semi-absolutists's account than I am to articulate a hopefully interesting form of relationalist interpretation which might lend itself to useful further discussion.

The Relationalist's Account of Both Nonempty and Empty Space-Times

It will be helpful to give the relationalist account by treating space-times in the following order: (A) NE-GQ space-times, (B) purely empty space-times and purely empty *regions* of NE-GQ space-times, (C) weakly empty space-times, and (D) NE-GE space-times. The rationale for this procedure will become clear from my discussion in (C) of the import of theorem 8.1 of John Winnie's essay in this volume.

A. NE-GQ SPACE-TIMES

What is otherwise just another symmetric, covariant, second-rank physical tensor field $g_{\mu\nu}$ does control the physical space-time trajectories of photons and freely falling ("free") massive particles, as well as the rates of atomic clocks and perhaps the world-lines of other (as yet unknown) entities as well. And *at least in NE-GQ space-times*, it is the idealized behavior (referred to in note 59) of this particular important class of external entities that is taken to be *canonical* for the metricality of these space-times in the sense that it is fundamentally *constitutive* ontologically for their metric structures. Thus what is otherwise just a physical tensor field $g_{\mu\nu}$ thereby *acquires* specifically geometric, additional, physical significance as being the metric tensor of the space-time manifold. And in NE-GQ space-times, $g_{\mu\nu}$ acquires this geometric significance because— and only because—the scalar $\left| g_{\mu\nu}\, dx^\mu\, dx^\nu \right|^{1/2}$ generated from that field is (ideally) *physically realized* by the behavior of that open-ended set of external devices that are taken to be canonical for the metric structure of space-time. Hence the physical field $g_{\mu\nu}$ also plays an ontologically constitutive role in the metric structure of space-time, but its role is ontologically a *derivative* or conditional one: it acquires this role only because it causally governs those external entities which play the more fundamental constitutive role ontologically. Were it not for this geometrically canonical role of photons, atomic clocks, etc., and for the "orchestrating" action of the $g_{\mu\nu}$ field on *them, what else* would confer on $g_{\mu\nu}$ the additional status of being physically the *metric* tensor of an NE-GQ space-time? What else would *compel* assigning to the solution of Einstein's equations the latter specific significance over and above its general significance as a second-rank, covariant, physical tensor field?

Can the conclusion drawn by these rhetorical questions be gainsaid on the basis that the external metric devices can properly be said to *explore* the physical $g_{\mu\nu}$ field by which they are *governed*, and hence to adjust to it? Photons, atomic clocks, infinitesimal rods, and free massive (gravitationally monopole) particles—hereafter the "family of concordant external metric standards" or "FCS"—*explore* the physical $g_{\mu\nu}$ field in the sense that they exhibit the latter's effects on them by concordantly *adjusting* to that field. But it does *not* follow that just because the $g_{\mu\nu}$ field can thus be explored and, *qua* physical field, *does indeed exist independently of being explored*, this tensor field *must* have the particular added physical significance of being the *metric* tensor field of a NE-GQ space-time without the ontologically constitutive mediation of the FCS. In other words, just because the FCS "adjusts itself" to the $g_{\mu\nu}$ field as a result of this field's action upon that family of external standards, it does not follow that the entity to which the FCS does adjust itself must be an *autonomously geometric* structure in whose metricality the FCS plays no ontologically constitutive role but which the FCS merely *manifests* by its behavior. Speaking analogically, to say that a conductor orchestrates the playing of his musicians and that they take directives from him is not to say that the directive motions of his hand alone would constitute a concert. In the "metrical concert," the physical $g_{\mu\nu}$ field of an NE-GQ space-time is the "conductor," but the members of FCS are the musicians who generate the music.

As emerges from my exegesis of Riemann above, the relationalist's view of the metricality of space-time derives its intellectual inspiration from Riemann's ontology of physical geometry, instead of being a holdover from the discredited homocentric ontology of operationalism or of the verifiability theory of cognitive meaning. I have always rejected the latter as conflating—to use Thomistic parlance—the order of knowing with the order of being. [67]

In the context of NE-GQ worlds containing the FCS, the relationalist distinguished betwen the primary physical status enjoyed by the $g_{\mu\nu}$ field qua second-rank, covariant tensor field, and its derived status as the metric tensor field of space-time, a further status which he deems to be ontologically mediated in these space-times by the metrically canonical role of the FCS and by the fact that the FCS are "orchestrally" governed by the action of the $g_{\mu\nu}$ field.

I shall advert to the distinction between the *single* duty and *double duty* status of $g_{\mu\nu}$ when I proceed to formulate the relationalist's interpretation further. In order to employ this distinction, I need a descriptive phrase that will refer to the physical $g_{\mu\nu}$ field of Einstein's equations *while still leaving open* the question whether, in a given case, that $g_{\mu\nu}$ field *also* has the additional physical significance of being the *metric* tensor field of the space-time manifold. Neither the term "graviton field" nor the unqualified term "gravitational field" seems to implement this objective. In the first place, "graviton field," as distinct from the more noncommittal "gravitational field," may have the restrictive connotation of particlehood, much as the term "photon field" has a more restrictive connotation than "electromagnetic field"; moreover, it may suggest the presence of gravitational radiation and would thus be useless to refer to the $g_{\mu\nu}$ fields in our *purely* empty space-times. Second, in some quarters, the term "gravitational field" has come to imply that the field to which it refers *necessarily* or automatically *also* has the status of being the *metric* field of space-time, an implication which I am here at pains to avoid. In order to distinguish the latter *metrically committed* usage of the term "gravitational field" from the usage which still leaves *open* the question whether $g_{\mu\nu}$ has metric physical significance as well, I shall do the following: when I intend to convey the metrically *uncommitted* sense just mentioned, I shall speak of "the pregeometric gravitational field." (In a *particular* case, the pregeometric gravitational field may, however, be characterized more specifically as a pregeometric *graviton* field.) But when the metrically committed meaning is intended, I shall use the term "gravitational-cum-metric field." It will be clear that far from begging the question and giving the relationalist an argumentative edge on the semi-absolutist, the terminology will help to pinpoint the agreement no less than the divergence between their respective accounts.

B. PURELY EMPTY SPACE-TIMES AND PURELY EMPTY REGIONS OF NE-GQ SPACE-TIMES

I remind the reader that the empty space-times that I call "purely" empty are *devoid* of even gravitational radiation, as are purely empty *regions* of a space-time.

1. GLOBALLY MINKOWSKIAN SPACE-TIME AND ONLY LOCALLY MINKOWSKIAN SPACE-TIMES

The relationalist exponent of OI views these purely empty space-times with complete equanimity when he appraises their pertinent physical significance as follows: standard GTR allows a purely empty world— devoid of even the *sourceless* gravitational radiation of the Ehlers and Kundt solutions—in which the space-time manifold carries a pregeometric gravitational field, which, however, does *not have to be* interpreted physically as gravitational-cum-metric. For after all, by what ontological edict of the GTR *must* the $g_{\mu\nu}$ tensor of its field equations be interpreted physically as gravitational-cum-*metric* in *every* space-time world allowed by that theory?

What if the semi-absolutist were to retort that the relationalist is trying to make gratuitous ontological capital out of the mere withholding of the *label* "metric" from the $g_{\mu\nu}$ field, but without genuine detriment to the latter's ontological geometricality? The semi-absolutist might say with Shakespeare:

> What's in a name? that which we call a rose
> By any other word would smell as sweet.[68]

The relationalist counters this retort by asking the semi-absolutist: *after having said that a purely empty vacuum world devoid of the FCS is one of the pregeometrically gravitational worlds allowed by Einstein's empty-space equations, just what is the *physical significance* of going on to say that this purely empty universe also has a metric space-time structure $g_{\mu\nu}$? If, in the case of such a purely empty world, the semi-absolutist eschews, as he does, an "ontological loan" from the putative behavior of *hypothetical "test*-particles" which are otherwise negligible, is it not devoid of additional physical significance—and thus merely *redundant* physically—to say that the $g_{\mu\nu}$ field is *also* the metric tensor field? Indeed, it would seem that in regard to playing games with words, the shoe is rather on the other foot: at least in the case of the purely empty world devoid of the FCS, it is the semi-absolutist who is engaging in verbal baptism when *insisting* on characterizing $g_{\mu\nu}$ as gravitational-cum-metric. For, as just explained, there is no difference in physical content *here* between the latter locution and the characterization of $g_{\mu\nu}$ as pregeometrically gravitational. Thus, by insisting that $g_{\mu\nu}$ *also* has *geometric* physical significance both in a purely empty world devoid of the FCS *and* in a world containing the FCS, the semi-absolutist is engaging in the stated redundancy at least for the case of the former world. In contrast, the relationalist bestows the additional label "metric" on the tensor $g_{\mu\nu}$ only in a physically nonredundant way: in an NE-GQ world containing the FCS, his bestowal renders the fact that the FCS *adjust to* what is otherwise a pregeometric physical $g_{\mu\nu}$ field, because this adjustment consists in physically realizing the scalar $ds = \left| g_{\mu\nu} dx^{\mu} dx^{\nu} \right|^{1/2}$ as well as the mathematical structures obtainable from that scalar (e.g., the curves satisfying $\delta \int ds = 0$). And insofar as, in the case of a purely empty world devoid of the FCS, the relationalist is also chary of an ontological loan from counterfactual "test-particles" which are otherwise negligible, he refuses to bestow the additional label "metric" on the tensor $g_{\mu\nu}$, because it would be physically redundant in that case.

In short, the relationalist poignantly turns the tables on the semi-absolutist's **Shakespearean** epigram by declaring with Goethe:

> Doch ein Begriff muss bei dem Worte sein. . . .
> Denn eben wo Begriffe fehlen,
> Da stellt ein Wort zur rechten Zeit sich ein.
> (Yet a word ought to represent a concept. . . .
> For just where concepts are lacking,
> A word comes to the rescue just in time.)[69]

I now remind the reader of the 1952 statement by Einstein which I used as a thematic quote at the very beginning of this essay. That summary statement by him would seem to espouse the Riemannian relationalist conception of a purely empty space-time which I have just tried to develop. But I must point out that such a construal does go beyond what is warranted by a conservative reading of the full text of the appendix to which Einstein's summary statement refers: in that appendix (5), he seems concerned to champion an avowedly *neo-Cartesian* (not necessarily Riemannian) *relationalism* which denies that there could be a geometrically-structured space-time devoid of *any* physical field: he tells us that if you remove all physical fields, including the gravitational one, then *"nothing* at all remains . . . an empty space, i.e., a space devoid of a field, does not exist." [70]

Clarifying comments by Howard Stein on an earlier draft of this essay have helped me reach the following tentative conclusion not necessarily endorsed by Stein: my relationalist account of purely empty space-times will bear scrutiny in the face of the following result of A. A. Robb discussed by Winnie [71] in his contribution to the present volume: [72] if one makes liberal use of physically *possible* (rather than only actual), primitive, "causal" relations among both actual *and* merely possible punctal physical events, [73] then Robb's work shows that the system C of primitive causal relations which yields the standard R^4-topology of Minkowski space-time suffices also to generate via purely nominal definitions, what is *formally* (modulo a constant factor) the Lorentz metric of that flat space-time *without* the antecedent imposition of the requirement of flatness.

In order to obviate *spurious* prima facie absurdities in the special case of *globally* Minkowskian space-time, a caveat concerning the relationalist's conception of the $g_{\mu\nu}$ tensor field in that purely empty space-time has to be made explicit. Within the framework of the GTR, the $g_{\mu\nu}$ of globally Minkowskian space-time is, of course, a solution set of the vacuum field equations and thus represents one of the purely empty worlds allowed by that theory. And in this GTR context, the relationalist's stated conception of the $g_{\mu\nu}$ fields in purely empty space-times applies, of course. On the other hand, if Minkowski space-time is taken to mean the space-time of the *pre*-GTR version of *special* relativity, then this space-time avowedly does not satisfy the emptiness condition of being a vacuum field; instead, it then contains ubiquitous material clocks, rods, and light rays. And in the latter construal of the Minkowski universe, the relationalist's conception of its $g_{\mu\nu}$ field would *not* be parsimoniously pregeometric.

2. PURELY EMPTY REGIONS OF NE-GQ SPACE-TIMES

As Allen Janis has rightly noted, consistency with my relationalist account of the $g_{\mu\nu}$ fields in purely empty worlds devoid of the FCS requires that the stated parsimonious, pregeometric gravitational interpretation of these fields also be given for *those portions* of NE-GQ space-times which are devoid of the FCS. Indeed, if the ten independent $g_{\mu\nu}$ values at *any one* world-point P of an NE-GQ space-time are to derive specifically geometric physical significance from the FCS, then the FCS must be suitably present in a sufficient number of space-time directions at P. Thus, elsewhere (*PPST*, pp. 732–3) I have recognized by reference to J. L. Synge's *chronometric* foundation of the space-time metric that on an ontologically relationalist rather than merely epistemic construal of this kind of physical foundation, generally ten atomic clocks are required on as many different world-lines which cross P. Hence, at any world-point P of a particular NE-GQ space-time where the FCS are insufficiently plentiful to confer added geometric physical significance on the pregeometric $g_{\mu\nu}$ field, the stated parsimonious, pregeometric gravitational interpretation of this field would also apply. Despite its prima facie oddity, I see no explanatory or physical loss in this consequence of the relationalist's conception but only an innocuous verbal one: What is lost *physically* by this conception but gained nonverbally by the semi-absolutist's insistence that in *every* $g_{\mu\nu}$ world allowed by the GTR, and in every portion of the space-time of any such world, the $g_{\mu\nu}$ field also unrestrictedly has metric significance? Let it be noted that at least in purely empty regions of NE-GQ worlds and in purely empty space-times, the relationalist confines himself to the parsimonious, pregeometric, gravitational interpretation if he foregoes resorting to counterfactual "test particles" from the FCS, which practicing physicists are wont to invoke. The renunciation of the hypothetical invocation of such "test particles" is motivated by the modification in principle, however slight and otherwise negligible, which the theory asserts to take place in the $g_{\mu\nu}$ field in virtue of the dynamical action of the FCS itself. Were it not for the modifying effects on the $g_{\mu\nu}$ ensuing from the putative added presence of the FCS, the relationalist would not need to deny himself an appeal to the corresponding counterfactuals.

Let me elaborate on this point briefly to indicate just why I think the relationalist should thus bend over backwards, as it were. In a physical theory which asserts that the physical geometry is *non*dynamical, "frozen," or "given once and for all," a *dispositional* account of metric interval ratios by means of counterfactuals is no less feasible than a corresponding dispositional account of, say, the solubility of sugar in water or of the brittleness of glass. Thus, in the *pre*-GTR version of the special theory of relativity ("STR"), for example, the four-dimensional metric ratio r of two timelike intervals I_1 and I_2 on the *physically possible* world-line of some hypothetical standard (atomic) clock C could be construed as follows: independently of how much matter-energy is present—and, in particular, independently of the amount of matter contained in the putative atomic clock C which *could* be applied mensurationally to both I_1 and I_2—the latter intervals have the disposition to accommodate respectively r_1 and r_2 "ticks" of C such that $r = r_1/r_2$. In short, in the nondynamical STR, the putative presence and application of metric devices like the hypothetical C no more affects the *truth* of the pertinent subjunctive conditional assertion which renders the chronometric dispositions of I_1 and I_2, than the putative or actual presence of a hammer affects the truth of a subjunctive conditional concerning the outcome of a vigorous hammer blow against a highly brittle piece of glass. But precisely this immunity of the truth of the pertinent subjunctive conditional concerning the putative metric application of the FCS no longer obtains in the dynamical context of the GTR, unless *approximations* and limiting processes can somehow nonetheless be legitimately invoked even in the context of the nonlinear GTR. Hence I fear that both prudence and intellectual honesty require the relationalist to "bite the bullet" here by denying himself resort to "test particles."

By asserting in OI that the $g_{\mu\nu}$ fields of NE-GQ worlds acquire geometrical significance through the mediation of external entities like the FCS, the relationalist's ontology can claim for itself the advantage of immediately being able to accommodate philosophically such "Two-Metric" theories as Dirac's proposed modification of GTR in the interest of accommodating quantum phenomena.[74] In Dirac's 1972 theory, macroscopic and atomic metric standards generate *incompatible* metric ratios among space-time intervals in the manner familiar from E. A. Milne's logarithmically related astronomical and atomic time scales.[75] But Dirac points out that only the macroscopic standards physically realize the scalar $\left| g_{\mu\nu}\, dx^\mu\, dx^\nu \right|^{1/2}$ corresponding to the $g_{\mu\nu}$ field of Einstein's equations. Now the relationalist's conception clearly countenances the existence of the $g_{\mu\nu}$ field qua pregeometric physical field. But it *denies* that this field also has ontologically unmediated, automatic, *geometrical* significance. Hence the relationalist allows Dirac's general relativistic-cum-quantum world to be *relationally* endowed with two *different* metric structures, each of which is physically realized and therefore significant. By the same token, the relationalist point of view accords ontological co-legitimacy to the aforementioned two different (nonlinearly related) time metrics of E. A. Milne's cosmology.

351

With respect to giving poignancy to the relationalist's ontology of metricality, Dirac's two-metric theory, in which the two sets of incompatible metric ratios are separately physically realized by important classes of external metric standards, is even more interesting than the most recent version of Nathan Rosen's "Bi-Metric Theory of Gravitation," in which only one of the two metrics is so realized. Rosen notes that "If the existence of black holes is confirmed, this will represent a brilliant success of general relativity."[76] But in the absence of convincing evidence that black holes do exist, Rosen regards it as permissible to take the view that a "black hole" would represent "a breakdown of the familiar concepts of space-time and hence is something unphysical."[77]

Thus his motivation for trying to develop a modified theory of gravitation springs from the fact that the formalism of the GTR can be interpreted as predicting the existence of black holes in nature. And Rosen's "Bi-Metric Theory of Gravitation" is intended to make available a theory of gravitation that is in agreement with known observations *without* permitting black holes.

He employs two tensors, each of which has the formal properties of a pseudo-Riemannian metric tensor: one of these, the familiar $g_{\mu\nu}$, "can be interpreted as describing the true gravitational field, that is, the field arising from matter or other forms of energy,"[78] while the other, $\gamma_{\mu\nu}$, is a *flat*-space metric tensor.

To my mind, the interesting thing about Rosen's bi-metric relativity theory vis-à-vis the relationalist's ontology is that, as he points out, the following physical interpretation of the two tensors $g_{\mu\nu}$ and $\gamma_{\mu\nu}$ is permissible though not required (unique): "one can regard the $\gamma_{\mu\nu}$ as describing the properties of space-time, which is now considered to be flat, while the $g_{\mu\nu}$ is interpreted as a gravitational potential tensor which determines the interaction between matter (or other fields) and gravitation."[79]

Rosen uses a tensor $\gamma_{\mu\nu}$ that is *not* realized physically by existing metric standards to impart a *flat* geometry to a space-time that does contain matter and various physical fields. This kind of imposition of a metric structure on space is akin to the more pedestrian cartographic device of using a metric tensor that is *not* realized physically by rigid rods to impart a *curved* two-dimensional geometry to an ordinary sheet of paper lying on a table-top: a practically useful case in point is furnished by Mercator's Projection in cartography.[80]

C. WEAKLY EMPTY SPACE-TIMES

Let us recall from the introductory part of the present reply (2) that the GTR allows empty space-times that contain gravitational radiation. I shall now consider this particular class of gravitational fields.

In my view, the first ontological moral to be drawn from the weakly empty space-times allowed by the GTR is this: although the name "gravitational" field traditionally denoted a field that requires generating material sources, *even* a $g_{\mu\nu}$ field *containing gravitational radiation* can exist *sourcelessly* and, at least to this extent, is a kind of gravitational field not dreamed of in prior field physics ontology. In regard to sourcelessness, Ehlers and Kundt stressed the analogy to the existence of source-free electromagnetic fields in electrodynamics.

But gravitational radiation has the important feature that its wave-fronts are null or "lightlike." Hence we must now consider the import of the latter fact for the divergence between the relationalist and the semi-absolutist in the light of the following consequence of John Winnie's theorem 8.1, stated in his essay elsewhere in this volume: if we are given *all* of the null space-time trajectories of a *weakly empty* space-time, then the tensor $g_{\mu\nu}$ is thereby determined up to a *constant* (conformal) factor.

But as Allen Janis has pointed out cogently, I believe, this result does *not* enable the semi-absolutist to adduce the nullity of the world-lines of gravitational wave-fronts to establish his thesis. Janis gives essentially three reasons for this conclusion:

(1) It is unclear that in a given weakly empty space-time, there exist gravitational wave-fronts in a sufficient number of null directions on the null cone at any given world-point to determine the ten independent $g_{\mu\nu}$ values at that point.

(2) In any event, *at best* the $g_{\mu\nu}$ are determined by null gravitational wave-fronts *only* at those world-points that lie on actual wave-fronts, but are *not* so determined elsewhere in the space-time; yet in at least some weakly empty space-times, there are such points and indeed whole regions that are devoid of gravitational radiation.

(3) The semi-absolutist could hardly invoke *hypothetical* or *counterfactual* gravitational waves ("gravitons") in the latter regions to vindicate his claim that the $g_{\mu\nu}$ field has gravitational-cum-metric physical significance *throughout* space-time: to assume hypothetical gravitational waves in those regions of an empty space-time actually devoid of them

is—by *definition*, as it were—to talk about a gravitational field fundamentally different from the one whose *geometrical* significance is here at issue.

It is true but unavailing to the semi-absolutist that whereas the introduction of hypothetical "particles" *belonging to the FCS* would in principle issue in an alteration of the $T_{\mu\nu}$ tensor, the introduction of hypothetical "gravitons" into a weakly empty world would not affect the vanishing of $T_{\mu\nu}$. After all, as explained in subsection B(2), the relationalist had to prudently "bite the bullet" by renouncing the resort to hypothetical members of the FCS because the latter would indirectly change the gravitational field as a matter of physical law (Einstein's field equations) via a change of the tensor $T_{\mu\nu}$. All the less can the semi-absolutist introduce a *direct* change in the very nature of the given gravitational field in order to confer geometrical status on it! Thus, despite Winnie's very interesting result, gravitational radiation does not help the semi-absolutist.

On the other hand, I must emphasize that the latter *may* be only cold comfort to the relationalist, since it *need* not redound to the credit of his position. For *if* it should turn out that in at least some weakly empty space-times, there is indeed sufficient gravitational radiation to determine the metric tensor (modulo a constant factor) for at least some world-points or even regions, then the relationalist would need to deal with a serious question put to me incisively by John Winnie to the following effect: in that putative case, doesn't the gravitational field autonomously and automatically have gravitational-cum-metric physical significance at these world-points or even in the regions in question? In the same vein, John Stachel has asked essentially: "If photon trajectories count as geometrically significant in the sense claimed by OI, why not also the wave-fronts of gravitational radiation?" I am inclined to think that under these posited circumstances of sufficient gravitational radiation, both the version of relationalism I have endeavored to articulate and semi-absolutism would be unsatisfactory: since they are contraries rather than contradictories, they can both be false. And indeed under the posited circumstances, I myself would take the moral of Winnie's and Stachel's challenging rhetorical questions to be that the following kind of alternative third position is reasonable: the gravitational field can acquire added geometrical and physical significance either from gravitational radiation itself or via the FCS. Thus under the posited conditions, the stated inclusion of gravitational radiation constitutes an enlargement of the FCS. Qua enlargement of the avowedly *open-ended* set of concordant metric standards, it is no less legitimate than the following: the inclusion of optical radar-ranging as a *spatial* metric standard for inertial systems along with rigid rods, once we are given the upshot of the null result of the Michelson-Morley experiment. But I tend to agree with John Winnie's suggestion that although such an enlargement would itself be altogether legitimate, it cannot be philosophically accommodated in my proposed relationalist framework without a kind of semantic skulduggery which would treat the gravitational field itself as no less external to the space-time manifold than, say, atomic clocks.

Why, it might be asked, do I *not* agree that the gravitational field of a *purely* empty space-time or a weakly empty one *devoid* of *sufficient* radiation is an adequate physical basis for constituting a space-time metric willy-nilly? In order to provide context for the reply which I am about to give to this question, I ask the reader to bear in mind what I already said in the present subsection (2) à propos of the following two facts: (a) Contraction of the *gravitational* field tensor g_{ik} with the tensor $dx^i\, dx^k$ yields a scalar having the same *mathematical* properties as the square of a suitable geometrical pseudo-Riemannian ds of space-time, and thus in particular, (b) for an appropriate set of space-time directions dx^μ, this contraction will yield *null* values. I shall speak of any trajectory on which this contraction everywhere yields a null value as "gravitationally null." With this preparation in mind, I now reply to the stated question along the following lines. In weakly empty space-times, the wave fronts of gravitational radiation *as such physically pick out special space-time trajectories*, and every such trajectory is *also* gravitationally null, i.e., is characterized by the fact that the scalar $g_{ik}\, dx^i\, dx^k$ is *null* along it, though not conversely. In other words, in a weakly empty space-time, a certain subset s of the gravitationally null trajectories can be picked out by gravitational radiation *independently* of that nullity. Now suppose that the thus independently characterized set s of trajectories is sufficiently rich for our impending purpose while its members are also taken to be the *geometrically* null lines of a pseudo-Riemannian space-time metric. Then at the points of the space-time which lie at an intersection of a sufficiently large number of members of s, the stipulated geometrical nullity of the latter will determine the metric tensor of that metric modulo a *constant* factor. In this way, the propagation of gravitational radiation serves as the physical metric standard for the given space-time manifold! But in a *purely* empty space-time, there is no corresponding *independent* physical picking out of a nonempty subset of the class Σ of gravitationally null trajectories, and indeed the set Σ is physically picked out *only* by the gravitational nullity of its members. Hence if this latter nullity were now declared to be *geometrical* as well with a view to establishing thereby the geometric physical significance of the g_{ik} tensor, this would be *physically* tantamount to no more than just *calling* this tensor "metric."

D. NE-GE SPACE-TIMES

The relationalist account of this class of space-times is evident from the account given of the NE-GQ ones under A above. But according as gravitational radiation itself can or cannot play the same geometrical role as the FCS in the sense just discussed, the corresponding assessment of the relationalist account given under the preceding section C will apply. Thus in a nonempty space-time no less than in a weakly empty one, gravitational waves may well qualify as a metric standard when sufficiently plentiful. And hence in a nonempty space-time, they enjoy parity with the FCS under the stipulated condition of sufficient plentifulness, a condition applicable alike to both.

This brings us to a statement of the semi-absolutist's account as part of the resumption of my reply to Stein. But I do *not* mean to imply that Stein is necessarily an exponent of the tenets of semi-absolutism as I shall formulate them. And I am glad that he will be speaking for himself on this point in his Reply in this same volume.

The Semi-Absolutist's Account

Not being an adherent of this account myself, I hope to do at least some justice to it by giving the most impressive formulation of it which I have encountered.

The semi-absolutist asserts—outright, as far as I can see, without giving any reason other than the *mathematical* tensor properties of $g_{\mu\nu}$—that ontologically every physical tensor field $g_{\mu\nu}$ allowed by Einstein's equations under specified conditions *autonomously* and automatically has the physical significance of being gravitational-cum-*metric*, i.e., of *also* being the *metric* tensor field of the space-time in which the $g_{\mu\nu}$ field exists physically. The semi-absolutist's thesis that the $g_{\mu\nu}$ field's metric significance is both autonomous and automatic asserts that this metric significance *never* depends ontologically on any mediation by the FCS.

Thus, with respect to the ontology of the metric structure of space-time, the semi-absolutist denies that the role of the FCS is ever constitutive and confines that role to being manifestational, probative, and epistemic. For brevity, let me introduce the acronym "TAM" (Thesis of Autonomous Metricality") to refer to the semi-absolutist's thesis that, in every Einsteinian $g_{\mu\nu}$ world, the physical significance of the $g_{\mu\nu}$ field is autonomously and automatically gravitational-cum-metric rather than pregeometrically gravitational. I have already remarked that within the framework of the *semi-absolutist's ontology*, I know of no reason given by him for his TAM other than that the pregeometric gravitational tensor $g_{\mu\nu}$ has the same *mathematical* properties as a physically bona fide metric tensor. And I explained at the beginning of the present reply section (2), by reference to the physical field tensors which I called "$f_{\mu\nu}$" and "$T_{\mu\nu}$," why the latter mathematical fact is insufficient as a reason for TAM. In contrast, for the cases of the NE-GQ and the NE-GE worlds, the relationalist offers an ontological reason for regarding the $g_{\mu\nu}$ field to be doing double duty as the gravitational-cum-metric field in regions of their respective space-times that contain the FCS in sufficient plenitude to determine the metric tensor. Thus he points to the fact that under the action of the pregeometrical gravitational field $g_{\mu\nu}$ of a NE-GQ space-time, any existing FCS physically realize and—if sufficiently plentiful—determine at least regionally the tensor representing that field, thereby enhancing its ontological significance, which then becomes gravitational-cum-metric, if only regionally.

Yet once we grant the semi-absolutist his TAM, his reasoning unfolds with impressive cogency. Subject to correction, of course, I take Stein to be affirming TAM in his subsection (a) when he speaks of the differentiable manifold of world-points and says that it "carries a physical field with the characteristics of the Einstein metric field." Having thus assumed TAM, Stein rightly feels justified in holding for any space-time that, *without* any ontological mediation by the FCS qua metric standards, certain space-time paths or curves have the bona fide geometrical status of being, say, null or time-like geodesics on the strength of the physical $g_{\mu\nu}$ field alone. And geometrically it is then just ontological "gravy" that the behavior of specified members of the FCS also *picks out* these *already* distinguished loci, since the latter's *geometrical* status then does *not* depend ontologically on the FCS qua metric standards. Instead, the FCS can manifest, explore, or probe the metrical structure by their behavior and can thus yield adequate geometric data for ascertaining that structure. Of course, qua matter-energy entities, the FCS contribute to the stress-energy tensor of Einstein's equations and thereby make their presence felt in contributing to the $g_{\mu\nu}$ field. But this does not make the metric field ontologically dependent on the FCS playing the role of metric standards. Small wonder, therefore, that any philosopher who assumes TAM is bound to conclude that the relationalist's OI is wrong-headed in its account of the role of the FCS.

On the other hand, by tacitly assuming TAM, the semi-absolutist can be driven unintentionally to put up *a straw man* as follows: he misportrays the relationalist, in effect, as *denying* that qua *pre*geometric gravitational field, the physical $g_{\mu\nu}$ field exists independently of being explored, manifested, or probed by the FCS in their role as metric standards. This would be a serious misrepresentation, since the relationalist, no less than the semi-absolutist, affirms rather than denies this independent existence of the $g_{\mu\nu}$ field *qua pregeometric physical field*. What the relationalist does deny instead is that the $g_{\mu\nu}$ field *also* autonomously and automatically qualifies as gravitational-cum-metric in any and every space-time. By tacitly assuming TAM, the semi-absolutist can be easily misled into reasoning thus: if someone denies that the $g_{\mu\nu}$ field of an NE-GQ or NE-GE world qualifies as gravitational-cum-*metric* independently of the metric role of the FCS, then he is thereby also necessarily denying the independent existence of the $g_{\mu\nu}$ field qua being pregeometrically gravitational. And, in this way, the semi-absolutist may innocently but nonetheless fallaciously erect a straw man by depicting the relationalist as denying altogether that in the context of standard GTR, the physical $g_{\mu\nu}$ field of Einstein's equations exists independently of the metric role of the FCS. I do not say that Stein means to attribute this denial to the relationalist. But I think it is important to guard against such a misattribution when reading the arguments of semi-absolutists, lest there be serious misunderstanding of the philosophical divergence of the relationalist from the semi-absolutist.

As I have been emphasizing, my relationalist account of the metric status of the $g_{\mu\nu}$ field has been within the framework of standard (1915) GTR. But as I stated in my reply (1) above, the relationalist's ontology of the metric structure is open to refutation in at least the following way: if standard GTR were to be supplanted by something akin to Wheeler's pre-1972 GMD program, then the putative success of the latter would furnish good grounds for *rejecting* the relationalist's ontology of the space-time metric as *false*. The justification for this claim of the putative refutation of OI becomes apparent by taking cognizance of the fact that in the monistic GMD ontology, the FCS are literally reduced ontologically to being parts of the $g_{\mu\nu}$ field, instead of existing independently of it as peers of the ontic realm. Being literally parts of the $g_{\mu\nu}$ field in GMD, the FCS *cannot* first externally and hence relationally *confer* any (geo)*metric* status on that field which it does not already have. And yet even in the GMD framework, the behavior of the now-reduced FCS can be held to furnish data attesting to the *metric* structure of space-time. Hence in the GMD ontology, any and all metric structure of the space-time manifold which may *appear prima facie* to be first generated ontologically by the FCS under the influence of the $g_{\mu\nu}$ field, is already vouchsafed ontologically qua metric structure by the very existence of that field. In short, by erasing, in reductionist fashion, the ontological line of demarcation which *standard* GTR had drawn between the metrically-structured space-time manifold and the FCS external to it, GMD precludes the relationality of any metric interval ratios yielded by the no longer "foreign" FCS.

Clearly, *my relationalist extrapolation of Riemann's OI* from pre-relativistic physical 3-space to physical space-time *is* not *an a priori philosophical claim; rather, my account is avowedly open to broadly empirical refutation by the putative success of an appropriate physical theory like GMD!* And it *may* have a genuine difficulty with weakly empty space-times for reasons set forth under subsection C above.

I hope to have shown that when I posed the question, "Whence does the metricality of space-time derive ontologically?" my question was *not* predicated on an a priori assumption of OI and thus did *not* preclude the kind of absolutistic answer given by GMD or by some other sort of absolutistic answer. Hence I do not see that my version of Riemann's OI is open to the complaint of illicit apriorism made by Stein when he said in his section (a): "whether we can ground the postulated metrical and differentiable manifold structures in yet deeper ones, for which we can find good evidence, is always again a legitimate matter for inquiry, but that we *must* do better and do so in a specified direction, is *never* a legitimate demand." In reply 3 below, I shall amplify the non-apriorism of my conception by reference to *non*metrical aspects of geometric structure.[81]

Moreover, note that the very strong assertion "[metrical] geometry [is] a mere bookkeeping for relations between particles"[82] relationalistically goes very far beyond Riemann's denial of *the metrical aspect* of Newton's absolutism. The just cited, very strong *contrary* of unbridled absolutism was entertained by Wheeler himself as part of his canvassing of various *alternatives* to his erstwhile GMD.

Thus we must heed John Stachel's admonition to specify the particular features or *aspects* of the geometrical structure that are held to be relational or absolute in a given physical theory, assuming that one has been initially specific in regard to what is to count as "geometrical."[83]

In order to state a concluding caveat along these lines, suppose that someone espouses the very strong relational view that "geometry is a mere bookkeeping for relations between particles" and, more specifically, maintains the following: the entire space-time structure is both ontologically and epistemologically a mere abstraction from the attributes and relations of physical things and events, such as causal relations. In this view, there is a physical substratum on which the entire space-time structure is parasitic ontologically. Thus it may be claimed that any geometrical world-point derives its very identity or individuality from an underlying, ideally punctal, actual, or merely possible, physical event. It should then be pointed out that in order to avoid logical havoc, even such a very strongly relational thesis must distinguish carefully between the abstracted space-time structure, on the one hand, and the physical substratum from which the abstraction is presumably made, on the other.

For consider any set A of ideally punctal, physical events which form an equivalence class in the sense that they all belong to the career of one and the *same* classical mass particle. The pertinent equivalence relation between any two event-members of A is called the relation of genidentity. Now consider two distinct classical particles P_1 and P_2 whose careers are such that they collide just once. Can we conceptualize that collision by saying that the equivalence class of events which constitutes the career of P_1 *intersects* in the set-theoretic mathematical sense with the equivalence class P_2? Such a conceptualization of the collision would be *inconsistent* with the status of genidentity as an *equivalence relation* among events belonging to the career of one and the same particle. For if we regarded the event classes P_1 and P_2 as intersecting in the mathematical sense, then the putative intersection event E_0 would be genidentical with some other event E_1 in P_1 and also with yet another event E_2 in P_2, but such that E_1 and E_2 are *not* genidentical with each other. Hence instead of conceptualizing the collision of our two particles as the intersection of two sets of genidentical events, we might more correctly regard the collision as a complex event constituted by the *coincidence* of two distinct, ideally punctal, physical events. So far, we have talked only about the physical substratum.

But our relationalist philosopher tells us that the space-time geometer has abstracted a geometrical entity called a world-line L_1 from the physical career of P_1, i.e., that the constituent world-points of L_1 collectively are the space-time trajectory of P_1. Similarly for P_2 and its corresponding world line L_2. Now note that on this level of geometrical abstraction, the relationalist can speak quite consistently about the intersection of L_1 with L_2 in the set-theoretic sense. But we saw that he could *not* with logical impunity similarly regard the careers of P_1 and P_2 as intersecting, even though he considers those physical careers to be the respective ontological substrata of L_1 and L_2. Thus even the aforementioned dyed-in-the-wool relationalist cannot talk interchangeably about world-points and the ideally punctal physical events from which they are held to be abstracted, or about a space-time trajectory L_1 and the event class P_1 from which it is presumed to be abstracted. Assume that the relationalist grants therefore the necessity of heeding the distinction between a world-point and the event occurring at that point, so to speak. Then he can agree with his absolutist opponent on the meaningfulness of this distinction, even though they disagree on whether the existence of a world-point depends ontologically on the physical event which occupies it. Hence the absolutist and the relationalist can each speak consistently of things and processes as populating space-time or of physical goings-on within space-time, but with a proviso: we are to understand that the relationalist gives a somewhat Pickwickian meaning to these locutions, because of his avowed abstractionism, whereas the absolutist blithely intends them to be taken literally.

I have illustrated the need to specify the aspects of the geometrical structure that are held to be relational or absolute in a given case. The preceding discussion of these illustrations will now enable me to deal with the last of the important questions put to me in Stein's comments and likewise asked by David Malament.

(3) *Is it legitimate to postulate a manifold topology and a differentiable structure upon the set of world-points as intrinsic—and as not "grounded" in anything prior—while asking, "Whence does the space-time manifold get its metricality and particular metric structure?"*

Stein's concluding question was: by what principle of dispensation does the underlying four-dimensional manifold structure escape the burden of justification that the metric field allegedly has to bear?

My reaction to this challenge is quite different depending on whether it is to be given *sub specie aeternitatis* or only relatively to the present state of physical theories.

Sub specie aeternitatis, let us ask: "Where does the four-dimensional differentiable manifold structure of space-time come from?" In line with some conference remarks by John Stachel and Jürgen Ehlers, let me suggest that the four-dimensionality of space-time *might* be discovered to be relational. Indeed, it might turn out that space-time is relationally endowed with two different dimensionalities in rough analogy to the way the universe of Dirac's two-metric theory is endowed with two interestingly different metric structures. Furthermore, as John Earman has pointed out to me, one might envision giving a *causal* underpinning not only for the four-dimensional manifold feature, but even for the differentiability. This merely illustrates how, in principle, one might envision a relationalist answer to the question we just asked, without thereby having precluded by the very posing of the question that the answer turn out to be an absolutistic one instead.

That the question is not a loaded one is shown by the history of the corresponding *metrical* question: as we saw in section 2 on Newton, the fourteenth-century exponents of *compositio ex punctis* asked, "Where do the metric interval ratios of Euclidean physical space come from?"—yet they gave an *absolutistic* answer by invoking their theory of the cardinalities of intervals.

In sum, *sub specie aeternitatis*, no aspect of the geometrical structure is exempt from the ontological scrutiny which I have asked the metric feature to bear in the tradition of Riemann and Weyl.

But in all of my writings on the ontology of physical geometry, I have been viewing things in the light of our present knowledge rather than *sub specie aeternitatis*. Specifically, I followed Riemann even a century after his death in still taking the *n*-dimensional manifold structure and the locally Euclidean topology of physical 3-spaces and various space-times as absolute or *given without relation to anything external*.[84] For, although that structure may not be here to stay in future science, at present we are presumably stuck with it as seemingly "rock bottom." Jürgen Ehlers has emphasized this bedrock status of the locally Euclidean topology by special reference to current quantum theory. But, as shown by the example of Dirac's Two-Metric Theory—to mention only one—no one metric structure is at all comparably "rock-bottom." For this reason, I focused my ontological question on the specifically *metrical* aspect of geometric structure and asked only, "Where does the *metricality* of space-time come from?" By the same token, in the perspective of the current state of our *physical* knowledge—a perspective which Stein himself adopted—it seems to me beside the point, qua criticism of my preoccupation with the ontology of the metric, to note with Stein that nowadays mathematicians know of many more abstract structures, topologies, or what have you that can be put on a bare point set than they were able to do at the time of Riemann.

On the other hand, *sub specie aeternitatis*, I reject a double standard of ontological scrutiny as between the *metrical* and nonmetrical aspects of geometric structure: in the latter perspective, the neo-Riemannian inquiry, "Where does the metricality of space-time come from?" opens the mind not only to the legitimacy but also to the possible fruitfulness of a corresponding ontological question concerning one or another *non*metrical feature of geometric structure. And therefore the remarks about conceivable alternative relational dimensionalities which I made above *sub specie aeternitatis* indicate that the receptivity of a bare point set to a plethora of alternative abstract structures lends poignancy to asking about the ontology of *all* features of the geometry. And in the case of the ontological question concerning nonmetrical features, my construal of the question no more loads the dice in favor of a relationalist verdict than it did in the metrical case.

Far from having become spurious or lost its pertinence to present-day theories of space-time structure, the distinction between absolute and relational properties of space-time now has considerably more sophisticated actual and potential relevance than envisioned by Newton and Leibniz. And thus, the philosophical rivalry suggested by the title of this paper is *not* an anachronism.

5. Epilogue: Time-Orientability and the Individuation of World-Points

Let me conclude by raising a question concerning (a) the theoretical capability of Einstein's space-time theory to individuate the world-points of any particular space-time manifold that falls within its purview, and (b) the bearing of that capability on the claim that actual space-time can always be held to be time-orientable with impunity. This question will be relevant alike to various absolutistic and relational construals of that theory. For example, it will be pertinent to Wheeler's early geometrodynamic kind of extreme absolutism, which asserts that empty curved space-time is everything. And it will be no less germane to the several versions of relationalism that deny this absolutism. We can go back briefly to Newton and Leibniz and illustrate how the problem of individuation confronts both the absolutist and the relationalist, although they each give a different twist to it. The problem of what makes one world-point different from another presents itself, because at least prima facie all world-points are alike with respect to their monadic properties. J. L. Anderson has overstated this alikeness in his book on relativity by saying: "The distinguishing feature of a particular point of . . . space-time is that it has no distinguishing features; all points of space-time are assumed to be equivalent." [85]

Turning back for a moment to Newton and Leibniz, we recall from a quotation from Howard Stein that Newton differed from Leibniz in regard to the *character* of the relations which Newton regarded as *individuating* the points of space: Newton held that purely *internal* or spatial relations themselves give the parts of space their individuality, whereas Leibniz believed that such individuation had to be grounded in relations involving nonspatial entities. But Stein also emphasizes that despite this difference concerning the ontological autonomy of space, Newton and Leibniz agreed on a cardinal philosophical principle of individuation, to wit, Leibniz's principle of the identity of indiscernibles. Says Stein: "The idea formulated by Leibniz as the principle of the identity of indiscernibles is obviously a familiar one to Newton; and he bases upon it a view of the standing towards one another of the parts of space that is strikingly similar to Leibniz's. [86]

We must be careful *not* to trivialize Leibniz's requirement that world-points must differ in some property or relation if they are to be distinct individuals. Thus one must *not* countenance as an individuating property of a point a the property of being identical to a, a property which the point a possesses uniquely in trivial, solitary splendor. For such a trivialization would really beg the question. Incidentally, Quine's uniquely instantiable predicates do not serve to abet such a trivialization. For as Hector-Neri Castañeda has explained, "In recent times, Quine has introduced, in the style of Boethius, uniquely instantiable predicates like Pegasizes and Hector-Neri Castañedizes. These predicates are, however, not intended by Quine to express an individuating property."[87]

Let me first give a very elementary example to formulate the point of view from which I shall ask my question. An ordinary hemisphere S sitting on a table in Euclidean 3-space can be turned into a model of *single* elliptic 2-space as follows: we *identify* antipodal equatorial points of that surface as being the same (or equivalent), so that any two great circles of S are now said to intersect only once, whereas before they intersected *twice* with the equatorial great circle.

This identification procedure is of formal mathematical interest. But I trust that at least some of us would react to this procedure somewhat as follows: (1) It is *meaningful* to say that in the physical world which surrounds us, the antipodal equatorial points that have been merged by the identification procedure are in fact ontologically distinct. To mention only one reason for their distinctness, note that they each coincide with different points of the table. (2) Not only is it meaningful ontologically to say that the antipodal points are distinct, but we are able *epistemologically* to ascertain that this assertion of nonidentity is true. For example, at a given time, two of us can respectively touch *disjoint* parts of the hemisphere that respectively contain these distinct points, even though they have been declared to be the same in the identification procedure.

In the much more interesting context of the De Sitter universe, Schrödinger has discussed the considerations that presumably warrant the identification of antipodal world points of the pseudo-spheric world model so as to give an elliptic interpretation of it.[88]

More generally, consider the case of those rival manifold topologies in cosmological models which are obtainable from one another by procedures of identifying points in a covering manifold or carrying out the inverse disidentification to generate the topologically different covering manifold. Here I am inclined to agree tentatively with Clark Glymour's view of the ontological status of these rival topologies. He says: "Considering these examples of alternative topologies, I find myself reluctant to grant that both of a pair of alternatives might in fact be true. The alternatives seem to be most naturally understood as contradictory and irreconcilable."[89] The reason given by Glymour in support of this view is that the alternative topologies depend on what basic individuals there exist in the universe, and that such ontological differences are matters of truth or falsity. More precisely, our actual universe may, of course, be *neither* of a particular pair of topologically different space-time models which are related by identification or disidentification of world-points. But it is claimed to be meaningful to assert on the ontological level that we live in at most one of the two models, and furthermore meaningful to ask in which one of the two, if either, we do actually have our being. It is then an additional difficult epistemological question whether we can discover empirically the answer to this ontologically meaningful question.[90]

Let these tentative assumptions be granted for now. And let us bring to bear the same kinds of assumptions on the case in which a given space-time (M, g) lacks the global conformal property of time-orientability but is turned into a corresponding time-orientable space-time by disidentifications which generate its double covering space.

Hawking and Ellis appeal to this disidentification procedure and write: "we shall assume that either [the space-time] (M, g) is time-orientable or we are dealing with the time-orientable covering space."[91]

This prompts my question about individuation of world-points, which is addressed to both the all-out space-time absolutist and to his relationalist critics of one stripe or another. The question is: what criteria of identity or distinctness for world-points and/or punctal events, if any, can give *physical meaning* to the required formal disidentifications at the ontological level of postulated space-time theory? This question is logically prior to the further hard question of whether such disidentifications can also be legitimated empirically at the epistemological level, rather than merely at the level of postulational meaningfulness.[92]

It seems to me that an adequate defense of Glymour's stated point of view, which I endorsed tentatively, depends on the provision of a viable criterion of individuation for world-points which are prima facie so much alike with respect to their monadic properties.

In casting about for such a criterion of individuation, I first turned hopefully to the Bergmann and Komar method of constructing "intrinsic coordinates" under sufficiently heterogeneous conditions. Typically such an intrinsic coordinate system is furnished by four independent scalar fields whose construction involves the use of the metric.[93] *If* this method were feasible globally as a means of individuating *all* of the world-points of a given space-time, then it might serve even within the all-out absolutistic framework of Wheeler's erstwhile vision of geometrodynamics.

But as Peter Bergmann kindly emphasized to me in personal correspondence, this method of intrinsic coordinates works only locally in patches of space-time, if at all. Even locally, it may fail as a sufficient condition for individuation, because existing continuous isometries would preclude the availability of four independently varying scalars.

Clark Glymour has argued that for the large class of "weakly distinguishing" space-times in which the two lobes of each light cone are disjoint, "we may then regard the entities of the manifold as individuated by their metric relations."[94]

Assuming the success of this criterion of individuation for the specified kind of space-times, I ask: to what extent can it serve to distinguish between a space-time and its covering space-time *as well as* to discredit the claim that actual space-time can always be held to be time-orientable with impunity?

John Earman has mentioned, without giving reasons, his view that the latter claim can indeed be discredited.[95] I raised my question partly to emphasize the desirability of the published availability of the views of Earman and others who have developed their ideas on defensible answers to the question.

Notes

1. J. Stachel, "The Rise and Fall of Geometrodynamics," in K. F. Schaffner and R. S. Cohen, eds., *PSA 1972* (Boston and Dordrecht: Reidel, 1974), pp. 51–52. This volume will be referred to hereafter as *PSA 1972*.

2. R. Penrose, "Angular Momentum: An Approach to Combinatorial Space-Time," in T. Bastin, ed., *Quantum Theory and Beyond* (Cambridge: Cambridge University Press, 1971), pp. 151–180, esp. p. 180. For a discussion of some open problems relating to the construction of three-dimensional geometry from the concept of solid body, see P. Suppes, "Some Open Problems in the Philosophy of Space and Time," *Synthèse* 24 (1972): 298–316; reprinted in P. Suppes, ed., *Space, Time, and Geometry* (Boston and Dordrecht: Reidel, 1973), pp. 383–401.

3. C. W. Misner, "Some Topics for Philosophical Inquiry Concerning the Theories of Mathematical Geometrodynamics and of Physical Geometrodynamics," *PSA 1972*, p. 26.

4. Anneliese Maier, *Die Vorläufer Galileis im 14. Jahrhundert*, 2nd enlarged ed. (Rome: Edizioni di Storia e Letteratura, 1966); translations into English of quotations from this work are mine.

5. *Ibid.*, p. 167.

6. *Ibid.*, p. 166.

7. *Ibid.*, pp. 166–167. In note 23 on p. 167, Maier claims that the scholastic exponents of *compositio ex punctis* did not associate their belief in differing transfinite cardinalities with the modern Cantorean distinction between denumerably and super-denumerably infinite sets.

8. *Ibid.*, p. 170.

9. Cf. Maier, *Die Vorläufer Galileis*, pp. 167–169, and John E. Murdoch, "The 'Equality' of Infinites in the Middle Ages," *Actes du XIe Congrès International d'Histoire des Sciences*, vol. III, pp. 171–174.

10. This quotation from Robert Grosseteste was kindly translated for me by Dr. A. George Molland from the Latin text: R. C. Dales, ed., *Commentarius in VIII Libros*

Adolf Grünbaum

Physicorum Aristotelis (Boulder: University of Colorado Press, 1963), pp. 91–94. I am also indebted to Dr. Molland for steering me to most of the other literature on the thirteenth- and fourteenth-century scholastics cited in this paper.

11. A cubit is a measure of length given by the length of the forearm from the elbow to the end of the middle finger, equivalent to 18 inches or 45.72 cm.

12. See note 10.

13. Cf. Maier, *Die Vorläufer Galileis*, p. 170.

14. Cf. J. E. Murdoch, *"Rationes mathematicae": Un aspect du rapport des mathématiques et de la philosophie au moyen âge* (Paris, 1962), p. 26; "Superposition, Congruence and Continuity in the Middle Ages," in *Mélanges Alexandre Koyré* (Paris: Hermann & Cie, 1964), Vol. 1, pp. 431–434.

15. Murdoch, *"Rationes mathematicae,"* p. 25.

16. Newton's 1692/3 letter to Bentley, in H. S. Thayer, ed., *Newton's Philosophy of Nature* (New York: Hafner Publishing Co., 1953), pp. 51–53. The words in square brackets are taken from H. W. Turnbull, ed., *The Correspondence of Isaac Newton*, vol. III, 1688–1694 (Cambridge: Cambridge University Press, 1961), p. 239.

17. I. Newton, *Principia*, ed. F. Cajori (Berkeley: University of California Press, 1947), p. 6.

18. *Ibid.*, p. 6.

19. Cf. H. Stein, "Newtonian Space-Time," *The Texas Quarterly* (Autumn 1967), fn. 9, pp. 199–200, for the nonexternality and monadicity of a property which Newton regarded as absolute.

20. My earlier discussion of the differences between various attributes which Newton regarded as absolute (*Geometry and Chronometry in Philosophical Perspective* (Minneapolis: University of Minnesota Press, 1968), p. 252, cited hereafter as "*GCPP*") suffers from the failure to separate Newton's *intention* to invoke a preferred rest system from his actual *success* in singling it out physically.

21. Newton, *Principia*, p. 6.

22. J. L. Anderson, *Principles of Relativity Physics* (New York: Academic Press, 1967), p. 466.

23. K. C. Clatterbaugh, "A Note on Newtonian Time," *Philosophy of Science* 40 (1973): 281–284.

24. Cf. I. Newton's "On the Gravity and Equilibrium of Fluids," in A. R. Hall and M. B. Hall, eds., *Unpublished Scientific Papers of Isaac Newton* (Cambridge: Cambridge University Press, 1962), pp. 133–136. I am indebted to Howard Stein for this reference.

25. B. Riemann, "On the Hypotheses Which Lie at the Foundations of Geometry," in D. E. Smith, ed., *A Source Book in Mathematics*, vol. II (New York: Dover Publications, 1959), p. 425.

26. A. Grünbaum, *Philosophical Problems of Space and Time*, 2nd ed. (Boston and Dordrecht: Reidel, 1973), pp. 500–501 and 545–546 (cited hereafter as *PPST*).

27. *"De Instantibus" Opera Omnia*, Parma edition (New York: Mesurgia Publishers, 1950), vol. 16, pp. 361–362. G. J. Massey also kindly furnished me with his English translations of the pertinent Latin passages.

28. Riemann, "On the Hypotheses," p. 413.

29. *Ibid.*, p. 422.

30. *Ibid.*, pp. 424–425.

31. A. Grünbaum, *PPST*, p. 527; italics in original. The most recent critic to have overlooked my repeated and emphatic characterization of RMH as a separate *empirical* hypothesis is Robert Weingard (*Journal of Philosophy* 72 (1975): 426–431), who depicts me as having uniformly claimed the *deducibility* of RMH from the assumed differentiable manifold structure of space-time.

32. For details, see my *PPST*, pp. 753–756, or A. Grünbaum, "Geometrodynamics and Ontology," *Journal of Philosophy* 70 (1973): 795–796.

33. P. Horwich, "Grünbaum on the Metric of Space and Time," *The British Journal for the Philosophy of Science* 26 (1975): 204–205.

34. H. Weyl, *Space-Time-Matter* (New York: Dover, 1950), p. 97. See also section 13 of the *sixth* edition of Weyl's *Raum-Zeit-Materie* (Berlin: Springer-Verlag, 1970).

35. Weyl, *Space-Time-Matter*, p. 98; italics in original. The corresponding original German passage is on p. 101 of the aforecited German edition. J. Alberto Coffa has called my attention to a passage in Weyl's 1922 paper "Die Einzigartigkeit der Pythagoreischen Massbestimmung" (*Mathematische Zeitschrift* 12 (1922): 114–146) where Weyl dissociates his own view of geometry to some extent from the one he attributes to Riemann in the italicized quotation above. After citing that Weylian passage (as I translate it into English) I shall comment on its merely *prima facie* incompatibility with Weyl's own seemingly explicit espousal of OI in other statements by him which I shall quote at the beginning of section 4 below. Weyl writes (Hermann Weyl, *Gesammelte Abhandlungen*, ed. K. Chandrasekharan (New York: Springer, 1968), vol. II, p. 266):

> I add an epistemological remark: it is not correct to say that space or the world in itself ["an sich"], prior to any material content ["vor aller materiellen Erfüllung"] is merely a formless continuous manifold in the sense of analysis situs; the *nature* of the metric [i.e., its infinitesimally Pythagorean character as codified in Weyl's *generalization* of a (pseudo-) Riemannian metric; A.G.] is peculiar to space in itself ["an sich"], only the mutual orientation of the metrics at the various points is contingent, *a posteriori* and dependent on the material content. (Of course the [infinitesimally Pythagorean] nature of space can present itself in reality only in a quite determinate quantitative particularization ["Ausgestaltung"], which is however to be regarded as contingent vis-à-vis space itself.

It appears from Weyl's various writings that he claims to be able to deduce a priori—in the manner of the presuppositional method employed in Kant's transcendental deduction of the categories—that physical space(-time) as such is endowed with an infinitesimally Pythagorean metric. But whether that metric will belong to the restricted (pseudo-)Riemannian species of Weylian metric or not will depend, in his view, on the material content of space-time (the absence or presence of electromagnetism). And indeed even if the metric is pseudo-Riemannian as required by the GTR, it is the contingently present material content and behavior that will determine to within a constant factor which one of an infinitude of incompatible metrics is physically realized.

Does Weyl's phrase "vor aller materiellen Erfüllung" indicate that he countenances an actually matter-*empty* space-time which is endowed with a determinate particular metrical structure and that he is thereby here contradicting his seeming explicit espousal of OI elsewhere (as documented in my section 4 below)? It seems to me that Weyl's phrase admits quite reasonably of a reading which avoids such a contradiction, if it is construed in its context as having the force of saying the following: *apart from* the contingent and a posteriori *specifics* of the material content, it is guaranteed a priori that the *general form* of any metric physically realized by matter will be infinitesimally Pythagorean. But this fully allows that in any one particular instantiation of this *general* form—which can be instantiated by any one of an infinitude of incompatible metric structures—material entities play the constitutive role assigned to external metric standards by Riemann's OI.

36. *Ibid.*, p. 97; italics in original.

37. *Ibid.*, pp. 97–98.

38. Anderson, *Principles of Relativity Physics*, p. 106.

39. *Ibid.*, p. 329; italics are mine. For a similar preoccupation with the nondynamicity component of Newton's absolutism, see A. Papapetrou, "General Relativity—Some Puzzling Questions," in R. Cohen and M. Wartofsky, eds., *Boston Studies in the Philosophy of Science*, vol. XIII (Boston and Dordrecht: Reidel, 1974), pp. 376–377.

40. For a definition of a Killing vector, see Anderson, *Principles of Relativity Physics*, p. 85.

41. B. Russell, *The Foundations of Geometry* (New York: Dover, 1956), p. 66.

42. Cf. A. Grünbaum, *PPST*, pp. 512–534.

43. Weyl, *Space-Time-Matter*, p. 101.

44. For a discussion of the rationale for possibly *excluding* solid measuring rods in the GTR, see my *PPST*, pp. 730–731.

45. Weyl, *Space-Time-Matter*, pp. 101–102.

46. L. Sterne, *The Life and Opinions of Tristram Shandy, Gentleman*, Introduction by Wilbur L. Cross (New York: Liveright Publishing Corp., 1925), p. 42.

47. Stein expressed these doubts to me in part in response to questions I put to him at the conference, "Absolute and Relational Theories of Space and Space-Time," held on June 3–5, 1974 in Andover, Mass. under the auspices of the Boston University Institute of Relativity Studies, and in part in subsequent correspondence. When reporting the *former* doubts below, I shall quote both my conference questions to him and his orally delivered responses.

48. See A. Grünbaum, *PPST*, chap. 16. Unfortunately the largely sympathetic discussion of my views in I. Hinckfuss, *The Existence of Space and Time* (Oxford: Oxford University Press, 1975) does not take account of the respect in which the *second* edition of *PPST* goes beyond the first of a decade earlier.

49. Stein, "Newtonian Space-Time," p. 194.

50. *Ibid.*, note 9, pp. 199–200.

51. This volume, chapter 1.

52. For a discussion of this point in regard to the ascription of the cardinality 2 to the set of Jack's feet, see my *PPST*, p. 846.

53. More specifically, in one of his letters (February 16, 1975), Stein elaborated on this point as follows:

> Turning, finally, to the globe sitting on your desk, I can, *prima facie*, entertain the question whether it "intrinsically" possesses the "metrical property of spherical shape," or the property of having a surface which is a Riemannian 2-manifold of constant positive curvature; despite the fact that to ascertain this property recourse is necessary to "external" instruments. I say "prima facie," because I don't think the question is a clearly posed one in this bald formulation (consider, e.g., Descartes's meditation upon what is "intrinsic" to the piece of wax); still, I can attach genuine content to the remark that, if the globe in question is a hollow glass one filled with water, sphericity is an "intrinsic" (although not unalterable) attribute of the glass, but is not intrinsic to the water. The rough sense of this statement is that, in the presently given state of the glass, its spherical shape is maintained by the interactive forces of the particles constituting the glass itself; but not so of the water. But when the analogous question is asked about space-time—whether it possesses, intrinsically, properties of a metric character—I am stymied from the outset. The difference between the two cases is this: whereas in both some clarification might be required of the meaning, or criterion, of "intrinsicality," at least a number of rough criteria suggest themselves; but in the latter case, unlike the former, there is a crucial need for *identification of the object about which the question is asked*. I know ("by acquaintance," as it were) what thing you mean when you speak of the globe on your desk; but I have no corresponding access, independent of *theoretical* identification, to space-time. The question "Is metric structure intrinsic to space-time *qua* Riemann-Minkowski-Einstein space?" or ". . . *qua* differentiable manifold?" is, as we all agree, purely mathematical—and trivial. The intended question is, "Is metric structure intrinsic to space-time *qua* space-time?" And my difficulty is that for space-time, unlike the material globe, I don't know any way to indicate the referent of the term without begging the question. I know, of course, that you have tried to give a general criterion of "intrinsicality" that might circumvent this problem; but, as I have said, I do not find your definitions clear.

54. See the quotation from Stein associated with note 49 above.

55. Cf. J. A. Wheeler, "Curved Empty Space-Time as the Building Material of the Physical World," in E. Nagel, P. Suppes, and A. Tarski, eds., *Logic, Methodology and Philosophy of Science: Proceedings of the 1960 International Congress* (Stanford: Stanford University Press, 1962), p. 361.

56. A. Einstein and N. Rosen, "The Particle Problem in the General Theory of Relativity," *Physical Review* 48 (1935): 73.

57. *Ibid.*, p. 76.

58. C. W. Misner, "Some Topics for Philosophical Inquiry," section 2, esp. p. 12.

59. This theory is "already unified" roughly in the sense that electromagnetic effects are described as particular characteristics of the curvature, which might not be found in other regions of space-time.

60. Wheeler, "Curved Empty Space-Time," p. 367.

61. Misner, "Some Topics for Philosophical Inquiry," section 3, B, pp. 19–20.

62. Cf. J. Earman, "Are Spatial and Temporal Congruence Conventional?", *General Relativity and Gravitation* 1 (1970): 155–156.

63. Cf. Weyl, *Space-Time-Matter*, p. 308, italics in original. See also the sixth German edition of this work as cited in note 32 above: on pp. 271 and 298–299, Weyl uses the German verb "sich einstellen" when speaking of the stated adjustment of metric standards.

64. For a detailed discussion of the important qualifying phrase "at least approximately and ideally," see my *PPST*, chap. 22, section 2, pp. 730–750.

65. J. Ehlers and W. Kundt, "Exact Solutions of the Gravitational Field Equations," in L. Witten, ed., *Gravitation* (New York: John Wiley, 1962), p. 61 and sec. 2–4, pp. 82ff.

66. *Ibid.*, p. 97.

67. See my "Operationism and Relativity" in P. Frank, ed., *The Validation of Scientific Theories* (Boston: Beacon Press, 1957), pp. 84–94.

68. *The Tragedy of Romeo and Juliet*, The Yale Shakespeare, R. Hosley, ed. (New Haven: Yale University Press, 1954), p. 37.

69. *Goethes Faust*, ed. Hans Gerhard Gräf (Leipzig: Inselverlag, n.d.), p. 189; the parenthetical translation into English is mine.

70. The German original of the pertinent pages of Appendix 5 can be found on pp. 100–101 of A. Einstein, *Über die Spezielle und Allgemeine Relativitäts-theorie*, 16th enlarged ed. (Braunschweig: Vieweg, 1954), pp. 84–85. I am indebted to J. Alberto Coffa for having drawn my attention to both the thematic quote from Einstein and to this Appendix 5. The translation of the quotation here into English is mine.

71. See J. A. Winnie, "The Causal Theory of Minkowski Spacetime," this volume, esp. section VIII. For a related detailed discussion, see D. Malament's doctoral dissertation (Rockefeller University, unpublished, 1975), chap. 2: "The Causal Structure of Minkowski Space-Time—What Robb Did and Did Not Prove."

72. Winnie and I completed our respective essays for this volume almost simultaneously. There was insufficient opportunity for me to study his text with the requisite care before going to press. Hence I ask the reader to refer directly to Winnie's paper in order to check and, if necessary, correct my impending statement of its import.

73. Cf. Winnie, "The Causal Theory," section 2, note 6.

74. The more recent version of Dirac's two-metric theory, in which the nonlinear relation between the two discordant metrics is logarithmic, is published in the section "The Two Metrics" (pp. 49–51) in P. A. M. Dirac, "Fundamental Constants and Their Development in Time," in J. Mehra, ed., *The Physicist's Conception of Nature* (Boston and Dordrecht: Reidel, 1973), pp. 45–54. For papers by him relevant to a different earlier version, see *Nature* 139 (1937): 323 and 1001, and *Proceedings of the Royal Society* A 165 (1938): 199.

There are other "*two-metric*" theories associated with more or less slightly modified forms of the GTR: for discussions of some of the earlier ones, see P. Havas, *Reviews of Modern Physics* 36 (1964): 961–962, and M. von Laue, *Die Relativitätstheorie* (Braunschweig: F. Vieweg & Sohn, 1953), vol. II, sec. 54, pp. 186–192. I am indebted to Peter Havas for the specifics of this earlier literature.

75. For a brief statement of the essentials of Milne's two scales, see *PPST*, pp. 22–23 *and* the emendation on pp. 557–558.

76. N. Rosen, *Annals of Physics* 84 (1974): 455.

77. *Ibid.*

78. *Ibid.*, p. 456.

79. N. Rosen, *General Relativity and Gravitation* 4 (1973): 435.

80. Mercator's projection provides a correspondence between the points of the (x, y)-

plane of the sheet of paper, and the points (θ, ϕ) of the globe (sphere) where θ is the angle of longitude and ϕ the angle of colatitude. The mapping is given by

$$x = k\theta, \; y = k \operatorname{sech}^{-1} (\sin \phi) = k \log \tan \frac{\phi}{2}$$

(James and James, *Mathematics Dictionary*, 3rd ed. (Princeton: Van Nostrand, 1968), p. 234.)

A metric that is bizarre in the sense that it is *not* yielded by rigid rods can now be imposed on the sheet of paper by transferring the globe's metric $ds^2 = a^2 d\phi^2 + a^2 \sin^2\phi \, d\theta^2$ (a is the radius in the chosen units) to the sheet of paper via the stated mapping. When the "transferred" ds^2 is then expressed in terms of the coordinates x, y of the points on the paper and of their differentials, the resulting form of the metric will *not* be $ds^2 = dx^2 + dy^2$ (to within a multiplicative constant) but will have a metric tensor which is *not* realized by rigid rods on the sheet of paper.

The map on paper and the bizarre metric imposed on it can serve to represent the metric interval ratios of the sphere on the sheet of paper as follows: (1) Parallels of latitude that are equidistant from one another on the globe (with respect to the usual metric there) are represented by lines on paper which—as judged by rigid rods applied to the paper—are increasingly *further* apart as the parallels of latitude approach the poles, but these lines on paper are interpreted as equidistant from one another with respect to the bizarre metric now imposed on the sheet of paper; and (2) Line segments between meridians (longitude lines θ = constant) which *decrease* on the globe from the equator toward the poles (with respect to the usual metric there) are represented on paper by line segments which are *equal* as judged by rigid rods, but the latter segments are interpreted as decreasing with respect to the bizarre metric now imposed on the sheet of paper. (For helpful diagrams, see "Map," subsection "Mercator Projection," *World Book Encyclopedia* (Chicago: Field Enterprises Educational Corp., 1967), vol. 13, pp. 141–142.)

81. I believe that, when coupled with chap. 16 and the appendix of my *PPST* as well as with reply 3 below, my articulation here of what I have called "the relationalist's position" above has also, in effect, dealt with some of the issues and objections raised by Michael Friedman's "Grünbaum on the Conventionality of Geometry," in Suppes, ed., *Space, Time and Geometry*, pp. 217–233.

82. J. A. Wheeler, "From Relativity to Mutability," in J. Mehra, ed., *The Physicist's Conception of Nature* (Boston: Reidel, 1973), pp. 233–234.

83. See the thematic passage cited from Stachel in section 1 of this essay, referenced in note 1.

84. Cases in which causal underpinnings of the topology *can* be given call for qualification of this statement about the presumably typical state of affairs.

85. Anderson, *Principles of Relativity Physics*, p. 4.

86. Stein, "Newtonian Space-Time," p. 194.

87. *American Philosophical Quarterly* 12 (1975): 136.

88. E. Schrödinger, *Expanding Universes* (Cambridge: Cambridge University Press, 1956), pp. vii–viii and 7–14.

89. C. Glymour, "Topology, Cosmology and Convention," *Synthèse* 24, nos. 1 and 2 (1972): 202. This paper is reprinted in Suppes, ed., *Space, Time and Geometry*; the pertinent page reference in the latter publication is p. 200.

90. For details on this epistemological question concerning the ascertainability of global space-time properties, see the previously cited paper by Glymour as well as his "Indistinguishable Space-Times and the Fundamental Group," in the present volume. See also David Malament's very valuable comments on the latter Glymour paper in the present volume.

91. S. W. Hawking and G. F. R. Ellis, *The Large Scale Structure of Space-Time* (Cambridge: Cambridge University Press, 1973), p. 181.

92. Concerning this important epistemic question, see the very helpful paper by David Malament cited in note 91 above.

93. See P. G. Bergmann and A. Komar, *Physical Review Letters* 4 (1960): 432, and P. G.

Bergmann, "The General Theory of Relativity" in S. Flügge, ed., *Handbuch der Physik IV* (Berlin: Springer, 1962), sections 21–29, pp. 247–271.

94. C. Glymour, "Physics by Convention," *Philosophy of Science* 39 (1972): 338, as amplified in a private communication.

95. J. Earman, "An Attempt to Add a Little Direction to 'The Problem of the Direction of Time,' " *Philosophy of Science* 41 (1974): 32. I have discussed critically some of the *other* tenets of this paper in *PPST*, chap. 22, sec. 4: "The Time-Orientability of Space-Time and the 'Arrow' of Time," as well as in my "Is Preacceleration of Particles in Dirac's Electrodynamics a Case of Backward Causation? The Myth of Retrocausation in Classical Electrodynamics," *Philosophy of Science* 43 (1976): 165–201.

——————HOWARD STEIN——————

On Space-Time and Ontology: Extract from a Letter to Adolf Grünbaum

I pass now from small-scale editorial points, and come to a more difficult task: the attempt to formulate in a clear way the issues that divide us, as I see them. We may roughly classify these under three heads: exegetical questions; substantive questions of the philosophy of science; and questions *about* (allegedly) substantive questions—about, in particular, the importance, or even the meaningfulness, of certain questions put forward as substantive. I do not suggest that a precise or a very satisfactory division of issues along these lines is really possible, but only that it is worth having these approximate headings in mind in our discussion.

I shall say relatively little about exegetical disagreements between us, with one important exception (important, at least, to me). We have, I think, some such disagreements—but not very sharp ones—about Newton; and we have plainly more serious ones about Riemann; but the one set of divergences over the interpretation of a writer that I cannot avoid dealing with in some detail has to do—(it seems so odd, I am actually experiencing some difficulty in setting down the words!)—with the interpretation of *me*. We shall come to this in a short while.

As to questions of a substantive kind (that is to say, the questions that I am myself inclined to regard as the ones that ultimately *matter*): I have been in one way very pleased, but also in one way very puzzled, to find in reading your paper that despite what appear to be serious disagreements between us, on what I consider the central substantive points we are evidently quite close together. At least, I have found nothing in your paper that contradicts, and several passages that seem to affirm, what you

NOTE: The letter was written in response to Professor Grünbaum's request for editorial criticism and for comments in reply; the extract comprises those comments. Changes occasioned by the subsequent revision of Grünbaum's paper are indicated by square brackets.

Acknowledgment is due the John Simon Guggenheim Memorial Foundation for support of this work.

374

quote from me on p. 331 ("I think I am only paraphrasing Newton—with whom I agree—when I say [etc.]"). For me, this is the main general point. And as to the interpretation and evaluation of relativity theory in particular, you hold (if I read you correctly) that the status of the tensor-field g as a legitimate physical field is independent of ontological theses about "absoluteness" versus "relationality" of metrical (or any other) attributes of space-time; the same, I presume, you would maintain of the laws governing this field and its relations to any other physical structures; you grant the admissibility of vacuum solutions in general relativity, and of vacuum regions in nonvacuum solutions; and you seem to concede the conceptual coherence of Wheeler's (old) program—which I had previously understood you to challenge—since you cite it (or, as you say, "an appropriate physical theory like GMD") as a theory that *might be empirically established*, when you characterize what you call "OI" as "open to broadly empirical refutation" (p. 359). With all this I agree; and I consider it as more or less exhausting the issues that are worth serious controversy in this domain—which suggests to me that the locus of our disagreement lies in my third class: i.e., that we differ seriously about what to consider a serious difference.

The question that seems to dominate your discussion (if we restrict its formulation to the context of the general theory of relativity) is this: *by virtue of what is the tensor-field g to be regarded as representing the metrical structure of space-time?* Now that seems to me a legitimate question, but not a clear and precise one: it seems to me (to use a distinction I appear to be growing fond of) "presystematic" rather than "systematic," and to pose a problem of explication. I should be inclined to reformulate the question as: "Just what do we *mean* when we say that this tensor-field describes *the metrical structure* of space-time?" You, however, appear not to see such need for explication, but to regard the question in its first version as clear, and of systematic "ontological" import. (For me, the word "ontological" itself presents seriously problematic aspects. You clearly do not use it in the sense advocated by Quine; and if you think Quine's usage not a very useful one for the philosophy of physics, I quite agree with you—but this surely suggests that there is not a generally accepted systematic use of the term.) You find that hard problems arise within this ontological domain: "[I]t is presumably very much *harder* to *discover* whether the metricality of space-time is ontologically absolute or relational than to discover whether a butcher is mistaken if he . . . regards the weight of a chicken . . . as . . . intrinsic to the chicken" (pp. 337–8). Even in the case of the butcher, I think some preliminary verbal clarification is needed, if only to determine whether he might distinguish in any way between what is measured by a spring balance and by a beam balance; but this is minor, of interest chiefly for the moral that verbal usage ought never to be taken quite for granted, except under the most closely controlled conditions. As to your *harder* problem, I think it may in one way be a quite *impossible* one, because ill-defined—but therefore no problem at all; and in another way (namely if, through adequate explication, it becomes well-defined), perhaps in principle not so hard. For consider: when the butcher's usage (and the intended sense of "intrinsic") has been fixed, the answer to the butcher question is logically determined by the body of knowledge codified in the theory of gravity (Newton's will serve the purpose). I ask, if the use of the phrases 'metricality of space-time' and 'ontologically absolute or relational' is made entirely clear, should we not expect that the answer to your question will also be determined by the content of the relevant physical theories—say, the general theory of relativity? Of course, the discussion of the theory—criteria for its application to phenomena, derivation of consequences, evaluation of evidence, etc.—may involve hard problems; but assuming these under sufficient control, it would seem prima facie that your ontological question, *if* it is clear, ought to be answered with comparative ease (unless, that is, contrary to what seems to me at all likely, the answer should turn out to depend upon further difficult *purely mathematical problems*; in which case these should at least be precisely posed). On the other hand, if the ontological question is not clear, there is no point in speaking of it as hard.

This complaint, lack of clarity, is just the one I have previously made—see, e.g., your quotation from me on pp. 328–9, and the end of my letter to you of February 16, 1975—against your attempt to explicate "intrinsic" in "Space, Time, and Falsifiability"; and in my letter, I excused myself from extended discussion of the matter, in part because, as I wrote, "I think it possible that you may have come to similar conclusions yourself." You have not confirmed or rejected this conjecture; but since you do not appear to rely, in your present discussion, on that earlier attempt, I shall still refrain from detailed comment on it. However, elaboration of my similar complaint against your latest discussion now seems obligatory. I have already indicated the general point, which can be put as follows:

Although I believe that in your quotation from *Faust* (p. 348) it is the student's sentiment you commend (*"Doch ein Begriff muss bei dem Worte sein"*), not that of Mephistopheles (*"eben wo Begriffe fehlen, da stellt ein Wort zur rechten Zeit sich ein"*), your practice in connection with certain crucial words—notably, 'ontological' or 'ontologically constitutive,' and 'metrical' or 'geometric'—seems to me rather of the devilish kind. I want to be unmistakably clear about this: I am not objecting to the fact that your use of these terms is different from mine—in such matters I am an extreme libertarian, believing not only that the utterer is entitled to any usage he finds apt, but also that the auditor has a certain obligation to be open and attentive, and to try to construe what is said in the way it is intended. (Just for this reason, I think it bad practice, except in connection with technical terms of quite established scientific usage, to assume that a word can be employed in systematic discussions without any need for elucidation.) What I object to is that, as it seems to me, you use the terms in question in a way that *is not clear at all*; or—to state a more modest claim, but with greater assurance—it not only "seems to me" but is indisputably true that you use those terms in a way that *is not clear to me*.

Perhaps it will be a help, in trying to give a more detailed account of what puzzles me about your point of view, if I first briefly sketch my own answer to the question raised a few pages back ("By virtue of what is the tensor-field g to be regarded as representing the metrical structure of space-time?" or: "Just what do we mean when we say that g describes the metrical structure of space-time?"). I do not think my answer differs very much, in substance, from yours: it has a good deal to do with the theory of the behavior of such things as measuring rods and clocks. But I would begin by making a preliminary, and in my opinion quite crucial, remark. This is that there is no "categorical" (in Kant's sense: derived from "categories" or "reine Verstandesbegriffe")—no innate, a priori, or (in your language) "canonical"—notion of *the metric of space-time*. Indeed, nobody before Minkowski employed such a notion at all. And when Minkowski invented, or discovered, this concept, what exactly did he do? He showed that the special-relativistic theory of space and time was tantamount to the statement that space-time has a particular structure, whose attributes are suggestively (although not perfectly) analogous to those of a Euclidean metric structure (thus, by the usual and useful liberty one takes in mathematics to *extend* or "generalize" a notion, a structure that may reasonably be called a "metrical" one), *and from which* (essentially by "orthogonal projection") *the Einstein geometric, chronometric, and kinematic relationships are determined.* So we have two main points: (1) a structure whose characteristics are, in a certain generalized sense, of that mathematical species which is called "metrical"; (2) a theory according to which the physical facts that belong, classically, to physical geometry (and chronometry and kinematics) are manifestations of that structure. Geometry having thus been "aufgehoben" into the Minkowski structure, one quite naturally refers to the latter as "the geometrical structure of space-time."

I shall not be so pedantic as to recite to you the story of the transition from special to general relativity, but only remark (a) that this transition reinforces the geometric analogy by the expansion from Euclid-Minkowski to Riemann-Minkowski, and (b) that it is with Minkowski, and the classification of world-directions into spacelike, timelike, and null, that *light rays* (or photons) and *free inertial particles* make *their first appearance as "probes" or "test-bodies" of the metric.*

Note, then, from this, that I fully agree with you (i) that the relationships of the metric tensor to various kinds of physical structures or systems that serve to "measure" it are of capital importance for the physical significance of that tensor, and (ii) that it is in virtue of certain of these relationships that it was natural for the creators of the theory to refer to this quasi-Riemannian metric structure as "the" metric, or "the" geometry, of space-time.[1] But note also (iii) that, as I see the situation and have just sketched it, the question to which (ii) is the answer is essentially one of *etymology*—hence essentially *historical*—rather than one either of physics or ontology. Why you (apparently) consider it both of the latter[2] is something I do not understand. I have tried, just above, to indicate how I use the word "geometric"—namely, rather flexibly: now in a broadly mathematical sense referring to purely structural characteristics of a certain (but not very sharply defined) type; now in a narrower sense having some reference to the historical root meaning, connected with "lengths (etc.) of objects." But when you speak (p. 345) of the question whether the structure to which measuring instruments adjust themselves is "an *autonomously geometric* structure," I simply and honestly do not know what the words mean. And as to "ontologically constitutive," I should have thought (whatever the finer analysis of the notions involved) that the application of that expression would be to situations in which some kind of "reduction" of one sort of entity to another was involved. Thus I have always taken Leibniz to be saying, not that spatial attributes *qualify as*—are "ontologically constituted as"—*spatial* by *their relationships to bodies*, but rather that those structures we call spatial *fundamentally consist in*—have no other meaning or being than as—*relations among bodies. This* is the doctrine that has made Leibniz's view fascinating to me; the other I should consider trivial. I might be wrong to consider it so—but only if the phrase "ontologically constituted as spatial" is being used with a meaning that I have so far utterly failed to grasp. And therefore—at least for me (*argumentum*, not *ad hominem*, but *ab homine!*)—some explication is required.

I must also say that I shrewdly suspect, from various signs, that such an explication is not to be had. I shall try to present a representative selection of these signs. But first, again, I emphasize that there is no question here of demonstrative refutations: at several points—and clearly important ones—your usage or your criteria *seem incoherent* to me; that they *are* incoherent I cannot prove, and I cannot be sure, because I do not understand them.

(1) Let me mention first your "family of concordant external metric standards," or "FCS," which in the view of your relationalist "ontologically mediates" the metrical status of the tensor-field g. Why do you include, besides the classical devices for measuring spatial and temporal quantities (measuring rods and clocks), also *photons* and *free massive particles* among these "metric standards" (p. 345)? As I have already remarked, before Minkowski these would not have been considered to be "geometrical" measuring devices.[3] Why are they now so regarded? From my point of view, the answer is plain: *because they can be used to explore the field g, which we now call "the space-time metric."* I—with my relaxed attitude towards the use of the term "metric"—find this entirely reasonable. But on the same grounds, I should be glad to consider as a geometric measuring device *any* instrument that could be made serviceable for exploring the field g. It is, as far as I can see, quite arbitrary to draw the line as you do, and to say in effect that those probes of the field g that are standard in the present stage of physical theory should be regarded—"canonically"—as "ontologically constitutive" of space-time geometry. [In your revised discussion, to be sure, you characterize the "set of external devices that are taken to be canonical for the metric structure of space-time" as "open-ended" (pp. 344–5). But this reinforces for me the

379

impression of a fluctuating—perhaps ambivalent—point of view. For "open-ended" seems hard to reconcile with "canonical": one would at least want to know how ordination is achieved! "Open-ended" and "taken to be" suggest conceptions of a dialectical and methodological order; "canonical" and "ontologically constitutive" suggest fundamental and unchanging objective principles.]

(2) Next: I wondered, on first reading your discussion of the "relationalist" 's view of vacuum solutions [(a discussion which has now been completely revised)], what you would say about vacuum regions in non-vacuum worlds. Then I found (pp. 349–50) that (the issue having already been raised by Allen Janis) you are prepared to deny the "existence of a metric" in such regions. This, to me, is astonishing. Let me quote from my own letter to you of February 16, 1975:

> The distinction between *constitutive* elements and "probative" ones ("test-bodies") seems to me a crucial one in these questions. Distance may be measured by measuring rods; it does not *consist in* (is not constituted of—or by—) measuring rods. (Descartes thought that space is constituted by extended substance—body—and so, quite consistently, argued that there can be no vacuum: i.e., that if there are no bodies between A and B then A and B must touch. But not even Bridgman would argue that if there are no measuring rods between the earth and the sun, the earth and the sun must touch.) In general, it's characteristic of probative elements (*a*) that alternative methods of probing are generally possible . . . , and (*b*) that the "field" or structure in question is presumed to exist independently of its being explored or manifested or probed.

But you seem willing to adopt something very like the view that I said "not even Bridgman" would. Consider your FCS closely for a moment: "Infinitesimal rods" in any strict sense clearly do not exist; so it is hard to see how they can be "ontologically constitutive" of anything, ever. "Free massive particles," taken in the strict sense, almost certainly also do not exist; in any case, they must be of extremely rare occurrence. "Atomic clocks": there can hardly be any such in the interatomic intervals. That leaves photons. Can one save the constitution of the metric by maintaining that photons are (quasi-)ubiquitous? Hardly—the photoelectric effect and the finiteness of total energy in a compact region pretty well exclude such a conception. Furthermore, it is hard to see how, if presence of members of the FCS is regarded as necessary for existence of a metric, one could justify considering the *mere* presence of *any* member of that family as *sufficient*—that is, without further conditions guaranteeing that a definite metric will be "determined"; and for this, light by itself is well known to be inadequate. It would seem to follow that the real world, viewed on a fine scale, is *rife* with "nonmetrical" regions—indeed, that an ordinary line connecting two ordinary points must *practically always* pass through such regions. And this leaves you almost in the position Descartes foresaw if a vacuum were admitted (namely, one of spatial "collapse"): you are not forced to conclude that two points *touch*, i.e., that their distance is *zero*; but you seem forced to conclude that distance, or the length of a line, hardly ever "exists" at all. It appears to me that a *reductio ad absurdum* lies dangerously near—or at least that the dialectics that may be necessary to avoid such a *reductio* are likely to cost more than the theory they are called upon to rescue is worth.

(3) You consider (pp. 358–9) "Wheeler's pre-1972 GMD program" to be incompatible with the "relationalist" view, or the thesis "OI." This, in the light of your preceding discussion, is quite baffling to me. If Wheeler's program were to succeed, it would quite certainly have to embrace all the physical structures, agencies, or instrumentalities, that make up what you call the "FCS." It is true that, because Wheeler's program was for a unified field theory, these things would all have to be characterized as states of the field *g*, and hence (I suppose), in your terms, as not "external" to it; nevertheless, when I try to adopt your way of speaking of and looking at these matters, it seems that they could perfectly well still be thought of as independent of—and in that sense "external to"—the "metrical status" of *g*. You have already conceded that fields, including *g*, can be regarded as "physical," and as "existing" (hence "ontologically constituted"), independently of any probing devices (FCS). If the field *does* "exist autonomously," and if the devices *can* be constituted by special states of the field, why should the role of such devices as "constitutive of the *metric* character of the field" have been undermined thereby? I admit that I do not understand, from the beginning, what you mean by this "constitutiveness"; still, it is a further perplexity for me that, having granted that the field *g* can "exist" independently of the FCS, but having maintained that it is "constituted as metric" by its relation to the FCS, you should then say that if the members of the FCS can be "reduced" to *g*, the power of the FCS (which, after all, still means, not just *g*, but some quite special configurations of *g*) to constitute *g* "as metric" has been lost. (Indeed, as I write this, my head swims.)

[(4) In your discussion of Winnie's new considerations (pp. 353–5) I believe I see the ghost of the same (as it seems to me) confusion that I remarked on in your earlier discussion of Winnie on Robb. In order to state the case clearly—and also because some of the substantive points I have made to you seem to me worth putting on record—I am going to allow my section on that earlier discussion of yours to stand, even though you have now deleted it from your paper.

[You said, then,] following Winnie, that "Robb's work shows that the system . . . of causal relations . . . suffices . . . to generate, via purely nominal definitions, the Lorentz metric . . . *without* the antecedent imposition of the requirement of flatness." First, there is a small technical point that ought to be clarified: it must be clearly understood that an infinite family of Riemann-Minkowski metrics is compatible with—or "exists over," in a mathematical sense—conformally flat spacetime, or "Robb space." *One and only one member of this family is metrically flat* (understanding that we identify Riemann structures whose metrics are related by a constant factor). There are, thus, fully ("canonically") determined by the Robb structure (conformality; "system of causal relations"), *both* (a) an infinite class of Riemann-Minkowski structures conformally equivalent to the Robb structure, *and* (b) a unique member of this class, which is metrically flat. (I am not sure your phrase "without the antecedent imposition of the requirement of flatness" quite does justice to this situation; although, indeed, it can be shown that in a certain sense the metrically flat member of the class mentioned in (a) is the only member of that class that can be distinguished from all others by *any* criterion that utilizes nothing but the assumed data.)

Given all this, and in the light of your general discussion, I should have expected you to ask, as on p. 342, "on the strength of just what" the structure singled out by Robb's "purely nominal definitions" has "the ontological status of being *the metric field* in the sense of endowing spacetime with the latter's particular metric structure." But instead, you accept the structure in question in this case without any "ontological mediation of the FCS." The reason you give is that this metric is definable in terms of ("arises from") a certain physical foundation—the Robb structure—which also allows one to define the (topological and differentiable) manifold structure of space-time; so: "It is philosophically gratuitous, or at best highly contrived, to single out the *metric* structure of this space-time as being relational in a way in which the underlying R^4-manifold structure of space-time is not. . . . And thus the metric structure is *not* ontologically dependent on relations to any *further* class of external entities or standards."

Now, first of all, a formally analogous statement can be made in the case of GTR. You have accepted the manifold structure as *somehow* "given," and as physical; and you accept the field g as physical, independently of the "FCS." This ensemble constitutes a structure that enables one to define all such notions as belong to the geometry of a four-dimensional Riemann-Minkowski manifold: the topological, differentiable, and metric notions all "arise from" this structure. Of course, you may argue that the Robb structure is "deeper": that Robb derives topology, for instance, from something physically more basic, whereas here it is cheating to say that the topology "arises from" our structure, since it has really been built into that structure explicitly and deliberately. This contention does certainly have merit; yet I feel uneasy about basing upon it a sharp ontological distinction between the two cases, since I do not see a sharp criterion for that distinction: it remains true that in both cases we have a body of structural data, of acknowledged physical and ontological legitimacy, sufficient to allow the definition of "*a* geometrical structure," and it is (again: to me!) obscure why in the one case but not the other something "ontologically more" is required to qualify this as "*the* geometrical structure."

But that does not yet hit the main point. The main point, from my perspective, is that whereas you and I agree, in the general case, that not the mere formal attributes of a single physical field, but also the ways that field affects particular "standard" physical configurations or "instruments," are involved in our choosing to characterize the field as describing the geometry of space-time, in this special case we unexpectedly part company on that issue: I do not think, as you do, that Robb's results have *any* bearing on the question why, in special relativity, the Minkowski structure (or "Lorentz metric") "deserves" to be called "the space-time metric." Indeed, any one of the infinite class of Riemann-Minkowski metrics compatible with the Robb structure might, without violating physical theory, be the field to which the members of your "FCS" "adjust themselves"; and if this should be one of the nonflat metrics, that nonflat metric would be the one to manifest itself in all spatio-temporal measurements, and would *undoubtedly* play the dominant role in the theory of the world. In this case, although the Lorentz metric would be "simpler" than the Einstein metric in the sense of requiring a smaller body of data for its determination (namely, it is determined by the Robb structure, whereas the Einstein metric would not be), yet the Lorentz metric and its mathematical distinction would be a mere curiosity; the Einstein metric would be preferred, just as it is preferred (in those contexts in which it makes a material difference) in the real world: because it dominates physical processes in general, and spatio-temporal measurement in particular.

How has it happened that we seem to have reversed sides here? My answer is that I maintain the same attitude to your "FCS" [4]—and to all other possible instrumentalities for probing a field, in all cases: they are essential to our knowledge of the field; physical laws of interaction adequate to determine how the field affects their behavior form an essential part of our concept of the field (without *some* such laws, we should have no adequate concept of a physical field at all); and special characteristics of that interaction influence us in giving "names," or brief descriptions, to fields—what we call "the electric field" is what exerts forces on what we call "charges" (and so it would still be, for instance, if the theory of Gustav Mie had been successful, according to which charges are "made out of" the field itself—a situation instructively analogous to that of Wheeler's GMD); what we call "the metric field" is what governs the "adjustment" of what we call "measuring rods (et al.)." But in no case do I consider the instruments, test-particles, etc., as "constitutive" of the field. You ascribe the latter virtue to the FCS (which implies, to me, a different use of the word "constitutive" from my own); but you withdraw the ascription in two cases—Wheeler and Robb—which differ very much from one another, but which have this in common: that in each of them there occurs, although in entirely different ways, an "ontological reduction" (or "constitution") in the sense in which *I* would use those words. For in the Robb case, it is true that the Lorentz metric can be "reduced" to the "causal structure"; and in the Wheeler case, the measuring instruments—the "FCS"—are reduced to the field. It seems fairly clear, then, that you boggle at saying either that the space-time geometry is "constituted" *by two different things* (Robb structure and FCS) or that it is "constituted" *by what it itself constitutes* (FCS in the Wheeler case). This would be quite reasonable if "constitutes" were being used in a single sense; but I submit that it is not—that the sense in which your "relationalist" ascribes "constitutiveness of the metric" to the FCS is *plainly* different from the sense in which, for Robb, the causal structure "constitutes" the Lorentz metric, or, for Wheeler, the metric field "constitutes" bodies—and thus that *your withdrawal of the relationalist ascription in these two cases exhibits a terminological confusion.*

[I take it now, from your revised paper, that you have accepted the foregoing comments *in part*—in particular, that you agree with what I have said about the bearing of Robb's theorem, but not with what I have said about the bearing of Wheeler's program.

[When we turn to the new points made by Winnie, it is to be noted first that the theorem cited by you (p. 353) is analogous to Robb's theorem: conformal structure together with a suitable "emptiness" assumption determines a metric structure. So one is led to ask, if this spirit was earlier exorcised successfully, whence does it now derive renewed power?

[The answer, evidently, is that "empty" is now a more powerful magic than before. Then we were in the context of "the pre-GTR version of the special theory of relativity," where in your opinion (p. 350) counterfactual appeal to hypothetical standards or test-bodies may be countenanced; now we are in the context of the GTR, where, you believe, appeal to such entities must be excluded, since their presence would modify the metric; and you refer (p. 351) to the *nonlinearity* of the GTR as especially pertinent here. Yet it is a standard point that this situation—modification of the field by test-bodies—exists in the classical linear fields as well. The crucial issue is not *linearity*, but *approximation*: can one suppose that "small" test-bodies "do not change things much"? If one could *not* make such an assumption in the GTR, then it is obvious that we could not apply the theory at all: for we do not know the true and exact impulse-energy distribution of the universe: in practice we ignore, not only mountains and planets, but—in cosmology—anything much smaller than a galaxy; yet you would have us reckon in, explicitly, each atomic clock.

[If, however, we envisage "weakly empty space-times" as *global approximations*—as universes in which "ordinary bodies" (and photons, etc.) are "gravitationally negligible" in comparison with gravitational radiation—then it would still be possible for you to appeal to the "FCS" as "constitutive" of the metric in such space-times. I remind myself that 'FCS' stands for "family of *concordant* . . . standards"; and conclude that *if the "standard" provided by gravitational waves did not agree with the behavior of particles, clocks, etc., it would not be allowed in as "concordant."* Is it not, then, a strange doctrine that admits gravitational waves (provided there are enough of them) as ultimate canonical ontological constitutors of "the metric" so long as space-time is weakly empty, but that deprives them of ultimate "constitutive" power if somewhere there is a neutron?[5] Let us, for instance, suppose that there is available a set of objects of (gravitationally) negligible mass belonging to the "FCS"; and that the behavior of these objects discloses a metric structure of space-time conformally, but not metrically, equivalent to the structure of a "weakly empty" world—one, moreover, in which gravitational radiation is abundant. We should undoubtedly conclude that we had thereby detected the presence of (gravitationally nonnegligible) nongravitational energy, and that the true gravitational field and associated Einstein metric are not here determined by the conformal structure. This case is quite parallel to the Robb one; and I assume that you (unlike me) would regard the objects in question as *ontologically constitutive* of "the physical metric." Now suppose that, holding everything else the same, we allow the objects of the "FCS" to vanish. Gravitational radiation will remain—sufficiently abundant, as we are supposing for the argument's sake, to determine the conformal structure. Does it now "constitute" a "physical metric," different from the one we had before (namely, a "weakly empty" one)? Note that the gravitational tensor g—to which you allow physical significance that is independent of the "FCS," and which continues as it was before—is *not* that of a "weakly empty" space-time, although its gravitational waves determine a conformal structure (that of g itself) to which Winnie's theorem applies, and thus determine a unique associated "weakly empty" metric tensor g^*. In the absence of the "FCS," does g^* acquire "metric status" that it did not have in their presence—or can gravitational radiation "constitute" a metric only on the added condition that $g = g^*$? Such a supplementary condition does not seem easy to justify. But if it is not required, then how is it that the *absence* of measuring standards allows gravitational radiation to assume an ontological burden it was unqualified to bear in their presence?—From a comment you have made in a letter to me, I infer that you might wish to resolve this issue by allowing, in the presence of the "FCS," *both* g and g^* as "ontologically constituted" metrics (in the spirit of your discussion of "two-metric" theories). This would indeed secure a consistent treatment of the "metrical constitutive power" of gravitational radiation (so far as we have gone—but see below!). However, it would seem to conflict with your response to my argument in the Robb case, where you did *not* conclude that we must accept two metrics (one flat, the other not so), but rather held the verdict of the "FCS" to be decisive, and your "relationalist" to win the exchange.

[In an important way, it seems to me, the discussion so far has been out of focus; for I do not believe that your requirement of sufficiently abundant gravitational radiation to determine all the null-lines of space-time—at least at some points—either *can* be realized (from this, if true, your relationalist will take comfort), or is a reasonable requirement to impose for the metric potency of gravitational radiation (so that I do not think that comfort is deserved). On the first point I speak with some diffidence, as insufficiently versed in the theory of the gravitational radiation-field; but I should expect, merely from the assumptions of continuous differentiability characteristic of this theory, that—strictly speaking—gravitational radiation can be propagated, at a given point in space-time, in no more than *one* direction. However that may be, the crucial point is the second. Here I take my stand, not on Winnie's theorem, but on what is implied by Stachel's question (quoted by you, p. 354): in effect, "If photons, why not gravitons?" In allowing light rays to play a role in relation to the metric, one surely does not restrict this role to space-times in which light alone is abundant enough to determine the metric fully (for it cannot, in principle, do so!). The most you require here is that *all members of the "FCS" taken together* are present in sufficient abundance to do the job. It seems, accordingly, when you speak of the "third position" (neither "relationalist" nor "semi-absolutist"—p. 354), that it would be fairer to the graviton—more in the spirit of your general treatment of the "FCS"—to allow metric power to gravitational radiation *whenever* it can be used to *supplement* other physical agencies in exploring the metric.

[But having said all this, and even assuming that you might be induced by these arguments to revise your stand—e.g., to withdraw your approval of my earlier discussion as a defense of "relationalism" against Winnie-Robb, advocating instead a "two-metric" view in that case, and to move further in the direction of the "third position" in the way I have just suggested—I am still left with an eerie feeling about these matters; a feeling that "Zeus has been dethroned, and Whirl is king." You cite with approbation three points of Allen Janis's, as counting against the appeal to gravitational waves alone to establish the space-time metric; and you conclude (p. 354) that Janis has thereby shown that "gravitational radiation does not help the semi-absolutist." Now, since I do not myself understand the "semi-absolutist" 's doctrine, I cannot quarrel with the conclusion against him. But I do note that Janis's three points have precise counterparts in respect of your "relationalists" 's doctrine. One of those points is an argument against any appeal to "hypothetical test-gravitons"; but you, of course, also reject appeal to "hypothetical" objects of the "FCS." The other two points amount to this: that gravitational radiation may not suffice to determine the conformal (hence a fortiori the metric) structure, at all points, or even at any point, of a weakly empty space-time. Again, you acknowledge the same possibilities in respect of your "FCS" in nonempty space-times. I have discussed partially, in (2) above, what seem to me the severe difficulties this makes for your "relationalist" 's position—namely, so far as the clause about "all points" is concerned; and have suggested that a *reductio ad absurdum* threatens. I think, however, that the threat is even more dire than I implied there: for to me it seems obvious that there *cannot* be, at *any* point of *any* general-relativistic universe, a "sufficient abundance" of objects of *any* of the kinds you have considered—members of the "FCS" and gravitational radiation taken together—to "constitute" a metric in the way you seem to require. Indeed, through any point of space-time there can be at most one particle world-line; and this is surely not enough to "constitute" a metric, if I understand your use of that term—and your endorsement of Janis's arguments—at all.[6]

[If the foregoing is right, it seems to me to follow that *all* the positions you discuss in your comments on Winnie (pp. 353–5) essentially fail, *including* the "third position": that "the gravitational field can acquire added geometrical physical significance either from gravitational radiation itself or via the FCS." Rather, whatever "added geometrical significance" means, from the conditions you lay down upon this notion it seems to follow that there is *no* way the gravitational field can acquire it. And yet this failure to achieve such added significance in no degree obstructs our normal use of the language of metrical relations, and of geometry in general, to characterize the physical world; and in no degree inhibits our usual procedures of measurement. Does it not follow generally—as you suggest (pp. 349–50) in a special case—that to abandon for the tensor field *g* the claim to "added geometrical physical significance" is "no explanatory loss . . . but only an innocuous verbal one"?

[Let me conclude this section with two further points illustrative of my sense of vertigo: my feeling of the presence, in all this discussion, of *Dinos—Vortex—Whirl*: (i) How does it come about that you are so concerned with the question of what "constitutes" the tensor g "metric," but that you are not concerned with what "constitutes" it "gravitational"? From what principled point of view is the notion of an empty universe with a metric structure *more* problematic than the notion of a universe devoid of matter but containing "gravitational radiation"? (After rereading your p. 304, I have tried to imagine the facial expression of Leibniz on contemplating the latter notion: too grim to describe!) (ii) Since, however, you are willing to concede not only that second notion, but the potency in such a universe (or at "weakly empty points" in a nonempty universe) of gravitational radiation—if only it is abundant enough to reveal all the null world-lines—to establish a "geometrical physical metric" for space-time, I wondered why you should draw the line here (that is, at gravitational *waves*): why not allow the *gravitational field itself* (waves or no waves) to "constitute" a metric? You now comment on this, saying that "whereas the wave-fronts of gravitational radiation pick out special space-time trajectories whose stipulated metrical nullity *then* might determine a metric tensor . . . , no corresponding determination exists in the gravitational fields of purely empty space-times: In the latter . . . , the gravitational field tensor g_{ik} would merely be *baptized ab initio* as *also* being physically metric in a physically redundant way."[7] Once again, you have not succeeded in making a distinction clear to me: I do not understand the principled difference between "stipulating the metrical nullity" of certain trajectories on the one hand, and "merely baptizing" a certain structure as metric on the other. (Does not Dinos seem to rule here?) As I see the two cases: (a) The "wave-fronts of gravitational radiation" are special space-time loci that can be constructed from the distribution of the gravitational tensor g; whatever "physical significance" they can claim derives from the "physical significance" of g itself, including its relationships with other structures that may have such significance. (b) Just as *some* space-time distributions of g may allow one in *some* regions to construct gravitational wave-fronts, and these in turn may allow one to construct *some* trajectories which are null-lines (of the tensor g!—see below); so *all* distributions of g allow one to construct *all* the space-time trajectories that are null geodesics of the tensor g, and all the timelike geodesics as well: an array of structures sufficient, in turn, to allow reconstruction (up to a constant factor) of "the metric tensor g." I still quite fail to grasp the basis

389

for your allegation of a deep difference in principle between these two constructions. Indeed, it seems to me obvious that there was never a philosophical, or a "principled" physical, reason to "stipulate the metrical nullity" of the world-lines of "gravitational rays": one is led to the possibility of using gravitational waves to explore the metric only by the *theorem* of general relativity that *such world-lines are necessarily null-lines of* g. In what I have above called the historical, or etymological, order, it is clear that the "metrical significance" of gravitational waves is derivative from, not constitutive of, the "metrical" character of g (even if the latter is regarded as a merely nominal affair). You maintain the existence, and importance, of a quite different relationship in the ontological order; but notwithstanding all the explanations you have given, the latter remains entirely dark to me.]

(5) One last point in this series (whose intent, I remind you, is to try to make plain *in what way your usage seems to me confused*—or *is* to me *puzzling*; not to demonstrate conclusively that your usage is in fact confused, or cannot in any way be coherently explicated): On pp. 345–6, you say that "the relationalist's view . . . derives its intellectual inspiration from Riemann's ontology of physical geometry" (as you construe the latter —we have some differences over it), "instead of being a holdover from the discredited homocentric ontology of operationalism or of the verifiability theory of cognitive meaning." I conclude that the Riemannian heritage—by contrast with the "discredited" sources you mention— counts as a recommendation (and on *this* we do not differ: in the matter of worshipping heroes, I would be loath to place any altar above Riemann's). But on p. 359, you make the (generalized) Riemannian "OI" (as you see it) a "broadly empirical" claim. Since Riemann does not offer a *shred* of empirical evidence, but only *conceptual argument*, in the passage that is your source for "OI," I am nonplussed.

[So I wrote before. You have now explained (p. 319) that you do not mean to invoke Riemann's authority to "lend credence to RMH." Yet you have allowed your statement on pp. 345–6, cited above, to stand—in which you contrast (I assume favorably) the intellectual inspiration derived from Riemann's ontology of physical geometry with the discredited ontology of the verifiability theory of meaning. This seems to me to leave things in a most unsatisfactory state. If you *reject* Riemann's own view, as aprioristic, in favor of an *empirical* hypothesis OI, why should you not dissociate the latter from Riemann's "discredited *a priori* ontology" just as much as from the "discredited ontology of operationalism"?

[You, I, and Riemann are in the following relationship here: You interpret Riemann's conclusion in one way; I in another. You believe that Riemann's argument for your version of his conclusion was utterly worthless (since in your version that conclusion is an empirical hypothesis, and since as I have said Riemann adduces no empirical evidence whatever); I believe that Riemann's argument for my version of his conclusion was essentially correct. Moreover, you distinguish in Riemann two hypotheses (OI, which appears at the beginning of his lecture, and DH, which appears at the end), and you argue that Riemann incorrectly inferred the second from the first; whereas in my reading, the application Riemann makes of his earlier remarks when, at the end of his lecture, he turns his attention to physical space, is entirely cogent. Under these circumstances, it is indeed altogether possible that you are right about Riemann and that I am wrong about him; but it is also certain that *if* you are right about Riemann, then Riemann was (in quite fundamental ways) *wrong* (and if I am right about Riemann, then he himself, on those fundamental issues, was *right*). It therefore seems to me very misleading—and, I venture to suggest, self-deceiving—for you to continue to claim to uphold a Riemannian philosophy of geometry.]

I should like now to try to clarify two aspects of my view of these matters, in connection with a pair of passages in which you comment on remarks of mine. First, you refer on p. 363 to my having said that mathematicians nowadays know of many more "structures" than were known in the time of Riemann; and you do not see how this is to the point. Let me try to explain what I had in mind:

In the ancient tradition of mathematics, its subject matter was conceived of as certain *particular "formal" aspects of the world*; most usually, under the two heads of "discrete" and "continuous" *quantity*. Riemann was one of the pioneers in breaking through to a freer and wider conception—one that is beautifully expressed in the saying of Riemann's devoted friend Dedekind, that the objects of mathematics (e.g., numbers) are "free creations of the human mind." In the older view, "space" was conceived, under the head of "continuous quantity," as a definite object of study,[8] having definite metrical attributes (cf. Kant, "Transcendental Aes-

391

thetic," section 2: "Space is represented as an infinite *given* magnitude").
One of Riemann's great philosophical contributions was the demonstration that, unlike the case of "discrete quantity" (where to determine a set is necessarily to determine its "quantity," or cardinal number), and contrary to all traditional expectation, in the case of a "continuous manifold" *the concept of such a manifold, and of* (roughly speaking) *its "continuity" or "smoothness" properties*, can be *separated* from *any* "metrical" determinations. *This* is what I believe to be the substance of the remarks of Riemann from which you distill "OI."

Now, it is in the older, the pre-Riemannian tradition that it appears appropriate and natural, if a "spatial manifold" is spoken of, say with various structures on it (like your flat paper with its map of the earth), to ask, "Which of these has the *added* significance of being *the* structure—topology, metric, etc.—of the spatial manifold?" In the newer conception, both of *what mathematics is about* and of *how mathematical concepts are brought to bear in physics*, such a question, even if it concerns a "physical manifold," seems to me inappropriate and even, in its tendency, obscurantist. A manifold as such *need* have *no* special, or distinguished, "metric" structure at all; on the other hand, it may have more than one such. If—as in several speculative theories you mention—physical space-time were to have more than one physically significant structure of a "metric" type, that would pose no special difficulty; one would only need to know how these several structures are related to physical processes and phenomena. To ask which of them has the *really geometrical* significance for space-time would be unreasonable, in a way whose connection with my remark about the multiplicity of mathematical structures may be made plainer by an analogy with a purely mathematical situation—and one from number theory (whose traditional domain was *"discrete* quantity"): In the theory of algebraic number-fields, the concept of the "whole numbers" or *integers* of a given algebraic field is introduced, *generalizing* the elementary concept of the (rational) integer. The generalized concept is "absolute" (not relative to a field), in the sense that if an algebraic number is an algebraic integer of any field, it is an integer in every field (algebraic over the rational numbers) to which it belongs. Now, $\sqrt{2}$ is an integer in this generalized sense of the term. Suppose someone objected to this rather scandalous statement by asking: "But is $\sqrt{2}$ an integer in the *ontologically arithmetical* sense of the word?"—or: "*By virtue of what* does a number qualify as *whole* in the *mathematically numerical* sense?"

The point is, again, that we must not be slaves to terminology. *Should* $\sqrt{2}$ be called an "algebraic integer"? Such questions have no answer: they are not well posed. The terminology of algebraic number theory is a very good one; it is clear, and it allows a rich abundance of relationships to be expressed perspicuously. But it is not *necessary*; one cannot even say that it is "optimal," or the "most natural" terminology: it is quite conceivable that some very different way of speaking might do just as good service in expressing the same relationships; and in the development of mathematics, good terminology *has* sometimes changed for the better. In any case, the question of terminology, although by no means negligible, is distinctly secondary: it has to be the *Begriffe* that control, not the *Wörter*. (Your "relationalist" and "semi-absolutist" quarrel—pp. 347ff.—over *which one* is "playing with words"; but I see them *both* as engaged in a battle over words without clear notions: *bilateral* logomachy!)[9]

The other passage I want to refer to is that on pp. 332ff., in which you use Jack, with his two feet and his unclehood, and Presidents Ford and Giscard, to dispute my skeptical comments on the notion "intrinsic to space." Here I feel distinctly to blame for certain deficiencies in my discussion (letter of February 16, 1975; referred to by you on pp. 327–8 and 337) of the difference between the cases of *space* (or space-time) and *an ordinary physical object*; and to some degree also in my conference remarks (passage quoted by you on p. 329)—which, as you know, were prepared somewhat hastily at the conference (and which utilized some material from an earlier letter to you). I think, now, that there are issues besides that of sheer individual "identification" that need to be emphasized. The phrase in my earlier statement to you (quoted by you, p. 329), referring to the difficulty "of giving an appropriate and adequate *characterization*" of the "object" in question, seems to me better chosen. The necessary point is also suggested in the earlier part of my paragraph about the globe: I said that I can, prima facie, entertain the question whether the globe has "intrinsically" the property of spherical shape (a question which seems to me, in its most natural "ordinary" interpretation, to have nothing to do with the fact that to test the globe's shape requires "external" instruments—if only, for a rough judgment, eyes and light). But I added that the qualification "prima facie" was there because "I don't think the question is a clearly posed one in this bald formulation (consider, e.g., Descartes's meditation upon what is 'intrinsic' to the piece of wax)." In other words, since I do not think you have succeeded in giving a satisfactory general explication of "intrinsic," I have to consider *what notion of "intrinsic" is appropriate to the context of the question*; and in the next portion of my paragraph, I suggested that in the case of the globe the relevant notion—the one that seems the most likely guess at the questioner's intent—is that of the distinction between the *internal physical constitution of* and *external constraints upon* a body: so that if a hollow glass globe is filled with water, one might reasonably say that spherical shape is an "intrinsic" property of that piece of glass (as constituted at that time), but an "extrinsic" property of the water.

Now, where did all this interpretation of "intrinsicness" in the given context come from? (Incidentally: it may be *mis*interpretation; the questioner may have something else entirely on his mind; and *you* may have quite a different view about what to consider "intrinsic" to the globe. You have not commented on this.) It came, in fact, from the circumstance that (as I have already said) I know what object you mean by the globe: your description serves to distinguish for me an object of whose "nature," or general characteristics, I have a conception quite independent of that description. But it is not "acquaintance" that is critically important here, and I was wrong to mention it: certainly not acquaintance with the *object* (in point of fact, I have never seen your globe); but also, not even acquaintance with *objects of a similar type* (although that is, in the present case, a principal source of the crucial knowledge); rather, what is really important is that I know enough about the thing you are talking about to have some kind of general view of the structure or attributes of such a thing. (Even this, I remind you, is not decisive: it merely makes prima facie understanding possible of what I have to consider a somewhat vague "presystematic" question—pending a satisfactory general explication of "intrinsic," which I do not really expect to see.) But clearly, in this respect "Uncle Jack" and the globe are entirely analogous—each belonging to a kind about which I possess an abundance of lore (or prejudice), namely *bodies* and *people* (antirespectively); and are quite unlike space-time, about which I happen to be deficient of such prejudice. Tell me exactly what *you mean* by 'space-time.' "I mean that structure," says Newton, "by virtue of which the notions of *length, duration, uniform rectilinear motion*, and *acceleration* are applicable to physical systems and processes." Ah, yes, I say, a metric is intrinsic to space-time for Newton.[10] "I mean," says X, "the mere four-dimensional differentiable manifold, independently of any further structure"; "I," says Y, "mean the smooth 4-manifold, with the distinction of directions at each point into *spacelike* and *causal* (timelike or null) directions." Ah, yes, I say, for X and Y the metric is extrinsic to space-time; whereas for Y, but not for X, the conformal structure is intrinsic. I not only have no criterion for adjudicating among these views; I have no *desire* to adjudicate, *see no point* in adjudicating, among them.

Again I feel it necessary to emphasize, so as to obviate any possible misunderstanding over it, the element of *personal judgment* in the position I have just expressed: I do not say that *there can be no criterion* for deciding among such views, or that there *is* no point in deciding; but that *I have* no criterion, and *see* no point in seeking one. It is certainly possible that I have overlooked something; but I cannot be persuaded that I have by a general argument, to the effect that such criteria are *conceivable* and *might* be of value, because that is what I have just admitted.[11]

But in reflecting further on this matter—on *why* you and I differ so much in our satisfaction/dissatisfaction with a position of openness on what is to be taken as "intrinsic"; why I am willing, and you are not, to consider this essentially a question of *façon de parler*—it has occurred to me that there is, after all, a point of "ontological perspective" involved, and one not without interest. You tend to think of the world in terms of "things"—"primary substances," in Aristotle's sense. I do not: I tend, rather, to think (Platonically?) of "structures" and "aspects of structure" ("Forms"?). Let us put aside modernity, and space-time with it; what do we mean—what did "we" *ever* mean—by "space"? Almost everyone who considers this thinks in the first instance of the question "Is there really *such a thing* as space?"—and the "absolutists" are ordinarily presumed to be the ones who answer affirmatively. But I find myself in profound agreement with Newton on this point (and I am exceedingly grateful to you for the stimulation of this discussion—because the significance of the text I am about to quote had never been clear to me until, just a few days ago, I was suddenly struck by its bearings upon our difference): Newton rejects the view that *extension* is either a "substance," or an "accident," or a "nonentity"; and asserts rather[12] that it

has its own way of existing, which fits neither substances nor accidents. It is not a substance both because it does not subsist absolutely of itself [*absolute per se*], but as it were as an emanative effect of God and a certain affection of every thing [or "every being": *omnis entis affectio*]; and because it does not stand under those characteristic affections that denominate a substance,[13] namely actions, such as are thoughts in a mind and motions in a body. . . . Moreover, since we can clearly conceive extension as existing without any subject, as when we imagine extra-mundane spaces or places void of any body [vacuum solutions and vacuum regions!] . . . , it follows that it does not exist in the way of an accident inhering in some subject. And hence it is not an accident. And far less is it to be called nothing, for in fact it is more a "something" than an accident, and comes nearer to the nature of a substance. Of nothing no Idea is given, nor has it properties; but we possess, of extension, an Idea the clearest of all, to wit by abstracting the affections and properties of a body, so that there remains only the uniform and unlimited stretching out of space in length, breadth, and depth. Moreover, its many properties are concomitant to this Idea. . . .

Having ruled out each item of what on the face of it seemed an exhaustive list of alternatives ("substance," "accident," and "nothing at all"), Newton propounds his own solution to the ontological problem of space in the following terms:

Space is an affection of a thing *qua* thing [or "of a being *qua* being"].[14] Nothing exists or can exist which is not in some way referred to space. God is everywhere, created minds are somewhere, and a body [is] in the space that it fills; and whatever is neither everywhere nor anywhere is not. And from this it follows that space is an emanative effect of the first existing thing;[15] because if anything is posited space is posited.

How is this quaint, even bizarre-sounding doctrine (and I confess that until recently it did seem so to me), to be understood: the doctrine that space is "neither substance nor accident" but *affectio entis quatenus ens*? I suggest that Newton is here expressing, in terms close to those that still prevailed in philosophical school-training, essentially the same view that I put to you when I spoke (your p. 329) of spatio-temporal attributes as "objective structural characteristics of the natural world": his doctrine is that *the fundamental constitution of the world*—its "basic lawful structure"—*involves the structure of space*, as something to which whatever may exist must have its appropriate relation. Notice that, contrary to a rather widely held interpretation of Newton (based upon his references to space as God's "sensorium"), he does not here represent space as an *organ* of God (which of course would be a kind of "thing"), but as what he calls an "emanative effect" of whatever thing first exists—because it is, as it were, *an aspect of the nature of "thinghood."* Transposing to a modern context, and assuming Einstein's GTR as "fundamental theory," we may paraphrase in this way: Whatever exists of a physical nature (let us leave out God and minds)—particles, fields—must be appropriately related to a space-time manifold with a fundamental tensor-field satisfying the Einstein equations. If, therefore, the real world is not the empty set(!), it must be characterized by the structure of such a manifold with such a field (Riemann-Minkowski metric): the latter is, in this sense, an "emanative effect" of the existence of anything. (I believe we may reasonably understand "emanative effect" to signify something that is not "created," or "produced by causal agency" (in accordance with "causal laws"), but that *is entailed by* (or "flows from") *the nature of something*.)

Perhaps I have been carried away into too long a digression upon Newtonian metaphysics. The essential point of view that I have been trying to explain is one that I believe Riemann also expresses—for instance, when he speaks of "the real"—or "actual"—"that underlies space" ("das dem Raume zu Grunde liegende Wirkliche") [16]: space, or geometric structure, is *an aspect of the structure of the world*. If, in this way, one does not think of space prima facie as a separate "thing" at all, you see that the question what is "intrinsic" to it can hardly mean anything else than, "Of the components or aspects of physical structure, which are the ones that we *call* 'spatial' or 'geometric'?" The point of view seems to me a fruitful and enlightening one; it puts the emphasis, I believe, on issues of genuine interest, and avoids (to use Riemann's term) "idle questions." Among the latter, in my opinion, one of the worst sorts—one that is not merely idle, but "reactionary," i.e., obstructive of scientific progress (and subversive of clarity)—is the demand, when some *fundamental conceptual change* occurs in scientific theory (some change in the basic characterization of physical structure), to be told—not just, as is reasonable and necessary, how the new conceptions "correspond" to the old, i.e., what the general connections are between them of agreement or disagreement in application—but what, in the structure posited by the new theory, *is the same* (or merits designation by the same term) as something posited by the old theory. (A recent example of this kind of unenlightening onomatologizing is the discussion of the question: When Newton spoke of "mass," was he referring to what we know as "rest mass" or to what we know as "relativistic mass"?) I believe that the effect of this kind of demand is the precise contrary of the admirable end proposed by Riemann, in a passage that I have quoted in my paper: Riemann, you will recall, said that investigations such as his own can serve to ensure that the work of

397

science shall not be hindered by a narrowness of conceptions, and that progress in the knowledge of the connections of things shall not be hampered by traditional prejudices. I know very well how far you are from wishing to defend narrow conceptions or traditional prejudices, or to trade on obscurity; but my most serious doubts about your point of view towards ontological questions of space-time lie in my suspicion that, quite counter to your intentions, that is their ("intrinsic") tendency: that to ask "Is—or *by virtue of what* is—the tensor-field g properly *geometric*?" is to ask a question of exactly the kind I have been deprecating.

Be it sae or be it nae, I hope at least that what I have now written will help to make it clearer to you (a) how I prefer to look at these questions, (b) how that way of looking influences my answers to them, and (c) what sorts of consideration recommend that way of looking to me. If I have been in the least successful in this, it should be apparent that the position you have assigned, in your paper (pp. 356ff.), to the "semi-absolutist," [which was presented, in an earlier version,] as an interpretation of my opinions, is very far from mine. I do not at all understand by what logic of exegesis you infer (p. 357), from the fact that I speak of the differentiable manifold of world-points and say that it "carries a physical field with the characteristics of the Einstein metric field" (a statement that seemed to me, and still seems, as simple and philosophically colorless a summary of the direct content of GTR as I can conceive), that I affirm what you call "TAM"—a verbal formula that seems to me grotesque, monstrous, and utterly devoid of sense. To try to avoid any possible misunderstanding on this issue, let me spell out more explicitly what my statement involves, and why I think it cannot have anything to do with TAM—*whatever* the latter may mean: (1) GTR postulates a four-dimensional differentiable manifold, ordinarily called "space-time"; I presume there is no controversy about this. Such a manifold, by its purely mathematical character, has associated with it various derived structures, in particular what is called its "tensor bundle"—and within the latter, the sub-bundle of "nondegenerate second-order symmetric covariant tensors of Lorentz signature." A tensor-*field* is a function or mapping of a particular sort from the manifold to its tensor bundle (a mapping of the sort called a "cross-section of the bundle"). There are also analogous notions connected with what are called "tensor-*densities*." (2) Physical theory—here speaking very generally—postulates the existence of various more special "physical" structures: i.e., among all the possible relations, functions, etc., mathematically conceivable (mathematically *existent*) within the context already laid down, certain particular ones (of particular types specified by physical laws), to be taken as "physically distinguished." Among these are (or may be), for example, "world-lines of matter-flow," "scalar-density field of mass," "tensor-density field of impulse-energy," "scalar-density field of charge," "electromagnetic tensor-field," etc. (3) More especially, Einstein in his GTR postulates a physical field which is a cross-section of the bundle of nondegenerate, second-order, symmetric covariant tensors of Lorentz signature; and he postulates certain relationships among this field and other physical structures. The field postulated by Einstein is often called the "metric field," or the "metric tensor," or indeed just the "metric." (4) We may turn now to my statement. Its reference to the differentiable manifold of world-points ought to be unproblematic, by (1). That this manifold *"carries"* a certain field means no more than that this field exists "on" the manifold—i.e., is a mapping from the manifold into the proper bundle (and, one should add, with the proper attributes: we want here a differentiable cross-section); surely, so far, no extra metaphysics. That the field is "physical": I have indicated what I mean by this in (2). Now, the critical point must somehow be the word 'metric'— since from my statement you infer "TAM," the "Thesis of Autonomous Metricality." But (a) my only use of that word to describe the field is an *indirect* one; my statement seems to me a very *model* of caution (not to say *excessively* cautious): it does not say that the field "is the—or a—metric," but only that it *"has the characteristics* of the Einstein metric field," i.e., the field g that occurs in Einstein's GTR and is there (often) called "the metric field." What those "characteristics" are I have indicated in (3) above: they include the structural attributes of the field considered by itself (namely, that it is a cross section of the bundle mentioned), and the laws relating this field to other physical structures (whether or not, I should add, these other structures might prove to be derivable from the Einstein field itself). I hope this is sufficient to establish my innocence of committing TAM. But in case I may seem to be constructing a lawyerly argument on the basis of my cautious language (although I assure you that the language was *deliberately* cautious, so that I was shocked to find it used against me!), let me now also point out (b) that even if I *had* meant to imply that the Einstein field g "is physically—or ontologically—metric," this *still* could not fairly be read as an endorsement of the view that its "metricality" is "*autonomous*," in whatever sense of that (to me obscure) assertion is obnoxious to you. For you, I believe, *do consider* the Einstein field to be "the metric of space-time"—although not "autonomously" so, but *by virtue of* something, namely some set of relationships to physical structures ("external metrical standards"). *But these relationships are, then, themselves among the "characteristics" of "the Einstein metric field"*; and, *ex hypothesi*, they *entitle* it to the designation "metric." Turn it which way you will, my statement *cannot* be made to yield the construction you have put upon it (and your p. 327 tempts me to add: "Q.E.D., Mr. Walter Shandy!").

Notes

1. It might, in fact, be urged that there is a motive for this usage independent of such considerations: namely, despite the correctness of your remark (p. 342) that there are other second-order, symmetric, physical tensor-fields on space-time (which in the purely mathematical sense qualify as "generalized metric" structures), the tensor-field g is nevertheless the only such physical field known that is *in principle* an *everywhere nondegenerate* metric; and, what is perhaps most telling, the only one that allows us to separate "spacelike" and "timelike" structures within space-time.

2. Since you ascribe (p. 342) to the tensor-field g both "the ontological status of being *the metric field* in the sense of endowing space-time with the latter's particular metric structure" (whereas I do not understand why this *is* an "ontological status"), and "the physical significance of being the metric tensor of space-time" (whereas I do not understand what this "physical significance" consists in); and similarly (p. 345) your "relationalist" speaks both of "the particular added *physical* significance," for the field g, of "being the *metric* tensor field of . . . space-time," and of "the *ontologically constitutive* mediation of the FCS."

3. It occurs to me that this can be challenged, so far as light rays are concerned—cf. the section on Riemann in my own paper for the Minneapolis and North Andover conferences. But (a) insofar as space was ordinarily conceived to be fully "measured out" by measuring rods, the law of the rectilinear propagation of light was usually regarded as an empirically established natural law, and not as in any way "constitutive" of spatial notions; and (b) for those who, like Riemann and Poincaré, did stress a fundamental role of light in geometrical determinations, that role had to do with the distinguishing of *spatial straight lines*—which is not at all the role played by light in relativistic measurement.

4. As objects; I am not enamored of their description, especially of the word "external."

[5. To prevent misunderstanding, it should be noted that this remark, in the sense in which it is intended, stands despite your acknowledgment to Winnie and Stachel that under suitable circumstances gravitational radiation may be admitted into the "FCS": for among those required circumstances is *concordance*, whereas I am here envisaging a possible *failure of concordance*.]

[6. That there can be no more than one particle world-line through a point is a clear consequence of the assumption of the existence, continuity, and differentiability of the impulse-energy tensor. In correspondence, you have questioned the point, saying that it "would make nonsense of particle collisions and various other kinds of coincidences." But what in practice we call "coincidence"—as of a pointer with a mark on a scale—is merely near juxtaposition. As to collisions, it is enough to remark (a) that the space-time point of a collision in the strict sense, involving an *instantaneous* interaction, must necessarily be a *singularity* of the world-lines involved (and therefore, presumably, on an ultimate analysis, of space-time itself)—so that "constitution" of the metric at such a point is out of the question; whereas (b) if "collision" is taken to involve, not an instantaneous, but only a *rapid* interaction, the bodies participating in it are, precisely during that process, not freely gravitating bodies—hence not qualified to "constitute" the metric.

[It might be asked how, if what I have here said is true, we ever actually determine ("measure") the metric. But clearly, in practice, our body of data is always *finite*; our conclusions about continuous structures are drawn partly by interpolation and extrapolation, and largely by inductive inferences guided by far more powerful theoretical principles—which, of course, are themselves based upon earlier instances of the same sort of process, so that an accurate genetic analysis of the whole construction would be extraordinarily intricate. (I must remark, autobiographically, that one of the crucial incidents in my own divorce from the verifiability theory of meaning in any strict sense of the latter occurred when, as a student, I read Reichenbach's *Axiomatik der Relativistischen Raum-Zeit-Lehre*, and saw that he postulates at each point of space-time an observer riding a particle in each possible state of motion, with all the observers sending and receiving light signals in all possible ways—somehow always recognizing the source of each signal! An "abundance" that is impossible to achieve—and is even, in a sense, inconceivable—cannot be a necessary condition for *knowledge that we actually have*; analogously, an abundance that is impossible cannot be a necessary *ontological* condition of a structure that we can attribute, without incoherence, to the real world.)]

[7. The passage is quoted from a letter, Grünbaum to Stein, dated 5 March 1976, which proposed the insertion of a new sentence at the end of part C of Grünbaum's section *The Relationist's Account of . . . Space-Times*, and invited comment on this sentence. In Grünbaum's final version, the inserted material has been revised and considerably expanded: it now forms the concluding paragraph of that part C (pp. 355-6). Instead of adapting my comments to this new exposition, I have decided to leave them as they were first formulated: Grünbaum's paragraph can thus be considered as—at least in part—a reply to the criticism made here.]

8. I here use 'object' in the sense of the older—medieval and Cartesian, not Kantian—distinction of "objective" and "subjective"; or of Locke's explanation of "idea" as "whatsoever is the *object* of the understanding when a man thinks."

9. There is what seems to me a similar mere verbalism in your discussion of the issue of "infinite numbers" (pp. 305ff.): You repeatedly speak, here, of "infinite *cardinalities*." Now, the word 'cardinality' no more possesses an "intrinsic" virtue or import than (say) the word 'integer' (cf. just above). The first useful definition, and of course the only one now used, of a notion of "cardinality" applicable to infinite sets, is Cantor's. On p. 367 (n. 7), you indicate that the doctrine you are discussing was not associated (according to Maier) with "the modern Cantorean distinction between denumerably and super-denumerably infinite sets." Now, if infinite sets have different "cardinalities" (in some sense); and if, among these, some are *larger* than others (as seems clearly implied); and if the set of the natural numbers is not maximal (as also seems clearly implied); then some sets will have "cardinality greater than that of the set of the natural numbers." The only sense I can make of a claim that "Cantor's super-denumerably infinite sets" are not involved here is that Cantor's *criterion of comparison* was not adopted. So far, well and good; except that it is then unclear (a) what criterion *was* assumed—or whether *any* clear notion really existed (Answer: almost certainly not!); (b) why the term 'cardinality' has any appropriateness here (was it actually used in the scholastic tradition?); and (c) why, if an honest *concept* of cardinality (or number) was in fact lacking, you should find any comfort here for your theory of the "intrinsic," or should conjecture an affiliation with Newton: is it a purely *verbal* tradition that you think Newton may have been following? But beyond this, real trouble—indeed, utter confusion—sets in for me when you say (p. 307) that the theory that different-sized point-sets have different infinite numbers of points was *refuted* by *the exhibition of one-to-one correspondences* between such sets; for this *is*, after all, Cantor's criterion of comparison! In this discussion, concepts seem to be in a

401

state of continual fluctuation, and only words to be constrained into a degree of stability. I can quite well believe that such confusions were present in the historical situation; my complaint against you is that you seem to see no need to explicate the issues, or to resolve the confusions.

10. I here speak somewhat loosely, without care for the distinction between the Newtonian and Minkowski space-time structures.

11. Similarly, when I said, in the passage you quote on p. 329, that "I see no way to confront the former question"—namely, whether a certain relation is involved in the structure of "the space-time manifold itself, considered apart from all other entities"—"independently of the latter" (how to explicate the notion of "the space-time manifold itself"—to draw a line, so to speak, between it and "all other entities"); and that yet the converse may also seem to hold, etc.: I was not claiming to offer a *proof* of vicious circularity in the enterprise under discussion. So your reply, in the Uncle-Jack-and-President-Giscard passage, aimed at refuting in general terms a charge of "necessary circularity," is in my opinion not to the purpose: *I still see* no way—you have certainly not shown me one—to confront the first question independently of the second, or to answer the second without begging the first.

12. *De gravitatione et aequipondio fluidorum*, in Hall and Hall, ed., *Unpublished Scientific Papers of Isaac Newton* (Cambridge University Press, 1962), pp. 131–132, 136; Latin on pp. 99–100, 103. Hall and Hall most irritatingly render *affectio* throughout as "disposition"—one among many seriously misleading mistranslations in their English version of Newton's Latin text.

13. Again the Halls give a really terrible mistranslation: they have "*it is not among the proper dispositions that denote substance.*" But it should be plain to anyone with a rudiment of philosophic discrimination that when Newton writes "*Non est* substantia *tum quia non absolute per se* . . . subsistit; *tum quia non* substat *ejusmodi propriis affectionibus quae substantiam denominant,*" the two verbs in the two dependent clauses—*subsistit* and *substat*—are deliberately chosen for their association with *substantia*: substance is what is self-subsistent, and is also the substrate or supporter of properties. The meaning of *substat* has to be, not "stands *among*," but "stands *under*": extension is not a substance because it does not support—underlie—stand under—the "characteristic denominations" (what Frege would call the "*Merkmale*") of a substance.

14. The article may, of course, be disputed, since Latin possesses no articles. Newton's text reads: *Spatium est entis quatenus ens affectio*; and the Halls render: "Space is a disposition of being *qua* being." The expression "being *qua* being" has indisputable standing in the metaphysical tradition—but not, I think, as denoting an individual subject of attributes, something that has "affections" (or "dispositions"). The translation I have given seems to me consonant with the sentences that follow.

15. Not, as the Halls translate, "of the first existence of being"!—The "first existing thing," of course, according to Newton, is God (of whom he has previously characterized space as "an emanative effect (as it were)"; but it is noteworthy that the *reason* he gives for his statement that space is "an emanative effect of the first existing thing" is *quite independent of what thing that may be.*

16. In what you call Riemann's "DH," it is this *Wirkliche*, rather than the spatial manifold itself, whose "binding forces" are said to give rise to the "Massverhältnisse" in the manifold: that is, Riemann does not speak of binding forces as *acting upon the spatial manifold* (a notion it is hard to make any sense of), but as *acting upon "the real."*

————MICHAEL FRIEDMAN————

Simultaneity in Newtonian Mechanics and Special Relativity

1. Introduction

Everyone will agree, I think, that the transition from Newtonian mechanics to special relativity taught us something of fundamental importance about time and simultaneity. Many philosophers have urged that there is a significant *semantic* lesson to be learned from this transition. For example, the following kinds of views have been expounded: Einstein was aided in his discovery of special relativity by an analysis of the "concepts" of time and simultaneity; the transition from Newtonian mechanics to special relativity resulted in a profound "change of meaning" of 'time' and 'simultaneous'—a change that was so extensive as to make any comparison of the two theories problematic; in a special-relativistic world the notion of simultaneity is in an important sense conventional—statements about distant simultaneity lack truth-value, they are mere "definitions."

Defenders of such views rarely provide explicit semantic theories within which their claims can be evaluated. There are, however, philosophical theories of meaning lurking in the background. Thus claims that Einstein analyzed the "concept" of simultaneity fit naturally into an operationalist account of meaning, since what Einstein did was to discuss ways of measuring time and distant simultaneity. Similarly, claims about the conventionality of distant simultaneity are supported (at least in Reichenbach's case) by a verificationist theory of meaning. Statements about distant simultaneity in a special-relativistic world lack truth-value, it is argued, because they are unverifiable in principle. On the other hand, views that make much of the noncomparability or "incommensurability" of the meanings of 'time' and 'simultaneous' in Newtonian mechanics and special relativity seem to involve some kind of "contextual" theory of meaning—meaning is to be identified with "role in theory" or the like.

In this paper I would like to see what light can be shed on these semantic issues using an approach to the theory of meaning that has been the subject of much contemporary discussion. The approach I have in mind takes reference rather than meaning as the central notion of semantics. According to this approach, the semantical properties of a sentence—its truth-value or lack of it, its inferential connections with other sentences, etc.—are determined by (a) the referential properties of its component words—the objects denoted by its singular terms and the sets (properties) determined by its predicates; and (b) the "logical form" of the sentence—how it is built up from its component words by means of grammatical constructions like truth-functions, quantifiers, etc. Let us call this kind of approach *referential semantics*. It is plausible to suppose that referential semantics can be an illuminating framework for discussing traditional semantic issues about the transition from Newtonian mechanics to special relativity, because many of these issues involve claims about truth. Thus the conventionalists hold that statements about distant simultaneity lack truth-value in a special relativistic universe. The "meaning change" and "incommensurability" theorists hold that we cannot apportion truth and falsity to the statements involving 'time' and 'simultaneous' made by a Newtonian physicist according to the truth and falsity of corresponding statements of special relativity, because such statements have different meanings in their different theoretical contexts. Similarly, according to the "meaning change" theorists, we cannot say that such statements made by a Newtonian physicist are approximately true, that some statements of Newtonian mechanics are logical consequences of special relativity, etc. From the point of view of referential semantics, all claims of this kind must depend on peculiarities in the referential properties of 'time' and 'simultaneous' (assuming that there is nothing problematic about the grammatical structure of the sentences in question).

A second feature of my approach is that I shall treat both Newtonian mechanics and special relativity as *space-time* theories. I view both theories as theories about a four-dimensional manifold, space-time, and the geometrical structures that characterize it. Where the two theories differ is with respect to the geometrical structures that space-time actually possesses. In particular, differences between the two theories as to time and simultaneity are to be understood as differences in the geometrical properties predicated of space-time. I adopt this view of the two theories because it seems to me to make their similarities and differences—their comparison—especially clear. However, I shall not argue directly for this view here (see, e.g., Earman, 1970, and Earman and Friedman, 1973). Nor shall I argue directly for referential semantics (see, e.g., Field, 1972 and 1973). Instead, I hope to show that the conjunction of these views provides a fruitful framework for the discussion of traditional philosophical issues relating to Newtonian mechanics and special relativity. Of course, if I am successful, this paper will constitute an indirect argument for the referential approach to semantics and the space-time approach to our two physical theories.

My argument will proceed as follows. In section 2 I shall briefly sketch four-dimensional formulations of Newtonian mechanics and special relativity, as such formulations will probably be unfamiliar to most readers. In section 3 I shall discuss the question of whether 'time' and 'simultaneous' underwent a "meaning change" in the transition from the former theory to the latter. I shall argue that with respect to the kind of meaning that is most relevant to questions about the truth of statements in the two theories—i.e., with respect to reference—it is plausible to suppose that there has been no change. In section 4 I shall discuss the issue of the conventionality of simultaneity in special relativity. I shall argue that conventionalists have not given us a good reason to regard statements of distant simultaneity as truth-valueless in the context of special relativity.

2. Four-Dimensional Formulations of Newtonian Mechanics and Special Relativity

According to the space-time point of view, the basic object of both our theories is a four-dimensional manifold. I shall use R^4, the set of quadruples of real numbers, to represent the space-time manifold. Both theories agree that there is a natural system of straight lines defined on this manifold. If (a_0, a_1, a_2, a_3), (b_0, b_1, b_2, b_3) are two fixed points in R^4, then a *straight line* is a subset of R^4 consisting of elements (x_0, x_1, x_2, x_3) of the form

(1)
$$x_0 = a_0 r + b_0$$
$$x_1 = a_1 r + b_1$$
$$x_2 = a_2 r + b_2$$
$$x_3 = a_3 r + b_3$$

where *r* ranges through the real numbers. A *curve* on R^4 is a (suitably continuous and differentiable) map $\sigma: R \rightarrow R^4$. Such a curve $\sigma(u)$ is a *geodesic* if and only if it satisfies

(2) $x_0 = a_0 u + b_0$
         $x_1 = a_1 u + b_1$
         $x_2 = a_2 u + b_2$
         $x_3 = a_3 u + b_3$

where $(x_0, x_1, x_2, x_3) = \sigma(u)$ and the a_i and b_i are constants. So if a curve is a *geodesic* its range is a straight line. Note that the geodesics are just the curves that satisfy

(3) $d^2 x_i/du^2 = 0$ $i = 0, 1, 2, 3$.

The importance of straight lines and geodesics is due to the fact that both theories agree that the trajectories of free particles are straight lines in space-time. So we can represent such trajectories as geodesics in R^4.

A *coordinate system* is a one-one (suitably continuous and differentiable) map $\phi: R^4 \rightarrow R^4$. A coordinate system is *affine* if and only if it is a linear transformation of R^4, i.e., it satisfies

(4) $y_i = \sum_{j=0}^{3} a_{ij} x_j + b_i$ $i = 0, 1, 2, 3$

where the a_{ij} and b_i are constants and $(y_0, y_1, y_2, y_3) = \phi(x_0, x_1, x_2, x_3)$. Affine coordinate systems are precisely those that preserve the condition

(5) $d^2 y_i/du^2 = 0$ $i = 0, 1, 2, 3$

for geodesics. As we shall see, such coordinate systems are a natural representation of the physicist's frames of reference.

So far, Newtonian mechanics and special relativity agree on the structure of space-time. But the two theories differ over what further structures exist on the space-time manifold, and, in particular, over the individual natures of space and time. In what follows I shall deal only with the kinematical aspects of our two theories, since these aspects are most relevant to the role of time and simultaneity. However, it should be noted that dynamics—i.e., gravitational interaction in the case of Newtonian mechanics, and electromagnetic interaction in the case of special relativity—can be easily dealt with within this framework as well (see Anderson, 1967, Earman and Friedman, 1973, Havas, 1964, and Trautman, 1966).

(a) Newtonian Mechanics

The central object that Newtonian kinematics postulates on the space-time manifold is an *absolute time*: a real-valued function $t: R^4 \to R$ defined by $t(x_0, x_1, x_2, x_3) = x_0$. Think of t as assigning a time to each point (event) in space-time. The hypersurfaces $t = $ constant are called *planes of absolute simultaneity*. Two points in R^4 are *simultaneous* if and only if they lie on the same $t = $ constant hypersurface. Furthermore, on each plane of absolute simultaneity Newtonian kinematics postulates a Euclidean metric, h, defined by

(6) $\quad h((t, x_1, x_2, x_3), (t, x_1', x_2', x_3'))^2 = (x_1 - x_1')^2 + (x_2 - x_2')^2 + (x_3 - x_3')^2$.

Now any geodesic curve $\sigma(u)$ satisfies $x_0 = a_0 u + b_0$, so if we use $x_0 = t$ as a parameter for σ it remains a geodesic: i.e., $\sigma(t)$ satisfies

(7) $\quad d^2 x_i / dt^2 = 0 \qquad i = 1, 2, 3$.

This is just Newton's law of inertia.

An *inertial* coordinate system is an affine coordinate system which is generated by a *Galilean* transformation; i.e., y_0, y_1, y_2, y_3 is inertial if and only if

(8) $\quad y_0 = x_0 = t$

$$y_i = \sum_{j=0}^{3} a_{ij} x_j + b_i \qquad i = 1, 2, 3$$

where the a_{ij}, $i, j = 1, 2, 3$ form an orthogonal matrix: $\sum a_{ij} a_{kj} = \delta_{ik} = 1$ if $i = k$, 0 if $i \neq k$. Inertial coordinate systems are just those that preserve the above form of the law of inertia and the above form of the spatial metric h. I shall say that an inertial coordinate system y_0, y_1, y_2, y_3 is *adapted* to a trajectory $\sigma(t)$ if and only if $\sigma(t)$ satisfies the equations $y_0 = t$, $y_i = 0, i = 1, 2, 3$. Thus one can think of σ as representing a particle at rest at the origin of y_0, y_1, y_2, y_3. There exists an inertial coordinate system adapted to σ if and only if σ is a geodesic. So if σ is a geodesic and ϕ is an inertial coordinate system adapted to σ, I shall call the pair $\langle \sigma, \phi \rangle$ an *inertial frame*. In inertial frames free particles satisfy Newton's first law.

(b) Special Relativity

In Newtonian kinematics time is represented by the function t, while space is represented by a $t = $ constant hypersurface, endowed with a three-dimensional Euclidean metric h. In special relativity we capture the roles of both time and space by a single object: a four-dimensional pseudo metric g defined by

$$(9) \qquad g((x_0, x_1, x_2, x_3), (x_0', x_1', x_2', x_3'))^2 =$$
$$(x_0 - x_0')^2 - (x_1 - x_1')^2 - (x_2 - x_2')^2 - (x_3 - x_3')^2.$$

g is called the *Minkowski* metric. Two points p, $q \in R^4$ have *timelike separation* if $g(p, q)^2 > 0$, *spacelike separation* if $g(p, q)^2 < 0$, *null separation* if $g(p, q)^2 = 0$. A curve is *timelike* if every point on it has timelike separation from every other point, and similarly for spacelike and null. Equivalently, a curve $\sigma(u)$ is timelike if and only if

$$(10) \qquad \sum_{ij} \eta_{ij} \frac{dx_i}{du} \frac{dx_j}{du} > 0$$

everywhere, and similarly for spacelike and null—where $\eta_{ij} = 1$ if $i = j = 0$, -1 if $i = j = 1, 2, 3$, and 0 if $i \neq j$. We require that the trajectories of free particles be timelike geodesics.

For any timelike curve $\sigma(u)$, we can define its *length* τ by the formula

$$(11) \qquad \tau(u) = \int^u \sqrt{\sum_{ij} \eta_{ij} \frac{dx_i}{du} \frac{dx_j}{du}} \, du.$$

τ is called the *proper time* of σ. On timelike curves we can use τ as a parameter, and if $\sigma(\tau)$ is a timelike geodesic it satisfies the law of motion

$$(12) \qquad d^2x_i/d\tau^2 = 0.$$

An *inertial* coordinate system is an affine coordinate system which is generated by a *Lorentz* transformation; i.e., y_0, y_1, y_2, y_3 is inertial if and only if

$$(13) \qquad y_i = \sum_j a_{ij}x_j + b_i$$

where

$$(14) \qquad \sum_{ik} a_{ij}a_{kl}\eta_{ik} = \eta_{jl}.$$

Inertial coordinate systems are just those that preserve the above form of the law of motion and the above form of the space-time pseudo metric g. Since in an inertial coordinate system a timelike geodesic $\sigma(\tau)$ satisfies $y_0 = a_0\tau + b_0$, we can use y_0 as a parameter on curves as well without disturbing the condition for timelike geodesics. The y_0 coordinate of an inertial system is called the *coordinate time* of the system. From now on I shall denote such a coordinate time by 't.' Thus in an inertial system the law of motion can be written in the form

(15) $d^2y_i/dt^2 = 0$ $i = 1, 2, 3$.

I shall say that an inertial coordinate system is *adapted* to a trajectory $\sigma(\tau)$ if and only if $\sigma(\tau)$ satisfies the equations $y_0 = t = \tau$, $y_i = 0$, $i = 1, 2$, 3—where τ is the proper time of σ. There exists an inertial coordinate system adapted to σ if and only if σ is a timelike geodesic. If σ is a timelike geodesic and ϕ is an inertial coordinate system adapted to σ, I shall call the pair $\langle \sigma, \phi \rangle$ an *inertial frame*. Relative to a given inertial frame we have hypersurfaces $t = $ constant, where t is the coordinate time of the frame. These hypersurfaces are spacelike (every point in one has spacelike separation from every other point) and are endowed with a Euclidean metric by g (if p, q have spacelike separation, define $h(p, q)^2 = -g(p, q)^2$). Two points p, $q \in R^4$ are *simultaneous with respect to the given inertial frame* if and only if they lie on the same $t = $ constant hypersurface.

Let us call a triple $\langle R^4, t, h \rangle$, where t is an absolute time and h is a Euclidean metric on the hypersurfaces $t = $ constant, *Newtonian space-time*; a pair $\langle R^4, g \rangle$, where g is the Minkowski metric, *Minkowski space-time*.[1] The basic claim of Newtonian kinematics is that our universe is a Newtonian space-time; the basic claim of special relativity is that our universe is a Minkowski space-time. Differences between the two theories over the roles of time and simultaneity turn on structural differences between Newtonian and Minkowski space-times. Thus in Newtonian space-time there is a unique global time determined by t, and a unique relation of simultaneity $\mathcal{S}$ such that $p\mathcal{S}q$ if and only if p and q lie on the same $t = $ constant hypersurface. Both time and simultaneity are independent of coordinate system or reference frame.

In Minkowski space-time, on the other hand, there is no such unique global time. Time is in the first instance a local property; the proper time of a particular timelike curve. Being local, proper time cannot be used to define a relation of simultaneity at all; it can be used only to compare the times of points lying on the same trajectory. However, relative to a particular inertial frame F there is a global time t_F—the coordinate time of the inertial coordinate system determined by F. Thus in Minkowski space-time there is a multitude of simultaneity relations. For each inertial frame F there is a simultaneity relation $\mathcal{S}^F$ such that $p\mathcal{S}^Fq$ if and only if p and q lie on the same t_F = constant hypersurface. So in special relativity (global) time and simultaneity are coordinate or frame dependent. It makes no sense to say that two events are simultaneous *simpliciter*, but only *relative to* this or that inertial frame or coordinate system.

3. The "Meaning" of 'Simultaneous' in Newtonian Mechanics and Special Relativity

If special relativity is true, Newtonian mechanics as a whole is false. Our world is a Minkowski space-time, not a Newtonian space-time; and neither a frame-independent global time nor a frame-independent simultaneity relation exists. Nevertheless, although the whole system of beliefs about time and simultaneity held by Newtonian physicists was false, we might plausibly (and perhaps naïvely) suppose that some of these beliefs were true. For example, we might suppose that when a Newtonian physicist uttered a sentence such as

(16)   Events e_1 and e_2 are simultaneous in frame F,

he said something true. On the other hand, when he uttered a sentence like

(17)   If e_1 and e_2 are simultaneous in frame F, then e_1 and e_2 are simultaneous in frame F',

he said something false. Our reasoning here is that (16) is true and (17) false because a relativistic physicist would accept (16) and reject (17), and we believe that special relativity is true.[2] Furthermore, although (17) is strictly false, we might plausibly (and perhaps naïvely) suppose that it is approximately true—as long as e_1 and e_2 are not widely separated in space and the relative velocities of F and F' are small. This is because of the following derivation in special relativity: Let events e_1 and e_2 have coordinates (x_0, x_1, x_2, x_3) and (x_0, x'_1, x_2, x_3) respectively in frame F. If y_0 is the coordinate time of event e_1, and y'_0 is the coordinate time of e_2 in frame F', it follows that their difference is given by

(18) $\left| y'_0 - y_0 \right| = v \left| x'_1 - x_1 \right| / \sqrt{1 - v^2}$

where v is the velocity of frame F' relative to frame F (I assume that F' is moving along the x_1-axis of F and that $c = 1$). This difference is small if v is small and $\left| x'_1 - x_1 \right|$ is small. Thus, if e_1 and e_2 are simultaneous in F, they will be approximately simultaneous in F' whenever they are spatially close and the velocity of F' relative to F is small.

410

However, a "meaning change" theorist would not be happy with this way of looking at the matter (see, e.g., Feyerabend, 1962 and Kuhn, 1962). He would deny that the fact that a relativistic physicist would accept (16) and reject (17) gives us a reason to think that a Newtonian physicist said something true when he uttered (16) and said something false when he uttered (17). For, according to the advocate of "meaning change," (16) and (17) do not express the same things when uttered by a Newtonian physicist and by a relativistic physicist; (16) and (17) have different meanings in their different theoretical contexts. Similarly, a "meaning change" theorist would deny that the fact that (18) is derivable in special relativity gives us a reason to think that (17) is approximately true in the context of Newtonian mechanics. The sentence that is derivable in special relativity is not an approximation to (17) as a principle of Newtonian mechanics, for the two have radically different meanings.

Now the first thing to notice is that the relevant issue here is not whether 'simultaneous' has different *meanings* in the two different theoretical contexts, but whether it has different *referents*. For, if there is anything right about the referential approach to semantics, truth-value is a function of the referents of the component words of the sentences in question. Thus, as long as 'simultaneous' has the same referent in our two theoretical contexts, (16) and (17) will have the same truth-values in the two contexts, whether or not they have the same meanings. As long as the reference of 'simultaneous' is preserved, our argument that (16) is true and (17) false in the context of Newtonian physics because (16) is true and (17) false in special relativity is correct. Similarly, if reference is preserved, we can regard (18) as an approximation to (17), and we can therefore regard (17) as approximately true. Thus, if the problem of "incommensurability" relates to the comparison of the truth-values of sentences in our two theories—e.g., if the problem is whether sentences in the two theories can *contradict* each other, whether sentences in one theory can be *derived* from sentences in the other, whether sentences in one theory can be *approximations* to sentences in the other, etc.—then the crucial issue is over the referents of words like 'time' and 'simultaneous,' not their meanings. The "meaning change" theorist must argue that 'time' and 'simultaneous' have different referents in their different theoretical contexts, not merely that they have different meanings.

How does the "meaning change" theorist argue for his view? Characteristically, he appeals to the radical differences in the theoretical princi-

411

ples involving time and simultaneity in the two theories. For example, in Newtonian mechanics time is an "absolute," nonrelational, frame-independent quantity; while in special relativity it is a relational, frame-dependent quantity. How could terms embedded in such radically different theoretical principles have the same meaning? This line of thought can support claims about the referents of 'time' and 'simultaneous' if we adopt the view that the reference of a theoretical term is determined by the theoretical principles containing the term. That is, we can use theoretical differences as an argument that 'time' and 'simultaneous' have different referents in Newtonian mechanics and special relativity if we adopt the view that a theoretical term refers to whatever satisfies the theoretical principles containing the term, or whatever satisfies a sufficient number of such principles, or the like. Thus if 'time' refers to anything at all in the context of Newtonian mechanics, it refers to an "absolute," frame-independent quantity; if 'simultaneous' refers to anything at all, it refers to an "absolute," frame-independent relation. On the other hand, in special relativity 'time' refers to a relational, frame-dependent quantity; 'simultaneous' refers to a frame-dependent relation. Therefore these terms cannot possibly have the same referents in the two different theories.

This view of how the reference of theoretical terms is determined—that a theoretical term refers to whatever satisfies (a sufficient number of) the theoretical principles containing the term—is closely analogous to the Russell-Searle account of how the reference of proper names is determined. According to the Russell-Searle account, the referent of a proper name is whoever satisfies (a sufficient number of) the descriptions we "associate" with the name. This account of the reference of proper names has been the subject of much recent critical discussion (see Kripke, 1972). It seems to me that the parallel account of theoretical terms has very similar flaws. In particular, if the account of theoretical terms in question is correct, it is hard to see how a theory can ever turn out to be false (or at least hard to see how the "central" principles of a theory can turn out to be false). If this account is correct, there either is an entity satisfying (a sufficient number of) the theoretical principles involving a given term or there is not. If there is, the principles are true; if there is not, the given term lacks a referent and the principles are truth-valueless. So, for example, if this account is correct, we cannot say that Newtonian mechanics represents a false view *of* time and simultaneity. Newtonian mechanics is not a theory *about* anything. The terms 'time' and 'simultaneous' have no referents, and, consequently, the theoretical principles involving these terms are not false but truth-valueless.

An obvious way out of this difficulty is to view the theoretical principles involving 'time' and 'simultaneous' of Newtonian mechanics as *existential* assertions; or, what amounts to the same thing, to view theoretical terms as analogous to definite descriptions, and to adopt Russell's rather than Frege's view of the truth-values of sentences containing nonsatisfied definite descriptions. That is, we construe Newtonian mechanics as containing assertions of the form

(19)   There exist a quantity t and a relation $\mathcal{S}$ such that _____ ,

where the conjunction of the various theoretical principles involving 'time' and 'simultaneous' is put in the blank. This construal allows us to say that Newtonian mechanics as a whole is false, since there exists no such quantity t and no such relation $\mathcal{S}$. However, it does not allow us to say anything about the truth-values of individual sentences of Newtonian physics. We cannot say, for example, that (16) is true and (17) false. Note, that it will not do to construe individual theoretical sentences again as existential assertions. We cannot, e.g., construe (16) as

(20)   There exists a relation $\mathcal{S}$ such that e_1 bears to e_2 in Frame F, and (17) as

(21)   There exists a relations $\mathcal{S}$ such that if e_1 bears to e_2 in frame F then e_1 bears $\mathcal{S}$ to e_2 in frame F'.

This makes (16) come out true, all right, but it also makes (17) true. For (21) is certainly true; there exist plenty of frame-independent relations, e g , the relation of having spacelike separation! This last move makes it far too easy for an individual theoretical sentence to be true.

These considerations suggest that it is a mistake to view the reference of theoretical terms as determined by the theoretical principles within which they occur. If we say that a theoretical term either refers to an entity that satisfies (a sufficient number of) the theoretical principles containing the term or to nothing at all, we make it too difficult for such theoretical principles to turn out false. On the other hand, if we construe theoretical terms as analogous to Russellian descriptions, and thereby construe theoretical principles as basically existential assertions, we make it too difficult for such principles to turn out true—for in this latter case, only the theory as a whole can be true or false. And note that this holds even if the theory as a whole is completely and exactly true—we still have no general method for apportioning truth to the individual sentences of the theory. However, if the reference of a theoretical term is not determined by the theoretical principles within which it occurs, how is it determined? In my opinion, so-called causal theories of reference are on the right track. That is, it seems to me that what a theoretical term refers to is not a matter of which entity (if any) satisfies the theoretical principles involving the term, but rather, a matter of which actual entities have the right sort of "historical" connection with the use of the term (see Kripke, 1972 and Putnam, 1973).

Now I grant that this way of talking is extremely vague, and I do not know how to give a precise account of what the right sort of "historical" connection is. Nevertheless, in my view, this way of looking at the reference of theoretical terms does not leave us at a total loss either. On the contrary, I think we have enough intuitive ideas about what the "right sort of connection" is to at least get plausible candidates for the referents of most theoretical terms. For example, such questions as: 'What actual quantities are being measured by the measuring procedures used to determine values for the quantities postulated by the theory?' and 'What entities are actually responsible for the phenomena explained by the theory?' seem highly relevant for determining which quantities and relations the theoretical terms of our theory actually refer to. Furthermore, although the "historical" connection view of reference does not have anything very precise to say about just what the reference relation is, it says enough to free us from the implausibilities of the satisfaction-of-theoretical-principles account. That is, it shows us how even the central principles of a theory can turn out to be false, and it allows us to attribute truth and falsity to the individual sentences of a theory in a plausible way.

The case of 'time' and 'simultaneous' in Newtonian mechanics provides a good illustration of these points. In determining the referents of these terms, we should not look for entities that satisfy the theoretical principles of Newtonian physics—there are no such entities! Rather, we should proceed as follows: given the entities—quantities, relations, etc.—that our best current theory postulates, we look for some among these which (a) give a plausible distribution of truth-values for the sentences involving 'time' and 'simultaneous' used by Newtonian physicists; (b) are actually responsible for the phenomena explained by Newtonian mechanics; (c) are actually measured by the measuring procedures used to test Newtonian mechanics. Supposing for a moment that special relativity is our best current theory, and using these (admittedly rough and incomplete) guides, I suggest we obtain the following results about the referents of 'time' and 'simultaneous' in Newtonian mechanics:

(i) In a context like 'time . . . in frame F,' 'time' refers to t_F—the coordinate time of frame F. In a context like 'simultaneous . . . in frame F,' 'simultaneous' refers to S^F—the relation of lying on the same hypersurface t_F = constant.

(ii) Where 'time' or 'simultaneous' occurs without explicit qualification as to reference frame, but other features of the context "attach" the sentence to a particular reference frame—e.g., the sentence is uttered within a particular laboratory frame on the surface of the earth—'time' refers to the coordinate time t_F of that frame and 'simultaneous' refers to S^F.

(iii) Where the context neither explicitly nor implicitly "attaches" the sentence to a particular inertial frame, 'time' and 'simultaneous' have no referents.

These suggestions accord with (a)–(c) above. We have the intuitively plausible consequence, for example, that (16) is true and (17) false; we are able to attribute truth and falsity to the individual sentences used by Newtonian physicists; and we make it neither too hard nor too easy for such sentences to come out true. The quantity assigned to 'time'—i.e., the coordinate time of a particular frame in a particular context—is the quantity actually responsible for the phenomena explained by Newtonian kinematics. The central explanatory principle of Newtonian kinematics is the law of inertia (7); and, according to special relativity, the correct form of this law is (15)—which determines the trajectory of a free particle as a function of *coordinate time*. Finally, the quantity assigned to 'time' is the quantity actually measured by (ideal) clocks. According to special relativity, (ideal) clocks measure the proper time along their trajectories. So a clock at rest at the origin of a particular inertial frame F measures the coordinate time of F.

If (i)–(iii) are correct, 'time' and 'simultaneous' have referential properties analogous to *indexical* words like 'I,' 'you,' 'here,' and 'now.' Just as indexical words refer to different things relative to different contexts— relative to different speakers, hearers, places, and times—'time' and 'simultaneous' refer to different things relative to different inertial reference frames. (And, as in the case of indexical words, the relevant context may be either explicit or implicit.) Just as the truth-values of sentences containing indexical words can vary with context, the truth-values of sentences containing 'time' and 'simultaneous' vary with inertial frame. Neither kind of sentence possesses a truth-value *absolutely*, but only relative to this or that context (reference frame). Thus, when the sentence in question is not "attached" to any context (reference frame) of the appropriate kind, it lacks a truth-value and its component words lack referents.

If I am right, the transition from Newtonian mechanics to special relativity *has* taught us a semantic lesson. In a special relativistic world the referents of 'time' and 'simultaneous' have to be taken as dependent on reference frame; 'time' and 'simultaneous' must be seen as possessing referential properties analogous to those of indexical words. If the world were Newtonian, this would not be necessary; 'time' and 'simultaneous' would have unique, frame-independent referents. However, it is not necessary to suppose that 'time' and 'simultaneous' have *changed* their referential properties in this transition. Since our world is and always was (so we believe—modulo note 2) a special-relativistic world, not a Newtonian world, the words 'time' and 'simultaneous' have and always had referential properties appropriate to a special relativistic world. Thus, when used by a relativistic physicist, 'time' and 'simultaneous' have the same referential properties as they did when used by a Newtonian physicist: i.e., (i)–(iii) still hold. (Of course, if a relativistic physicist is careful, case (iii) will never occur!) One is able to argue for a significant semantic change in the transition from Newtonian mechanics to special relativity only by employing wildly implausible theories about the reference of theoretical terms.

4. The Conventionality of Simultaneity in Special Relativity

The problem of the conventionality of simultaneity is typically introduced in the following way: we are asked to imagine two points, p_0 and p_1, in a given reference system. Situated at each of the points is a clock. A light signal is sent from p_0 to p_1, where it is reflected back to p_0. The light signal leaves p_0 at t_1—as determined by the clock at p_0—and returns to p_0 at t_2. Our problem is to synchronize the clock at p_1 with the clock at p_0; to say when, according to p_0-time, the light signal arrives at p_1. We must determine which event between t_1 and t_2 at p_0 is simultaneous with the event E at p_1. According to the conventionality thesis it is a matter of definition which event between t_1 and t_2 is simultaneous with E; no choice is any "truer" than any other. Of course, if we assume that the velocity of light is the same from p_0 to p_1 as it is on the return trip, the p_0 time of E would be unambiguously determined as

(22) $t = t_1 + \frac{1}{2}(t_2 - t_1)$

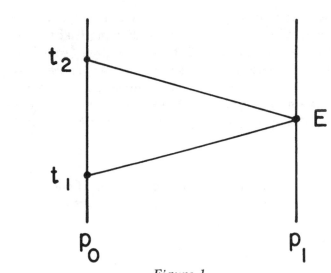

Figure 1

However, conventionalists argue that any claim about the one-way velocity of light—as distinct from its round-trip velocity—is just as conventional. They argue that

(23) $t = t_1 + \epsilon(t_2 - t_1)$

is just as good as (22) for determining the p_0-time of E, where ϵ is any real number such that $0 < \epsilon < 1$. Only computational simplicity can favor the choice $\epsilon = \frac{1}{2}$ over any other admissible value of ϵ. There are no facts that make (22) true and (23) false.

I think this problem can be greatly clarified by looking at special relativity from the space-time point of view of section 2. Our discussion will be facilitated if we consider Minkowski space-time as a two-dimensional manifold—i.e., as R^2 instead of R^4. This device simplifies the algebra without essentially changing the conceptual situation. Our theory remains

417

the same, except that the Minkowski pseudo-metric takes the simpler form

$$g((x_0, x_1), (x_0', x_1'))^2 = (x_0 - x_0')^2 - (x_1 - x_1')^2$$

in R^2. Thus, inertial coordinate systems y_0, y_1 are characterized by the condition

$$g(p, q)^2 = (y_0 - y_0')^2 - (y_1 - y_1')^2$$

where p has coordinates (y_0, y_1) and q has coordinates (y_0', y_1'). To set up the problem in this framework, consider a given inertial frame associated with the time-like geodesic $\sigma(\tau)$. Let there be given two null geodesics (light rays) which intersect $\sigma(\tau)$ at $y_0 = \tau_1$, and $y_0 = \tau_2$ respectively, and intersect each other at E. Since null geodesics have constant unit velocity in inertial systems (I have set $c = 1$), it is clear that if we fix the time at point E according to the synchronization role (22)—i.e., if we let the time of E be

$$t = \tau_1 + \tfrac{1}{2}(\tau_2 - \tau_1)$$

we are merely adopting the coordinate time $t = y_0$ of our given inertial frame as our global time. That is, the rule (22) amounts to fixing the time

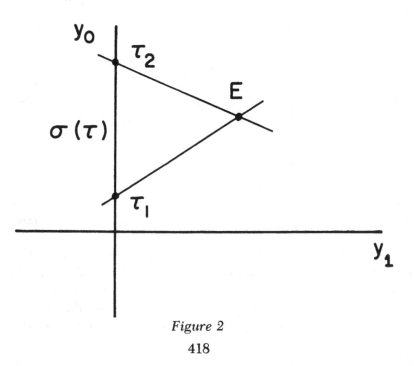

Figure 2

of events not on the trajectory $\sigma(\tau)$ by means of the coordinate time of an inertial coordinate system adapted to $\sigma(\tau)$.

What are we doing if we use (23) instead of (22) to fix the time of E—i.e., if we let the time of E be

$$(23) \qquad \bar{t} = \tau_1 + \epsilon(\tau_2 - \tau_1)$$

with $\epsilon \neq \frac{1}{2}$? This latter procedure can be viewed as using the coordinate $\bar{t} = z_0$ of a *noninertial* coordinate system $z_0\, z_1$ as our global time. It amounts to fixing the time of events not on the trajectory $\sigma(\tau)$ by means of the $z_0 = \bar{t}$ coordinate of a noninertial coordinate system adapted to $\sigma(\tau)$ (in the sense that $\sigma(\tau)$ satisfies $z_0 = \tau$, $z_1 = 0$ in $z_0,\, z_1$).

The relation between the noninertial system $z_0,\, z_1$ and our original inertial system $y_0,\, y_1$ is easily seen to be

$$(24) \qquad \bar{t} = z_0 = t + 2\delta y_1$$
$$z_1 = y_1$$

where $\delta = \epsilon - \frac{1}{2}$. The inverse relation is of course

$$(25) \qquad t = y_0 = \bar{t} - 2\delta z_1$$
$$y_1 = z_1$$

Thus, using $\epsilon \neq \frac{1}{2}$ in (22) amounts to performing the coordinate transformation (24) and using the "coordinate time" $\bar{t}$ of the new system to define simultaneity. Using (25) we find that the Minkowski metric takes the form

$$(26) \qquad g(p,\, q)^2 = (z_0 - z_0')^2 - 4\delta(z_0 - z_0')\, (z_1 - z_1') + (4\delta^2 - 1)\, (z_1 - z_1')^2$$

in our new system $z_0,\, z_1$. Now a minimal condition for a z_0-coordinate to be a temporal coordinate is that the curves $z_0 = $ constant be spacelike. It follows from (26) that this is the case if and only if

$$(27) \qquad 4\delta^2 - 1 < 0.$$

If we substitute $\epsilon - \frac{1}{2}$ for δ in (27), this minimal condition becomes

$$(28) \qquad \epsilon(\epsilon - 1) < 0.$$

(28) implies that $0 < \epsilon < 1$. So the "coordinate time" of an ϵ-*system*—a system in which (26) holds—is a suitable temporal coordinate only if $0 < \epsilon < 1$.

Some useful facts about ϵ-systems are the following. First, if a trajectory has velocity $v = dy_1/dt$ in an inertial coordinate system and "velocity" $\bar{v} = $

419

$dz_1/d\bar{t}$ in an ϵ-system, it follows from (24) and (25) that the two are related by

(29) $v = \bar{v}/(1 - 2\delta\bar{v})$.

Second, the relation between the coordinate time t and the proper time τ of a trajectory in an inertial system is given by

(30) $d\tau = \sqrt{1 - \bar{v}^2}\, d\bar{t}$.

In ϵ-systems (30) becomes

(31) $d\tau = \sqrt{1 - 4\delta\bar{v} + (4\delta^2 - 1)\, \bar{v}^2}\, d\bar{t}$

or (32) $d\tau = \sqrt{(1 - \bar{v}(2\epsilon - 1))^2 - \bar{v}^2}\, d\bar{t}$.

Finally, we know that two inertial systems, t, y_1 and t^*, y_1^* are related by a Lorentz transformation

(33) $t^* = (t - vy_1)/\sqrt{1 - v^2}$

$y_1^* = (y_1 - vt)/\sqrt{1 - v^2}$

where v is the relative velocity of the two systems. How are two different ϵ-systems related?

Let there be given two ϵ-systems, I and II, with coordinates $\bar{t}$, z_1 and $\bar{t}^*$, z_1^* respectively. Let the respective values of ϵ in the two frames be ϵ_1 and ϵ_2, and let frame II move with "velocity" v with respect to frame I. We can use the following procedure to find the transformation connecting the two frames: (1) use (25) to transform I into an inertial frame t, y_1; (2) use (28) and (32) to transform t, y_1 into a second inertial frame t^*, y_1^* moving with velocity v with respect to the first; (3) use (24) to obtain the frame II. This procedure results in some tedious algebra and

$$\bar{t}^* = z_0^* = \frac{(2\bar{v}(1 - \epsilon_1 - \epsilon_2) + 1)\bar{t} - (2(\epsilon_1 - \epsilon_2) + 4\bar{v}\epsilon_1(1 - \epsilon_1))z_1}{\sqrt{(1 - v(2\epsilon_1 - 1))^2 - v^2}}$$

(34)

$$z_1^* = {}^! \frac{z_1 - \bar{v}t}{\sqrt{(1 - \bar{v}(2\epsilon_1 - 1))^2 - \bar{v}^2}}$$

Note that when $\epsilon_1 = \epsilon_2 = \frac{1}{2}$, $\bar{v} = v$ and (34) reduces to a Lorentz transformation (33).

John Winnie (1970) derives the above transformations from a completely different point of view.[3] He calls the relations (34) the ϵ-*Lorentz transformations*. The purpose of Winnie's paper is to argue that special relativity as formulated using the standard synchronization rule (22) is "kinematically equivalent" to a formulation using the nonstandard rule (23) with $\epsilon \neq \frac{1}{2}$ —thus vindicating, according to Winnie, the thesis of the conventionality of simultaneity. In the present framework, Winnie's claim is that special relativity as formulated in ϵ-systems is equivalent to special relativity as formulated in inertial systems. It seems to me that there is one sense in which this claim is obviously true, but completely trivial; and there is a second sense in which it is not at all obvious, and completely unsupported by Winnie's arguments.

The sense in which the equivalence claim is obviously true is that Minkowski space-time can be described equally well from the point of view of ϵ-coordinate systems as from the point of view of inertial coordinate systems. Formulations of special relativity in ϵ-systems say the same thing about Minkowski space-time as formulations in inertial systems. Indeed, they are nothing but different coordinate representations of the same theory (the theory expressed in coordinate-independent form in note 4). Thus the two formulations cannot disagree about the behavior of light— light follows null geodesics independently of coordinate system; nor about the behavior of free particles—free particles follow timelike geodesics independently of coordinate system; nor about the behavior of clocks— (ideal) clocks measure the proper time along their trajectories independently of coordinate system; etc. But note that in this sense of 'equivalence' there is no need to restrict ourselves to ϵ-coordinate systems. Minkowski space-time can be equally well described from the point of view of *any* coordinate system; our theory can be represented in *arbitrary* coordinate systems. (This is especially obvious in the formulation of note 4.) Thus the equivalence of ϵ-systems and inertial systems in this sense reveals no deep facts about Minkowski space-time or special relativity. Newtonian space-time can be represented in arbitrary coordinate systems as well; Newtonian kinematics can be formulated in systems that are not inertial with no change in theory. In fact, of course, any theory expressible in *tensor form* will have this property.

Thus, if the equivalence claim is to be nontrivial, it must amount to something more than the assertion that ϵ-coordinate systems and inertial coordinate systems are equally good representations of the basic facts about Minkowski space-time hypothesized by special relativity. Let us look a little closer. According to special relativity there is no unique global

421

time defined on space-time. However, special relativity in its usual $\epsilon = \frac{1}{2}$ formulations associates a unique global time with every state of inertial motion. For every timelike geodesic $\sigma(\tau)$, there is a unique (up to a linear transformation) way of extending its proper time to a global coordinate time t—the y_0-coordinate of an inertial coordinate system adapted to $\sigma(\tau)$. Now a defender of the equivalence claim can be construed as asserting that there are other, equally good, ways of extending the proper time of a time-like geodesic to a global time—namely, the z_0-coordinates of ϵ-systems adapted to $\sigma(\tau)$. That is, he is claiming not merely that ϵ-systems and inertial systems are equally good coordinate representations of Minkowski space-time, but that the z_0-coordinate of an ϵ-system is an equally good candidate for the global time associated with a given state of inertial motion as the y_0-coordinate of an inertial system. The $\overline{t}$ of an ϵ-system is an equally good representation of physical time as the t of an inertial system. This explains why a defender of the equivalence thesis considers only ϵ-systems with $0 < \epsilon < 1$, and not *arbitrary* coordinate systems. For only the z_0-coordinate of an ϵ-system with $0 < \epsilon < 1$ satisfies minimal conditions for representing physical time: the hypersurfaces $z_0 = $ constant being spacelike.

If this is correct, arguments like Winnie's, which simply amount to showing how special relativity as formulated in inertial systems can be translated into a formulation in ϵ-systems, do not support a nontrivial version of the equivalence thesis. Such translation procedures merely prove that ϵ-systems and inertial systems are equally good coordinate representations of Minkowski space-time, a fact that is obvious in a tensor formulation of special relativity. In support of a stronger version of the equivalence claim, we must be given some reason to think that the z_0-coordinate of an ϵ-system is an equally good representation of physical time as the y_0-coordinate of an inertial system. Clearly the condition $0 < \epsilon < 1$ is a necessary condition for a z_0-coordinate to represent physical time—but is it sufficient? Are there any plausible additional conditions that narrow the choice of ϵ further?

The advocates of so-called *slow-transport synchrony* (see Ellis and Bowman, 1967) may be understood to propose a further such necessary condition for a z_0-coordinate to represent physical time. Consider again the problem of synchronizing two clocks in a given reference frame, one at P_0 and the other at P_1. The two are said to be in slow-transport synchrony

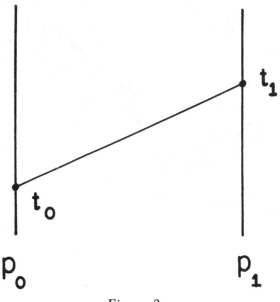

Figure 3

if a clock synchronized with P_σ-time at t_0 is transported "infinitely slowly" to P_1 and is in agreement with P_r-time at t_1. (We consider only "infinitely slow" transport to avoid the velocity-dependent relativistic time-dilation effects.) More precisely, let there be given an arbitrary ϵ-system z_0, z_1 adapted to a given timelike geodesic $\sigma(\tau)$.

Consider a timelike geodesic $\rho(\tau^*)$—representing a clock transported with constant velocity—which intersects $\sigma(\tau)$ at τ_0. Consider two events E and E' on $\sigma(\tau)$ and $\rho(\tau^*)$ respectively, and let the proper time τ^* of ρ equal that of σ at their intersection. i.e., let $\tau_0 = \tau_0^*$—the two clocks are synchronized. Finally, let the "velocity" of ρ in z_0, z_1 be $\bar{v} = dz_1/dz_0$. E and E' are *slow-transport simultaneous* if and only if $\lim_{r \to 0} (\tau_1 - \tau_1^*) = 0$, where τ_1 and τ_1^* are the respective proper times of E and E' (Fig. 4).

Now I take the advocates of slow-transport synchrony to be imposing the further condition on a z_0-coordinate that it agree with slow-transport simultaneity; i.e., that two events are simultaneous according to z_0—they have the same z_0-coordinate—if and only if they are slow-transport simultaneous. It is not hard to show that this requirement fixes ϵ at $\frac{1}{2}$; only the y_0-coordinates of inertial systems satisfy this condition. For suppose that

423

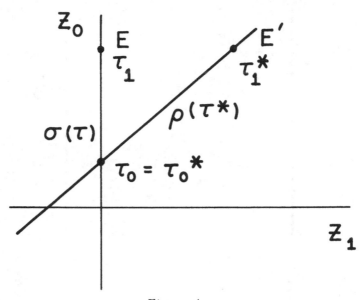

Figure 4

E and E' are z_0-simultaneous in our diagram: the z_0-coordinate of E' is just τ_1. It follows from (31) that

(35) $(\tau_1{}^* - \tau_0{}^*) = (\tau_1 - \tau_0)\sqrt{1 - 4\delta\,\bar{v} + (4\delta^2 - 1)\bar{v}^2}$

Expanding the "dilation term" in a binomial series we obtain

$$(\tau_1{}^* - \tau_0{}^*) = (\tau_1 - \tau_0)\,(1 - 2\delta\bar{v} - \tfrac{1}{2}(4\delta^2 - 1)\,\bar{v}^2 - \ldots$$

where the rest of the series consists of second and higher powers of $\bar{v}$. Since $\tau_0{}^* = \tau_0$ we have

$$(\tau_1 - \tau_1{}^*) = (\tau_1 - \tau_0)\,(2\delta\bar{v} - \tfrac{1}{2}(4\delta^2 - 1)\,\bar{v}^2 - \ldots$$

But $(\tau_1 - \tau_0) = z_1/\bar{v}$ where z_1 is the "spatial" coordinate of E'. So we have

$$(\tau_1 - \tau_1{}^*) = z_1 2\delta - \tfrac{1}{2}(4\delta^2 - 1)\,\bar{v} - \ldots$$

Letting $\bar{v} \to 0$ and substituting $\epsilon - \tfrac{1}{2}$ for δ we finally get

$$\lim_{\bar{v}\to 0} (\tau_1 - \tau_1{}^*) = z_1\,(2\epsilon - 1)$$

So z_0 simultaneity agrees with slow-transport simultaneity everywhere if and only if $\epsilon = \tfrac{1}{2}$, if and only if z_0 is the coordinate time of an inertial coordinate system adapted to $\sigma(\tau)$.

424

What does this show? It shows that *if* a necessary condition for something being a representation of physical time is that it agree with slow-transport simultaneity, *then* the z_0-coordinates of ϵ-systems with $\epsilon \neq \frac{1}{2}$ are not equally good representations of physical time as the y_σ-coordinates of inertial systems. However, this refutes the conventionalist only if he concedes that this requirement—agreement with slow-transport simultaneity—is not itself conventional. And the conventionalist does not have to (nor does he in fact: see Grünbaum, 1969 and Salmon, 1969) concede this. He can maintain that just as choosing $\epsilon = \frac{1}{2}$ is not any "truer" or more "factual" than choosing $\epsilon \neq \frac{1}{2}$; so, requiring agreement with slow-transport simultaneity is not any "truer" or more "factual" than not requiring it. Both choices may have the advantage of simplicity over their alternatives, but not the advantage of truth. But now the debate over conventionalism begins to look hopeless. The conventionalist asserts that a certain system of description is not "factual," and produces alternative descriptions which he claims are "equally good"; the anti-conventionalist points to various asymmetries between the original system and the conventionalist's alternatives; the conventionalist replies that these differences are not "factual" either, they are merely differences in simplicity; etc. If this debate is to have any point we need some kind of *independent* characterization of the difference between "factual" and conventional statements or descriptions.

Now, if we look at the conventionality thesis from a semantic point of view, it is clear that one important difference between conventional statements and "factual" statements is that the former are supposed to have no determinate truth-value, while the latter are either determinately true or determinately false. Therefore, one possible source of an independent characterization of the difference between "factual" and conventional statements is a semantic theory that is capable of dealing with sentences that lack determinate truth-value. As I suggested earlier, I think that so-called *referential semantics* is the most promising theory of this kind. According to referential semantics, there are at least two ways in which a (grammatically well-formed) sentence can lack a determinate truth-value: (1) it can contain words that pick out *no* referents; or (2) it can contain words that have a *multiplicity* of referents. In this latter case, the sentence (like sentences containing indexical words) is neither true nor false *simpliciter*, but has different truth-values relative to different choices from among the multiplicity of referents in question. With this in mind, I would like to turn to what I think are the most important arguments for the conventionality thesis.

It seems to me that there are at bottom only two arguments for the conventionality of simultaneity in the literature: Reichenbach's and Grünbaum's. Reichenbach argues from an epistemological point of view; he argues that certain statements are conventional as opposed to "factual" because they are unverifiable in principle. Grünbaum argues from an ontological point of view; he argues that certain statements are conventional because there is a sense in which the properties and relations with which they purportedly deal do not really exist, they are not really part of the objective physical world. Thus, Reichenbach's and Grünbaum's arguments depend on two different characterizations of the difference between conventional and "factual" statements. According to Reichenbach, the "factual"/conventional distinction is just the verifiable/unverifiable distinction. According to Grünbaum, the "factual"/conventional distinction rests on a prior distinction between properties and relations that are objective constituents of the physical world and those that are not.

How does Reichenbach argue for the conventionality thesis? He considers various methods for *determining* distant simultaneity in a given reference system—various methods of *verifying* statements of the form 'Events e_1 and e_2 are simultaneous with respect to the given state of inertial motion M'—and tries to show that none of these methods furnishes an unambiguous answer in a special-relativistic world. Thus, for example, if there were no upper limit to the velocity of signals, we could determine which event at a given place P_0 is simultaneous with a given event E at P_1 by considering arbitrarily fast signals that are sent from P_0 and are reflected back from P_1 at event E. In a special relativistic world, on the other hand, there is an upper limit to the velocity of signals. Consequently, we can use signals to determine simultaneity only if we know their velocities; and knowledge of (one-way) velocity presupposes knowledge of distant simultaneity:

Thus we are faced with a circular argument. To determine the simultaneity of distant events we need to know a velocity, and to measure a velocity we require knowledge of the simultaneity of distant events. The occurrence of this circularity proves that simultaneity is not a matter of knowledge, but of a coordinative definition, since the logical circle shows that a knowledge of simultaneity is impossible in principle (1958, 126–127).

Of course, just because one method of determining simultaneity involves circularity, it does not follow that they all do; so Reichenbach considers, in addition, the possibility of determining distant simultaneity by transporting clocks from one place to another. About this method he makes two points: (1) in a special-relativistic world it does not determine a unique simultaneity relation, because the rate of clocks depends on their velocity; (2) even if the relation so determined were unique, it would still only constitute a definition, because it would depend on unverifiable assumptions to the effect that if two clocks are seen to run at the same rate when together they continue to run at the same rate when spatially separated (1958, pp. 133–135).

I think Reichenbach's treatment of the clock-transport method is not so convincing as his treatment of the signal method. First, the method of "infinitely slow" clock transport avoids problem (1). Slow-transport simultaneity is a unique simultaneity relation. Second, while it is true that slow-transport simultaneity depends on assumptions about the rates of spatially separated clocks, these appear to be *additional* assumptions. That is, we do not appear to be faced with the same kind of obvious circularity as in the signal method, in which the determination of simultaneity depends on assumptions about velocity, and assumptions about velocity depend on the determination of simultaneity. Let me try to be more precise. The uniqueness of the slow transport method—its agreement with $\epsilon = \frac{1}{2}$ simultaneity—depends on assumptions about the proper time metric. That is, we assume that the proper time metric in a particular ϵ-system is given by (31), i.e.,

$$d\tau = \sqrt{1 - 4\delta\bar{v} + (4\delta^2 - 1)\,\bar{v}^2}\, d\bar{t}$$

If we assume instead a different proper time metric, e.g.,

$$(37) \qquad d\tau = \sqrt{1 - 4\delta\bar{v} - (4\delta^2 - 1)\bar{v}^2}\, d\bar{t} - 2\delta\bar{v}\, d\bar{t}$$

we can eliminate the uniqueness of slow-transport simultaneity. Thus, according to the metric (37)

$$(38) \qquad \lim_{\bar{v}\to 0} (\tau_1 - \tau_1{}^*) = 0$$

in *all* ϵ-systems. Therefore the method of slow-transport depends on assumptions about the temporal metric. However, these assumptions seem to be independent of assumptions about the value of ϵ—even if we fix the value of ϵ we are still free to choose between (31) and (37) as our proper time metric. One can argue that such assumptions about the temporal metric are themselves conventional, but this requires an *independent* argument.

In any case, the main problem with Reichenbach's argument is this: whether or not statements about distant simultaneity are in some sense unverifiable in the context of special relativity, we have been given no reason to suppose that unverifiability implies lack of determinate truth-value. It would seem that sufficient conditions for a sentence's possession of a truth-value are: (1) that it be grammatically well-formed, and (2) that its component words pick out determinate referents. If (1) and (2) are satisfied, the sentence has a determinate truth-value, regardless of its epistemic status. Thus it seems to me that Reichenbach's approach to the problem of conventionality is vitiated by his reliance on bad semantics— his reliance on the verifiability theory of meaning. Note that Reichenbach himself was perfectly explicit about his reliance on this theory. For example, in his comments on the significance of Einstein's views on simultaneity—understood as a version of the conventionality thesis, of course—Reichenbach writes:

The physicist who wanted to understand the Michelson experiment had to commit himself to a philosophy for which the meaning of a statement is reducible to its verifiability, that is, he had to adopt the verifiability theory of meaning if he wanted to escape a maze of ambiguous questions and gratuitous complications. It is this positivist, or let me rather say, empiricist commitment which determines the philosophical position of Einstein (1949, pp. 290–291).

Grünbaum's approach to the conventionality thesis is very different. Unlike Reichenbach, he does not rely on the verifiability theory of meaning; he does not use verifiability as a criterion for possessing a truth-value. Instead, he argues that in a special-relativistic world there is no objective simultaneity relation at all, there is no genuine physical relation for 'simultaneity' to refer to. Grünbaum's argument proceeds as follows: Let us say that two events, at P_0 and P_1 respectively, are *topologically simultaneous* just in case they are connectible by no causal signal. In a Newtonian world, in which there is no upper bound to the velocity of causal propagation, there is a unique event at P_0 topologically simultaneous with a given event E at P_1. In such a world, the relation of topological simultaneity uniquely determines the relation of *metrical simultaneity*. In a special-relativistic world like our own, on the other hand, in which there is a finite upper bound to the velocity of causal propagation, there are a multitude (in fact an infinity) of events at P_0 which are topologically simultaneous with E. In this kind of world, therefore, the relation of metrical simultaneity is not uniquely determined by the relation of topological simultaneity (see 1973, pp. 28ff; pp. 345ff.)

If this is correct,[4] in a special-relativistic world it is impossible to define a relation of metrical simultaneity solely on the basis of causal relations between events, while in a Newtonian world such a definition would be possible. But why *should* the relation of metrical simultaneity be definable solely on the basis of causal relations? Why should we take the indefinability of metrical simultaneity on the basis of topological simultaneity as a reason for concluding that there is no objective physical relation of metrical simultaneity? Why can't metrical simultaneity stand on its own feet, as it were?

The answer, in Grünbaum's case, is that he holds a *causal* theory of time. He believes that all objective temporal relations are constituted by causal relations between events; the only temporal relations that objectively exist are those determined solely by causal relations:

By maintaining that the very *existence* of *temporal* relations between non-coinciding events depends on the obtaining of some *physical* relations between them, Einstein espoused a conception of time (and space) which is *relational* by regarding them as systems of relations between physical events and things. Since time relations are first constituted by the system of physical relations obtaining among events, the character of the temporal order will be determined by the physical attributes in virtue of which events will be held to sustain relations of "simultaneous with", "earlier than", or "later than". In particular, it is a question of physical fact whether these attributes are of the kind to define temporal relations *uniquely*. . . . (1973, pp. 345–346).

So in a world in which metrical simultaneity is not definable solely on the basis of causal relations, there is no such physical relation. Note the similarity between Grünbaum's argument here and his argument for the conventionality of congruence. He argues that on a *continuous* set of spatial or temporal points there is no objective ("intrinsic") congruence relation, because on such a set congruence is not definable solely on the basis of topological properties (like cardinality) and order relations. Thus this argument depends on the claim that the only objective physical relations on a set of spatial or temporal points are those constituted by topological and ordinal relations[5]—just as the argument for the conventionality of simultaneity depends on the claim that the only objective temporal relations are those constituted by causal relations between events.

Grünbaum's argument, unlike Reichenbach's, has the advantage that *if* it were correct, we *could* draw semantic conclusions about the truth-value of sentences containing 'simultaneous' on the basis of the referential properties of their key terms. For, if Grünbaum's argument is correct, it follows that 'simultaneous' has no referent—there is no objective physical relation for it to refer to. And this would make the conventionalist contention that sentences like 'Events e_1 and e_2 are simultaneous with respect to state of inertial motion M' lack determinate truth-value highly plausible. However, it seems to me that Grünbaum's actual argument is much less persuasive than Reichenbach's. Reichenbach has given some plausibility to the claim that statements about distant simultaneity may be unverifiable within the context of special relativity. As far as I can see, Grünbaum has given us no reason to accept the view that the only objective temporal relations are constituted by causal relations. Indeed, how could one possibly support such a view? Our only grip on which properties and relations are objective constituents of the physical world is via our best theories of the physical world. The properties and relations that we hold to exist objectively are those that our best physical theories postulate. And since our best theories do not merely postulate the kind of *ordinal* (causal) temporal relations favored by Grünbaum—they postulate *metrical* relations as well—we have no reason to grant such ordinal (causal) relations the privileged ontological status that Grünbaum wants to ascribe to them.

In sum, it seems to me that we have not been given a basis for the "factual"/conventional distinction on which (a) conventional statements turn out to lack determinate truth-values, and (b) statements about distant simultaneity turn out to be conventional. Reichenbach has given a criterion for conventionality—i.e., unverifiability—which statements about distant simultaneity in a special-relativistic world can be held to fulfill with some plausibility. The verification of such statements is at least much more complicated in a special-relativistic world than it is in a Newtonian world. But there is no clear connection between Reichenbach's criterion and the lack of a determinate truth-value. Reichenbach's argument for the conventionality thesis rests on a dubious semantics. On the other hand, Grünbaum has given a criterion for conventionality—i.e., having constituent terms with no objective physical referents—which has a plausible connection with the lack of a determinate truth-value. However, Grünbaum's argument that 'simultaneous' indeed lacks an objective referent depends on an unsupported, and seemingly unsupportable, a priori judgment as to which relations are objective. Grünbaum's argument for the conventionality thesis rests on a dubious ontology.

Notes

1. It is worth noting that both Newtonian mechanics and special relativity can be formulated within a more general point of view by starting with a four-dimensional C^∞ manifold M instead of R^4 (cf. Anderson, 1967, Earman and Friedman, 1973, Havas, 1964, and Trautman, 1966). In this framework, a *Newtonian space-time* is a quadruple $\langle M, \Gamma^i_{jk}, t_i, h^{ij} \rangle$, where Γ^i_{jk} is a symmetric affine connection, t_i a C^∞ covector field, and h^{ij} a C^∞ symmetric tensor field of type (2, 0) and signature (0, 1, 1, 1). These objects satisfy the field equations

(1) $R^i_{jkl} = 0$
(2) $h^{ij}_{;k} = 0$
(3) $t_{i;j} = 0$
(4) $h^{ij}t_it_j = 0$

where R^i_{jkl} is the curvature tensor of Γ^i_{jk}.
Our law of motion is

$$(5)\ \frac{d^2x_i}{du^2} + \Gamma^i_{jk} \frac{dx_i}{du} \frac{dx_k}{du} = 0.$$

A Minkowski space-time is a triple $\langle M, \Gamma^i_{jk}, g_{ij} \rangle$, where Γ^i_{jk} is a symmetric affine connection and g_{ij} is a C^∞ symmetric tensor field of type (0, 2) and signature (1, −1, −1, −1). Our field equations are just

(6) $R^i_{jkl} = 0$
(7) $g_{ij;k} = 0$

and our law of motion is again (5). This more general framework facilitates the comparison of these two theories with general relativity. In this context a *general relativistic space-time* is a quadruple $\langle M, \Gamma^i_{jk}, g_{ij}, T^{ij} \rangle$, where Γ^i_{jk} and g_{ij} are as in special relativity and T^{ij} is a C^∞ tensor fields of type (2, 0) representing the mass-energy density. Our equation of motion remains the same, and we have one field equation

(8) $R^{ij} - \frac{1}{2}g^{ij}R = -8\pi kT^{ij}$

where R^{ij} is the Ricci tensor of Γ^i_{jk}, R is the contracted Ricci tensor, and k is the gravitational constant. (The notions from differential geometry used here are explained in Hicks 1965.)

2. Of course, we really think that special relativity is only *approximately* true. However, my discussion will be much simpler if I ignore this. If I were to take account of the actual situation, I would have to change 'inertial frame' everywhere to 'approximately inertial frame,' etc.

3. Compare (33) with the relations in Winnie, 1970, p. 234, remembering that I have set $c = 1$. Note that at the end of his paper Winnie briefly alludes to the possibility of obtaining his transformations in something like the above manner—cf. pp. 236–237.

4. (Added in proof) Even this much seems actually incorrect. David Malament has recently shown that the standard $\epsilon = \frac{1}{2}$ simultaneity relation *is* (in a natural sense) uniquely definable in terms of causal relations in Minkowski space-time. See Malament, "Causal Theories of Time and the Conventionality of Simultaneity," forthcoming.

5. See Friedman, 1972 for such an interpretation of Grünbaum's argument.

References

Anderson, J. L. (1967). *Principles of Relativity Physics*. New York: Academic Press.

Earman, J. (1970). "Space-Time, or How to Solve Philosophical Problems and Dissolve Philosophical Muddles Without Really Trying," *Journal of Philosophy*, vol. 67, pp. 259–277.

Earman, J. and M. Friedman. (1973). "The Meaning and Status of Newton's Law of Inertia and the Nature of Gravitational Forces," *Philosophy of Science*, vol. 40, pp. 329–359.

Ellis, B. and P. Bowman. (1967). "Conventionality in Distant Simultaneity," *Philosophy of Science*, vol. 34, pp. 116–136.

Feyerabend, P. K. (1962). "Explanation, Reduction, and Empiricism" in H. Feigl and G. Maxwell, eds., *Minnesota Studies in the Philosophy of Science*, vol. 3. Minneapolis: Minnesota Press, 1962.

Field, H. (1972). "Tarski's Theory of Truth," *Journal of Philosophy*, vol. 69, pp. 347–375.

Field, H. (1973). "Theory Change and the Indeterminacy of Reference", *Journal of Philosophy*, vol. 70, pp. 462–481.

Friedman, M. (1972). "Grünbaum on the Conventionality of Geometry," *Synthese*, vol. 24, pp. 219–235.

Grünbaum, A. (1973). *Philosophical Problems of Space and Time*. 2nd enlarged ed. Dordrecht: Reidel.

Grünbaum, A. (1969). "Simultaneity by Slow Clock Transport in the Special Theory of Relativity," *Philosophy of Science*, vol. 36, pp. 5–43.

Havas, P. (1964). "Four-Dimensional Formulations of Newtonian Mechanics and Their Relation to the Special and the General Theory of Relativity." *Reviews of Modern Physics*, vol. 36, pp. 938–965.

Hicks, N. J. (1965). *Notes on Differential Geometry*. Princeton: Van Nostrand.

Kripke, S. (1972). "Naming and Necessity," in D. Davidson and G. Harman, eds., *Semantics of Natural Language*, Dordrecht: Reidel.

Kuhn, T. (1962). *The Structure of Scientific Revolutions*, Chicago: University of Chicago Press.

Putnam, H. (1973). "Meaning and Reference," *Journal of Philosophy*, vol. 70, pp. 699–711.

Reichenbach, H. (1958). *The Philosophy of Space and Time*. New York: Dover.

Reichenbach, H. (1949). "The Philosophical Significance of the Theory of Relativity," in P. A. Schilpp, ed., *Albert Einstein: Philosopher-Scientist*, Evanston: Library of Living Philosophers.

Salmon, W. (1969). "The Conventionality of Simultaneity," *Philosophy of Science*, vol. 36, pp. 44–63.

Trautman, A. (1966). "Comparison of Newtonian and Relativistic Theories of Space-Time," in B. Hoffman, ed., *Perspectives in Geometry and Relativity*. Bloomington: University of Indiana Press.

Winnie, J. A. (1970). "Special Relativity Without One-Way Velocity Assumptions," *Philosophy of Science*, vol. 37, pp. 81–99, 223–238.

On Conventionality and Simultaneity—
Another Reply

1. Introduction

In "Conventionality in Distant Simultaneity," Brian Ellis and I (1967) discussed the position Reichenbach and Grünbaum had taken on this issue. That article received considerable comment (Grünbaum and Salmon, 1969; Winnie, 1970; Feenberg, 1974), much of the critical part of which Ellis answered in "On Conventionality and Simultaneity—A Reply" (1971). Here I shall reformulate, extend, and supplement his answer to some of the critiques (Grünbaum, 1969; Salmon, 1969; van Fraassen, 1969). Elsewhere I treat the topic in a less polemical manner (Bowman, 1974 and 1976).

The conventionality of distant simultaneity, as maintained by Reichenbach and Grünbaum, is after all this commentary so widely known that it can be stated very briefly. Let us consider two points A and B which are separated from one another in an inertial frame K. For a light signal emitted from A and reflected at B back to A, we compare the time interval for the outgoing trip to that for the round trip. This ratio is called "epsilon" (ϵ). In formulating the special theory of relativity, Einstein effectively took ϵ to be $\frac{1}{2}$; thus we may use $\epsilon = \frac{1}{2}$ in defining what is now called "standard signal synchrony." Reichenbach views ϵ as restricted *only* by the causal relations involved in the signaling process. That is, the reflection of the light ray at B must take place after the ray's emission at A but before its return to A. These considerations require us to restrict ϵ between zero and one, but Reichenbach insists that within these limits values of $\epsilon = \frac{1}{2}$ "could not be called false" (1958, p. 127). He claims that there are no facts that would mediate against using these values in definitions that are now called "nonstandard signal synchrony." This allegedly

NOTE: This paper follows subsection J.1 of my dissertation (1972) with only minor expository changes except for the last page of the present subsection 2.c, which is a substantive revision.

physical possibility of choosing ϵ between zero and one is "the conventionality of distant simultaneity (or, *mutatis mutandis*, synchrony) as determined by signals." Grünbaum also argues for this thesis, making clear that it obtains within a single inertial frame.[1] In this paper I shall consider a nonsignaling definition of synchrony and discuss its implications for simultaneity in Newtonian mechanics.

The Reichenbach-Grünbaum approach establishes the basic concept of special relativity, distant simultaneity, through the transmission of signaling processes; e.g., electromagnetic, gravitational, or matter waves and particles. The approach I wish to take is characterized by the transport of actual clocks; e.g., mechanical, light, or atomic clocks. Since a clock is defined as any physical system which passes through the same process periodically, and since on some construal a signal might possess this property, it would not be possible to distinguish sharply between a clock and a signal in that sense.[2] However, for the purpose of my characterization of a signal, it has the salient feature of being an infinitesimal disturbance or a point mass which, except for its own presence and absence, carries no information.

2. Simultaneity in Newtonian Mechanics

Elsewhere (1972, section F.5; 1974, section 3; 1976, section 3), I have taken the position that distant simultaneity as defined by the transport of clocks is conventional in the same sense as any other quantitative equality at a distance. In doing so, I have followed the position taken by Ellis and myself in our joint article (1967, p. 134). Grünbaum (1969, pp. 26–27), Salmon (1969, p. 56) and van Fraassen (1969, p. 67) have argued in reply that this is not the case in a world like that described by Newtonian mechanics, in which distant simultaneity can be established as a matter of temporal fact through the use of arbitrarily fast causal chains. Ellis (1971, p. 179, section 2) rejoins that this conclusion is false if "the basic time-ordering relationships in the Newtonian world are taken *e.g.* to be those given by the *local entropic order*," i.e., by what I call "clocks." In the third part of this section I shall reinforce this argument; but in the first two parts I shall put aside temporarily the assumption that simultaneity can be defined by the transport of clocks in Newtonian mechanics, in order to show what happens when Grünbaum excludes the latter, fully legitimate procedure from his account.

a. Without the Use of Transported Clocks

Let us imagine a point-mass P being sent as a signal (physical causal chain) between point-masses A and B at rest in an inertial frame (see figure). The departure of P from A we shall call E_1; its arrival and reflection at B, E'; and its return to A, E_2. Even though Newtonian mechanics allows arbitrarily fast causal chains, there is a unique event E on the world-line of A between E_1 and E_2 which cannot also be on the world-line of P if negative transmission times are precluded. For we may accelerate P as much as we wish, thereby making the transmission time approach zero, but we can never make P depart and return at the same instant: we can always find some instant E between P's departure and return. In other words, P is not a first signal in Reichenbach's sense[3] because we can always accelerate some other point-mass Q so that it will return before P. Still, there is a greatest lower limit, zero, to the transmission times of P or Q if negative transmission times are ruled out. Since it is impossible to connect the events E and E' in either direction by P used as a signal, they

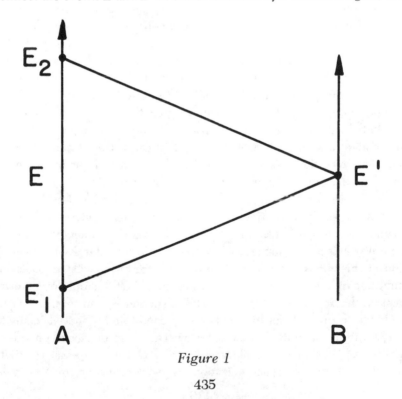

Figure 1

are indeterminate as to time order and hence simultaneous according to definitions given by Reichenbach (1958, pp. 144–45) and Grünbaum (1973, pp. 28–29). If the events were connectible, they would be temporally ordered and, with regard to a third event, they would exhibit temporal betweenness.

Now in Grünbaum's most recent treatment of simultaneity in Newtonian mechanics (1969, pp. 19–20), clocks play the roles of the point-masses A, B, and P. However, it does not follow from the above situation, as he would have us believe, that "the non-metrical, purely ordinal character of the unique simultaneity relation between E and E' [is] furnished by Newtonian clock transport." (1969, p. 21.) Rather, on the usual understanding of clocks and signals which was delineated in section 1, the simultaneity relation has been established by a signaling procedure. Although clocks could be substituted for the point-masses (as they were in Grünbaum's description), they do not have to be, since no clock was used as such.

Grünbaum goes on to provide a second characterization of "the relations of *temporal* betweenness and simultaneity in Newton's world on the basis of the causal betweenness defined by the following subclass K of its physically possible causal chains: K is the set of all those genidentical chains which are not spatially self-intersecting in at least one inertial frame." (1969, p. 22.) Once it is realized that the latter are infinitely fast influences (e.g., gravitational ones) and that those influences were also precluded from the first characterization, it becomes apparent that the first and second characterizations are identical in this respect: they are both based on the possibility of arbitrarily fast causal chains; neither is based on the possibility of obtaining consistent readings on transported clocks. Therefore the latter possibility, which was a legitimate physical procedure for synchronization in Newtonian mechanics according to Grünbaum's original position (1973, p. 370), has been effectively deleted from his revised position (1969, pp. 27–28). Concomitantly, the assumption of the impossibility of obtaining such readings, what he calls "assumption (i)," no longer characterizes for him the transition from Newtonian mechanics to special relativity (1964, section 1), and that assumption no longer represents for him a necessary condition for special relativity.

Rather than admitting that the slow transport of clocks provides a legitimate physical means of synchronizing clocks in special relativity, Grünbaum excludes synchronization by clock transport in Newtonian mechanics. If he is to be consistent, he has to put reliance on a clock's periodic processes as durational measures either in both theories *or* in neither, and he takes the latter implausible option. Grünbaum acknowledges that he modifies his position in that the absoluteness of simultaneity is no longer a sufficient condition for its nonconventionality (1969, p. 28). But he does not admit that even in Newtonian mechanics he must exclude clock transport as a physical means of synchronization distinct from signaling, because in that theory as well as in his "quasi-Newtonian universe"[4] the assertion that E and E' are simultaneous according to the readings of transported clocks (as distinct from signaling) rests on a convention in what he would consider a nontrivial sense: the assertion makes "a metrical appeal to the *durational congruence* of intervals on different world-lines" (1969, p. 25).

While Grünbaum concludes that simultaneity in a "quasi-Newtonian" universe is conventional in this sense, Putnam comes to a conclusion the statement of which is diametrically opposed to Grünbaum's, yet apparently similar to that of Ellis and myself:

In the quasi-Newtonian world, the customary correspondence rule for simultaneity involving the transported clocks does not lead to inconsistencies that cannot be explained as due to the actions of differential forces upon the clocks. There is then, in my view, no reason to regard simultaneity as a notion needing a definition except in the trivial sense in which every notion requires a definition (TSC). It is, in my view, an empirical fact in such a world that the to and fro velocities of light are equal relative to anything else. We *could*, of course, define the to and fro velocities of light to be equal relative to some system in motion and put up with the incredible complication that would result in the statement of all physical laws. This is, however, just an instance of the possibility of altering the nomenclature in connection with any physical magnitude and as such is an instance of TSC [= "trivial semantic conventionalism"] (1963, pp. 237–238).

Our conclusion is more restricted in that it refers to any physical magnitude that depends on local comparison. This is necessary to preclude quantities such as those depending on fundamental charge e and Planck's constant h, which are "physical magnitudes" but which are based on manifolds having intrinsic properties that provide a built-in measure. In contradistinction to such magnitudes there are magnitudes like spatial and temporal congruence that represent continua devoid of such intrinsic metrics; these magnitudes that have to have a metric provided for them by fiat or convention Grünbaum calls "Riemann-conventional" (1969, pp. 25, 33). Ellis and I [5] accept the existence of the latter conventionality, while pointing out that in certain theoretical contexts there are legitimate physical procedures (like slow-transport synchronization in special relativity) that override or subjugate it. Conversely, in the absence of such procedures, there are contexts in which the Riemann-conventionality of continuous manifolds prevails. [6]

However, as for the *grounds* for conventionality, we seem to have more in common with Grünbaum. In view of the theoretical-context-dependence of our version of Riemann-conventionality, we must distinguish it from a fact-dependent formulation like Grünbaum's. But even more we must separate it from Putnam's "trivial semantic conventionalism" (TSC), which is based on purely linguistic considerations. We can, like Grünbaum (1970, pp. 471, 476), imagine theories other than Newtonian mechanics (about which we disagree) in which simultaneity would not be Riemann-conventional. For example, we can conceive of physicists, in trying to reconcile quantum mechanics with relativity theory, devising a notion of time which would be quantized like charge and action and would not be based on a continuum of instants. Presumably, Putnam would say that such a magnitude can be altered so as to be a continuous one (cf. n. * on his p. 222). However much this may *simplify* (rather than complicate, as in Putnam's statement above) the mathematical representation, physicists would regard it as an approximation just as they do quantized magnitudes whose quantum numbers become very large (correspondence principle). The underlying reality, they would say, is nonetheless discrete, and an exact treatment must reflect this fact.

b. Without the Use of Gravitational Influences

Clock transport is not the only physical procedure for synchronizing clocks in Newtonian mechanics that Grünbaum eliminates. He also precludes gravitational influences from his characterization, evidently for the reason that "the K-defined relations of temporal betweenness do not allow the deduction of the paradoxical conclusion that either of two simultaneous events belonging to a gravitational chain is temporally between the other and some third event." (1969, p. 23.) Nevertheless, except for creating problems for Grünbaum's philosophical account, both of these procedures are perfectly legitimate alternative criteria for distant simultaneity in Newtonian mechanics. And he ought to admit them into his account if he really sees theoretical terms in physics as "open multiple-criterion concepts." He criticizes "the crude operationist claim that any particular 'definition' chosen by the physicist exhaustively renders '*the meaning*' of spatial congruence in physical theory." (1973, p. 15.) However, in practice, Grünbaum is doing this himself by restricting his characterization of distant simultaneity in Newtonian mechanics to the "subclass *K* of its physically possible causal chains."

Not only is he methodologically inconsistent in this respect, but he also weakens his argument for the nonconventionality of distant simultaneity in Newtonian mechanics beyond the point of collapse. As it now stands that argument rests merely on the lack of a limit on Newton's second law, i.e., the physical possibility of particles being accelerated without limit (1973, p. 350n). To be sure Newton himself believed that his law was limitless. And, of his famous Rules of Reasoning in Book III of the *Principia*, Rule III provided him sufficient grounds for such a belief: "The qualities of bodies, which admit neither intensification nor remission of degrees, and which are found to belong to all bodies within the reach of our experiments, are to be esteemed the universal qualities of all bodies whatsoever."[7] However, it is doubtful that he would have sustained his belief in the face of an empirically based challenge to the unlimited speed of gravitation. Rule IV provides just this necessary condition: "In experimental philosophy we are to look upon propositions inferred by general induction from phenomena as accurately or very nearly true, notwithstanding any contrary hypothesis that may be imagined, till such time as other phenomena occur, by which they may either be made more accurate, or liable to exception." (1966, p. 400.) Thus the second law is to be regarded as unlimited until such time as there is discovered some actual phenomenon which seems to limit it.

Now Newton was well aware from Roemer's observations of Jupiter's moon that the speed of light was finite. But he had no similar grounds for believing that gravitational propagations were limited. Thus he was justified in holding that particles could be accelerated without limit. However, the lack of a limit was by no means essential to Newtonian mechanics, and as J. D. North shows, there were attempts to modify Newtonian mechanics in this respect both from inside and from outside the theory (1965, chap. 3, section 5). First he generalizes the sort of external criticism of Newtonian mechanics that, as a matter of historical fact, Poincaré gave: "anyone who hoped to cast a new theory of gravitation in the form of Maxwell's electromagnetic field equations was committed to forces which could not be supposed to act simultaneously over finite distances."[8] What needs to be brought out explicitly here is that the modifications electromagnetic theory forced one to make in the laws of gravitation and acceleration were physically rather than logically grounded: the modifications were being made to obtain empirical agreement between two different theories (or three, if Newton's three laws are regarded as a theory of inertia distinct from his law of gravitation); logically, there is no reason why the theories could not disagree.

It is true that the foregoing modifications were made in the context of an electron-theoretic modification of Newtonian mechanics. But, as logical possibilities, they could have been made on the basis of some theory other than the electron theory, or they could have been made for reasons internal to Newtonian mechanics. North argues secondly that, at one time, Newton's gravitational law appeared unsuitable for the explanation of what were perceived as "anomalies" in the motions of the planets. Thus, in order to account for an apparent secular acceleration of the moon, Laplace considered the possibility of a finite velocity for gravitation, one with a lower limit of 7×10^6 times the velocity of light (North, 1965, p. 44). Thus, from within Newtonian mechanics, Newton's second law could have been altered prior to the electron theory if there had been no overweighing reason to believe that gravitational influences propagated faster than light and probably with infinite speed. Indeed, if such influences were unknown to the Newtonian, then he would have been even freer to require, in response to some perceived anomaly, that the actually known speed of light serve as a limit for the possible speeds of particles. Just as Poincaré felt obliged on physical grounds to modify Newton's second and gravitational laws to make them agree with electromagnetic theory, so also the Newtonian would have felt obliged under these hypothetical circumstances to modify the second law.

What are the implications of this discussion for Grünbaum's account of distant simultaneity in Newtonian mechanics? If he precludes faster-than-light gravitational propagations as a criterion for distant simultaneity, as he apparently does, then hypothetically one could insist that the acceleration of particles be limited by the speed of light and, concomitantly, that the criterion for distant simultaneity making use of accelerated particles agree with the light-signaling criterion. Under these conditions, then, Grünbaum's truncated version of the Newtonian world (not that world itself) can be identified with what he calls a "quasi-Newtonian world." And the Riemann-conventionality which he takes to characterize the latter is also found in the former. In particular, the Reichenbach-Grünbaum definition of simultaneous events as being those that are indeterminate as to time order is seen to be compatible with this interpretation of Grünbaum's Newtonian world. Or, by transposition, any two events belonging only to the career of the fastest causal chain (light) must be time-ordered and hence nonsimultaneous (Grünbaum, 1969, pp. 21–22.) This being the basis of Grünbaum's coordinative definition of nonsimultaneity, it is available in his version of the Newtonian world "to permit the inference that E and E' are paradoxically nonsimultaneous," despite his claim to the contrary (1969, p. 23).

Having just adduced one argument which purports to show that the Newtonian world as described by Grünbaum is Riemann-conventional, I shall now consider a second. In his rejoinder to Grünbaum, Salmon, and van Fraassen (abbreviated "G. S. & F." below), Ellis argues against their[9] attempted demonstration of the contrary conclusion by trying to show that

. . . if time in a Newtonian world is a quantity for which the basic ordering relationships are considered to depend for their determination upon local comparisons, then distant simultaneity in such a world is Riemann-conventional, and the conclusions of G. S. & F. are false.

If, on the other hand, the time order is defined by the causal order and not say the *local* entropic order, then in the Newtonian world, the basic time-ordering relationships do *not* depend only on local comparisons, and the argument is irrelevant. Our claim was that distant simultaneity is conventional in the way that any relationship of quantitative equality *which depends upon local comparisons* is conventional. It is irrelevant to show that it is conventional in a way different from that of some relationship of quantitative equality which in some possible world would not depend on local comparisons (1971, p. 179).

In section 2 of his paper Ellis explicates "the temporal order at any place A" in terms of "the local entropic order for certain classes of closed and isolated systems at A." Since a "clock" in the sense in which I have been using the term provides an instantaneous measure of the entropy of such systems, I can incorporate Ellis's explication. Then I can define the *distant* entropic order in terms of transported clocks.

This allows me to clarify the role of my conclusions in (a) and (b) above.

441

In (a) I have shown that Grünbaum does not base temporal order on *distant* entropic order. I have said nothing there about *local* entropic order, because it is unclear whether or not Grünbaum bases temporal order on it. Thus Ellis's irrelevancy argument quoted above complements my argument in (a): he argues that if Grünbaum does *not* base temporal order on local entropic order, his rebuttal is irrelevant to our position. These two arguments together allow me to conclude that, in regard to entropic order, the position taken by Ellis and me in our joint article (1967, p. 134) is not affected by Grünbaum's most recent argumentation. The argument in (b) presupposes the one in (a) and shows that, if Grünbaum *further* restricts the temporal order of Newtonian mechanics so as not to be based on the distant causal order of gravitational influences, then the temporal order is not unique.

c. With the Use of Transported Clocks

Finally, it is essential to know whether or not Grünbaum's characterization of the temporal order would lead to a conclusion in opposition to our own if he were to eliminate his self-imposed restrictions on it. In other words, the supplementary issue before me now is, does the local entropic order conjoined with the distant causal or entropic order lead to a unique temporal order? It is to this issue that Ellis directs his attention. He argues that, although the round-trip time of a signal can be made to approach zero in a Newtonian world, this is not true of the one-way time in either direction "unless we insist that a one-way signal time *shall not be negative*" (1971, p. 190). He then asks, "Is there then any reason why we must insist that a one-way signal not be negative?" Admitting there are reasons of descriptive simplicity, he points out that "to invoke the causal theory of time to justify the requirement . . . is to render the example of distant simultaneity in the Newtonian world irrelevant to our thesis." (*Ibid.*) What must be shown is "that it is a matter of 'ordinal temporal fact' that one-way signal times are non-negative." For this seems to be the criterion to which Grünbaum, Salmon, and possibly van Fraassen adhere for what would constitute a good physical reason. "If it is a factual claim in the required sense, then presumably we must have some independent criterion for distant simultaneity which we accept, and which we can use to test this claim." (Ellis, 1971, pp. 190–191.) Finally, taking as that criterion distant entropic order, he adapts the initial argument to "the one-way transport time relative to K . . . as determined in K . . ."

442

(1971, p. 191). Thus this duration can also be interpreted as being negative.

If I understand Ellis's argument correctly, it attributes to the positions of each of our three critics a need for a *testable* criterion for distant simultaneity. But certainly Grünbaum and possibly van Fraassen would dispute this. Grünbaum explains at length how his basis for the Riemann-conventionality of temporal congruence (and hence distant simultaneity as determined by transported clocks) is not the testability on which Reichenbach and Carnap grounded their position (Grünbaum, 1973, pp. 81–82). Van Fraassen apparently acquiesces to what he takes to be an argument on Poincaré's part that the choice of a congruence standard is purely conventional for a continuous manifold (1969, p. 65). Salmon follows Reichenbach, interpreting testability along the lines of physical rather than logical possibility.[10] Although Ellis argues properly and cogently in his response to Salmon, his use of testability may vitiate his rejoinder to Grünbaum and van Fraassen. However, it seems to me that his argument is just as cogent if it is reinterpreted without any reference to testability: no matter which criterion one chooses (arbitrarily fast particles, gravitational influences, transported clocks), distant simultaneity in Newtonian mechanics is Riemann-conventional unless some high-level fact (e.g., an ordinal temporal one) of that theory precludes negative (one-way) signal or transport times; but there seems to be no such fact in Newtonian mechanics.

Ellis establishes the former clause perfectly well, it seems to me; what remains to be shown convincingly is the latter. Ellis merely states it without discussion (1971, p. 192). He does, it is true, describe without contradiction the situation in which negative times are used, but this merely shows that it is logically possible. Whether the situation has a stronger kind of possibility, that which is called "conceptual,"[11] has been debated by Weingard (1972), but I am inclined to agree with Earman that it is a "somewhat nebulous question" (1967). Like Ellis (1971, p. 190) and Grünbaum (1973, p. 351), I do not put much stock in ordinary usage. The more pertinent question, it seems to me, is whether negative (one-way) signal or transport times are physically possible in Newtonian mechanics. And the scant treatment which I have seen of this question seems mistaken.

Weingard argues, in effect, that negative times are impossible because "in classical physics we have only the notion of time simpliciter, i.e. there is simply a single universal time" (1972, p. 118). But anyone who has read Newton's *Principia* knows that, as a description of his mechanics, this is at best misleading and at worst false. For in addition to what he calls "absolute, true and mathematical" time (also called "duration"), he allows that there exists "relative, apparent, and common time, [which] is some sensible and external (whether accurate or unequable) measure of duration by the means of motion, which is commonly used instead of true time. . . ." (Newton, 1966, p. 6). Thus, contrary to what Weingard (1972, p. 119) would have us believe, we already have in Newtonian mechanics the "notion of time [with] respect to a frame of reference." The difference vis-à-vis special relativity is that this time, if accurate, agrees with universal time; in special relativity, there is, of course, no such notion as universal time. At a practical level the problem exists just as much for the Newtonian as for the (special) relativist to find a proper frame of reference in which to record the time: for the Newtonian it would be the true or universal frame, whereas for the relativist it would be an inertial frame. (One cannot simply choose "a frame of reference whose space axes are at rest with respect to the physical system," as Weingard (*Ibid.*) seems to think, as that frame might be noninertial.)

Finally, let us examine the implications of all this for the possibility of negative (one-way) signal or transport times in Newtonian mechanics. In the situation in which something travels from event b to event a according to the universal frame, one cannot say in Newtonian mechanics, as Weingard tries to, that an observer who judges by the experiences of his rest frame that events at b occur after those at a "would be wrong since events at a are simply after those at b" (*Ibid.*) To say this one would have to know that his frame is equivalent to the universal frame. But this is a proposition which one might not know (much less know simply) in Newtonian mechanics, just as in special relativity one might not know whether his frame is inertial. Therefore, Weingard's objection is obviated, and Ellis can assert in Newtonian mechanics, just as well as in special relativity,[12] the possibility of negative times in the following sense: if, with respect to the time in an observer's rest frame, events in that frame are temporally ordered one way (with respect to earlier-later) while they are ordered the reverse way with respect to a second observer's frame of reference, then with respect to the second observer's frame of reference the first is going back in time (paraphrase of Weingard, *Ibid.*). Indeed, I do not see how Weingard can consistently deny for Newtonian mechanics the possibility of going backward in time when he later says that one can divorce the ideas of backward time travel from the particular space-time structure of special relativity (1972, pp. 119–120).

3. Conclusion

Ellis's answer to Grünbaum, Salmon, and van Fraassen on the simultaneity of Newtonian mechanics was complete and cogent only in response to Salmon. In reformulating it so as to make clear its applicability to Grünbaum and van Fraassen, I have no doubt rendered it incomplete. However, the reformulation is easily completed in the way I have just indicated. Further, I have argued that Grünbaum's position on simultaneity in Newtonian mechanics, insofar as it excludes the determination of simultaneity through the use of gravitational influences and transported clocks, suffers difficulties of its own: that simultaneity can be interpreted as being Riemann-conventional, notwithstanding his claims to the contrary.

Notes

1. 1973, p. 353. Ellis and Bowman (1967, section 1) derive an expression for one kind of nonstandard signal synchronization of the clocks of any inertial frame.

2. The construal mentioned above occurs in Grünbaum (1969, p. 21), who uses a clock as a signal. This conflation produces a certain difficulty, as I point out in subsection 2.a below.

3. A "first-signal" is, according to Reichenbach, "the fastest message carrier between any two points in space" (1958, p. 143).

4. A "quasi-Newtonian universe" is a world described by Newtonian mechanics except that the speed of light limits all causal chains; thus initially synchronized clocks continue to give the same readings despite being separated and rejoined.

5. In my case, at least, this is what I shall call "preanalytic acceptance," i.e., acceptance based on the way scientists actually operate. It is to be distinguished from acceptance based on the rational reconstruction of what scientists do. Since the behavior of scientists is sometimes based on mistaken judgments, it obviously cannot be the ultimate criterion for what is correct. However, it can serve as an interim criterion in the absence, for whatever reason, of an ultimate one. In other words, scientists behave rationally until they are proved to be mistaken.

6. I can mention two simple contexts in which the transport of clocks fails to provide a legitimate physical means of synchronization. The first is Reichenbach's "static gravitational field" (1958, p. 259), a static non-Euclidean space with an orthogonal time axis; i.e., the components of the field, $g_{\mu\nu}$, are functions of spatial coordinates alone (they are independent of time) and $g_{\mu 4} = 0$ for $\mu = 1, 2, 3$. The second is a special case of Reichenbach's "stationary gravitational field" (1958, pp. 261–262). The latter differs from the static in that the time axis is not orthogonal; i.e., $g_{\mu 4} = 0$ for $\mu = 1, 2, 3$. The special case results from making the space Euclidean, i.e., letting $g_{\mu\nu} = 0$ for $\mu \neq \nu = 1, 2, 3$. For this case Eddington (1924, p. 15) shows that even synchronization by infinitely slow transport along a straight line is impossible.

7. 1966, p. 398. In his elaboration he mentions some such qualities: gravitation (i.e., the mutual attraction of bodies) and what are now called the "primary qualities" of Newtonian mechanics (extension, hardness, impenetrability, mobility, and inertia).

8. 1965, p. 44. Cf. also Bowman, 1972, subsection B.4, part a.

9. Ellis can direct his argument at all three of our detractors, since it applies equally to slowly transported clocks or arbitrarily fast accelerated particles or infinitely fast gravitational influences. In contrast, my argument applies only to particles (it assumes the absence of the other methods), so I have had to restrict its effect to the position of Grünbaum, the only one of the three who seems to preclude gravitational signals. Salmon remains agnostic on them (1969, p. 52n); and van Fraassen does not mention them, although he may imply that consideration of them is not necessary (1969, p. 67, note 1), in which case his position is the same as Grünbaum's.

10. 1969, p. 61. Cf. Bowman, 1972, subsection K.3.

11. Apparently, a situation is a "conceptual possibility" if it can be described "in terms of our present concepts of time, travel, and change," i.e., without "a change in usage or meaning from the ordinary way of speaking" (Weingard, 1972, p. 120).

12. This is true of special relativity only insofar as the present problem is concerned. There are other problems in saying in that theoretical context that things can go back in time, as Earman (1967, section IV) has shown.

References

Bowman, Peter A. (1972). "Conventionality in Distant Simultaneity: Its History and Its Philosophy." Ph.D. dissertation, Indiana University.

Bowman, Peter A. (1974). "Simultaneity as a Theoretical Term." Paper read to Annual Meeting of the American Philosophical Association, Eastern Division, 28 December 1974, at Washington (D.C.) Hilton.

Bowman, Peter A. (1976). "The Conventionality of Slow-Transport Synchrony." In *PSA 1974*, ed. A. C. Michalos and R. S. Cohen, pp. 423–434. Boston Studies in the Philosophy of Science, vol. 32. Dordrecht: Reidel.

Earman, John (1967). "On Going Backward in Time," *Philosophy of Science*, vol. 34, pp. 211–222.

Eddington, Arthur S. (1924). *The Mathematical Theory of Relativity*. 2nd ed. Cambridge: Cambridge University Press.

Ellis, Brian (1971). "On Conventionality and Simultaneity—A Reply," *Australasian Journal of Philosophy*, vol. 49, pp. 177–203.

Ellis, Brian and Peter Bowman (1967). "Conventionality in Distant· Simultaneity," *Philosophy of Science*, vol. 34, pp. 116–136.

Feenberg, Eugene. "Conventionality in Distant Simultaneity," *Foundations of Physics*, vol. 4, pp. 121–126.

Grünbaum, Adolf (1964). "The Bearing of Philosophy on the History of Science," *Science*, vol. 143, pp. 1406–12. (Reprinted in Grünbaum, 1973, pp. 709–727.)

Grünbaum, Adolf (1969). "Simultaneity by Slow Clock Transport in the Special Theory of Relativity." In Grünbaum and Salmon, 1969, pp. 5–43. (Reprinted in Grünbaum, 1973, pp. 670–708.)

Grünbaum, Adolf (1970). "Space, Time and Falsifiability: Introduction and Part A," *Philosophy of Science*, vol. 37, pp. 469–588. (Reprinted in Grünbaum, 1973, pp. 449–568.)

Grünbaum, Adolf (1973). *Philosophical Problems of Space and Time*. 2nd ed. Dordrecht, Holland: Reidel (1st ed. New York: Knopf, 1963).

Grünbaum, Adolf and Wesley C. Salmon (1969). "Introduction: The Context of these Essays," pp. 1–4 of "A Panel Discussion of Simultaneity by Slow Clock Transport in the Special and General Theories of Relativity," *Philosophy of Science*, vol. 36, pp. 1–81. (Includes Grünbaum, 1969; Salmon, 1969; van Fraassen, 1969; and Janis, 1969.)

Janis, Allen I. (1969). "Synchronism by Slow Transport of Clocks in Noninertial Frames of Reference." In Grünbaum and Salmon, 1969, pp. 74–81.

Newton, Isaac (1966). "Principia." In *Sir Isaac Newton's Mathematical Principles of Natural*

Philosophy and His System of the World, vols. 1 and 2, trans. A. Motte, rev. F. Cajori. Berkeley: University of California Press.

North, J. D. (1965). *The Measure of the Universe*. London: Oxford University Press.

Putnam, Hilary (1963). "An Examination of Grünbaum's Philosophy of Geometry." In *Philosophy of Science: The Delaware Seminar*, ed. B. Baumrin, vol. 2, pp. 205–255. New York: John Wiley and Sons.

Reichenbach, Hans (1958). *The Philosophy of Space and Time*, trans. M. Reichenbach and J. Freund. New York: Dover Publications.

Salmon, Wesley (1969). "The Conventionality of Simultaneity." In Grünbaum and Salmon, 1969, pp. 44–63.

Van Fraassen, Bas C. (1969). "Conventionality in the Axiomatic Foundations of the Special Theory of Relativity." In Grünbaum and Salmon, 1969, pp. 64–73.

Weingard, Robert (1972). "On Travelling Backward in Time," *Synthèse*, vol. 24, pp. 117–132.

Winnie, John (1970). "Special Relativity Without One-way Velocity Assumptions," parts 1 and 2, *Philosophy of Science*, vol. 37, pp. 81–99 and 223–238.

INDEXES

Name Index

451

Name Index

Subject Index

CPSIA information can be obtained
at www.ICGtesting.com
Printed in the USA
LVHW042226210819
628468LV00025B/1095

9 780816 657520